AF204192

Reality and Hermeneutics

Bonn Studies in the New Humanities

Editors

Markus Gabriel · Marion Gymnich
Tobias Keiling · Birgit Ulrike Münch

Advisory Board

Jocelyn Benoist (Paris) · Alice Crary (New York)
Günter Figal (Freiburg i.Br.) · Monika Kaup (Washington)
Paul Kottman (New York) · Irmgard Männlein-Robert (Tübingen)
Jürgen Müller (Dresden) · Takahiro Nakajim (Tokio)
Jessica Riskin (Stanford) · Xudong Zhang (New York)

2

Maurizio Ferraris

Doc-Humanity

Translated by
Sarah De Sanctis

Mohr Siebeck

Maurizio Ferraris, born 1956; Professor of Theoretical Philosophy at the University of Turin and president of Labont (Center for Ontology). He is also the director of "Scienza Nuova", an institute of advanced studies – dedicated to Umberto Eco and uniting the University and the Polytechnic of Turin.
orcid.org/0000-0002-3027-1691

Original title: Documanità. Filosofia del mondo nuevo by Maurizio Ferraris.
Copyright © 2021, Gius. Laterza & Figli, All rights reserved.

The series *Reality and Hermeneutics* is supported by the *Faculty of Arts and Humanities, University of Bonn*, in cooperation with *The Institute for Philosophy and the New Humanities, New School for Social Research, New York*, and the *East Asian Academy for the New Liberal Arts, University of Tokyo*.

ISBN 978-3-16-161666-2 / eISBN 978-3-16-161667-9
DOI 10.1628/978-3-16-161667-9

ISSN 2751-708X / eISSN 2751-7098 (Reality and Hermeneutics)

The Deutsche Nationalbibliothek lists this publication in the Deutsche Nationalbibliographie; detailed bibliographic data are available at *http://dnb.dnb.de*.

© 2022 Mohr Siebeck Tübingen, Germany. www.mohrsiebeck.com

This book may not be reproduced, in whole or in part, in any form (beyond that permitted by copyright law) without the publisher's written permission. This applies particularly to reproductions, translations and storage and processing in electronic systems.

The book was typeset by Gulde Druck in Tübingen using Times typeface. It was printed on non-aging paper and bound by Hubert & Co. in Göttingen.

Printed in Germany.

Table of Contents

Prologue: The New World

The short century, which began in 1918 and allegedly ended in 1989, actually lasted a little longer than usual: 102 years. Indeed, the 20th century really ended in 2020, and the 21st century started this year. Or, if you prefer, the short century was followed by a *very* short one, lasting thirty-one years, and now we are entering the 22nd century. However you want to look at it, this is a new century, a new world, and also a new humanity – in fact, humanity is not something that has been defined once and for all, but an open project. What entitles me to such seemingly misplaced optimism, you may ask? Am I blind to the looming dangers and the ongoing crisis? Of course not, but for that matter an illustrious colleague like Augustine of Hippo wrote *The City of God* while under siege by Genseric, as the entire world was collapsing all around him. Rest assured, I have a different project in mind here, not only because unfortunately I am no Augustine, but especially because transitional periods, like unhappy families according to Tolstoy, are each unhappy in their own way. The point is to understand where this unhappiness comes from, and to find ways of translating it into confidence in the future, instead of letting pessimism or nostalgia paralyse our efforts. Now, however little we may think of ourselves, and however justified we may be in judging the crooked timber of humanity, the fact remains that signs of progress abound, and neglecting them would be a crime against ourselves and future generations. These signs may not be blatant, but this is only natural, since no age has ever been able to read, from the beginning, the tables of the law that would govern it. But precisely for this reason it is necessary to try and grasp the traces of the future in the confused fabric of the present.

First sign: the documedia revolution. For some decades now, we have been witnessing a revolution. This is not a political and ideological revolution like those that took place around the First World War, but a technological revolution, which is much more radical because it does not depend on people's beliefs, but on the tireless labour of machines. Specifically, the core of this revolution is the Internet, which, regardless of any short-term enthusiasm or execration, has introduced a degree of automation destined, in an indefinite period of time, to finally free humanity from the burden of toil and alienation. And this is not because of some form of industrial philanthropy, but simply because, unlike humans, machines have no rights, no dreams, no nightmares, no neuroses, no hunger, no thirst, no fatigue, no boredom, no divorces, no shared expenses, no retirement, and no ideas – and if they die, they can rise again.

So, there is nothing wrong in thinking that artificial intelligence was invented to take away our jobs: that is exactly right (on the other hand, as we shall see, it is not true that machines will eventually take over in general). Robotics and domotics are experimenting in home care and the military, but the real goal is the automation of all production. Indeed, it is hard to deny that automation is the destiny of a process that may take many years, but which has already been underway since the first hominid used a bone as a club or a lever, enhancing its strength with a rudimentary mechanism. There has been no turning back since then. So let's take this for granted – after all, it is a comforting thought, given that no one regrets mines or assembly lines, or even just the office manager strolling among the typists, tapping his watch. And let us consider that, just as those jobs have disappeared or are disappearing, so too will the delivery people, the on-demand workers, and all those who do the micro-jobs required by artificial intelligence insofar as it is not yet sufficiently developed, although very ready to learn. This is a dramatic and potentially hopeless situation, which heralds a crisis incomparable to any other known to humanity. Such a crisis cannot be addressed by mourning the past, but requires understanding the present in its singularity and originality.

At the end of the 18th century began the world of industrial capital: it produced goods, generated alienation and made much noise, that of factories. Then came financial capital, which produced wealth, generated adrenalin and made a modicum of noise, this time in stock exchanges. Today, a brand new capital is coming forward, thanks to a technology that did not exist before (though it manifests the essence of the capitalisation that constitutes the social world): this new capital produces documents, generates mobilisation through automation, and makes no noise. Compared to financial capital, it is richer and even more influential on the creation of value, on social relations and on the organisation of people's lives – and I am not just talking about their professional existence, of course, as it is prone to blur the distinction between life and work.

This capital is revolutionary and is linked to a *technological* transformation, the kind that took us from the chipping of the first flint to the Bronze and Iron Ages, to the exploitation of steam. I call the ongoing change "documedia revolution" because it is based on the intersection between the increase in documentality, i.e. the production of documents as a constitutive element of social reality, and the growth of the media, which today no longer work as one-to-many but as many-to-many. While the environment of the industrial revolution involved factories and working-class towns, the milieu in which production takes place today is the Web, i.e. a potentially ubiquitous place. This transformation (which implies a generalisation: factories are somewhere, the Web is everywhere) easily explains the radicality of the changes that have taken place in the last few years. This revolution is the largely accidental result of the dizzying increase in means of recording, comparing and profiling humanity – that is, the capital of acts that humans perform in the world.

When watching a video online one may feel like one is dealing with a more interactive television, but in reality there is a whole Copernican revolution between using an analogue and a digital medium. In the first case, we passively watch the video, so much so that we often fall asleep. In the second case, so to speak, it is the video that watches us, keeping track of our habits and preferences, the comments we make, the people to whom we send the link, the frequency with which we revisit it, etc. The Web then pushes us to take action, so much so that I don't think anybody ever fell asleep in front of the mobile phone, unless they used it as a TV set – but even then, unlike the TV set, the mobile phone impassively noted down the time, the day, and many other things. The fact that 90% of all documents currently stored in the world have only been generated in the last two years[1] is thus difficult to prove, but intuitively reasonable. In Europe, in the United States, and progressively all over the world – with an enormous competitive advantage for China, which has almost one and a half billion inhabitants and, above all, one billion mobile phones – everyday acts that until very recently would have disappeared into thin air are now being recorded and, as I will demonstrate at length in this book, turned into capital.

Second sign: from prod-humanity to doc-humanity. This is another sign that may not be apparent at first, but constitutes the starting thread of the tangle that I will try to unravel in this book. To those who, with legitimate concern, speak of the end of labour and therefore of employment, and to those who fear the dystopia of machines taking over as humanity falls, I propose a simple reflection: What is the point of machines without humans? What is the point of production without consumption? The growth of automation has brought about a revelation of something that had hitherto remained hidden in the workshops of *homo faber*. That is, there are very few functions in which a machine cannot replace a human being, but not *consumption*, be it material (machines need energy, but they can also do without it, whereas organisms, including humans, die if deprived of it) or spiritual (one can imagine a machine producing symphonies, but not a machine enjoying them).

---

[1] SINTEF, *Big Data, for better or worse: 90% of world's data generated over last two years*, in "Science Daily" (2013, May 22): www.sciencedaily.com/releases/2013/05/130522 085217.htm [12/11/2020].

If production was at the heart of industrial capital, which was made of matter, then consumption is at the heart of documedia capital, which is made of memory. Machines can produce infinitely more and better than humans, but no machine can ever consume *in the place of* a human. Instead, every human is capable of consuming, and indeed must necessarily do so to stay alive. For this very reason, consumption, with its physiological urgencies and the social mobilisation it entails, provides a purpose for the entire productive system, which would otherwise be meaningless. So, consumerism is not just the dirty word that has been deprecated, I am not sure how sincerely, over the last few centuries, but should be seen as the great engine of human growth. Reciprocally, the non-automated functions of consumption, which cannot be automated precisely because they are exclusively human and organic, are the future of value production. Mechanisms, in fact, consume only in a metaphorical or at least derivative sense: a car without fuel does not die, a human without food eventually does. As for non-human organisms, they satisfy their needs through processes that have nothing to do with consumption, as they are not part of a technical and economic cycle (when they are, it is again with a view to human consumption: farms, pets, etc.).

But why should consumption be equated with production? This is the crux of the whole argument, and the answer is simpler than it may seem: because it generates documents. For a long time, production demanded physical effort and entailed alienation, but this is no longer necessarily the case. Prodhumanity, the humanity that toils in fields and workshops, as crucial to the industrial age as it was for the ten thousand years before it, is still with us. But we are already seeing the rise of doc-humanity, a humanity whose greatest function is to produce documents about itself – be they minor documents such as the data about our online activities, or major documents such as spiritual or cultural products. All these documents are essential for the profiling of needs and behaviours, the resulting automation of production, and the rationalisation of distribution, with the consequent lowering of prices.

If understood in its true structure, the documedia revolution thus brings about a virtuous circle between the automation of production generated by the collection of documents produced by humans, and the world of life where human acts, which feed the production of documents, take place. On the one hand, automation enables the recording, archiving, profiling and reuse of human life forms. What we call "artificial intelligence" is just that. On the other hand, the whole process receives nourishment, meaning and purpose from the fact that there are human agents, mortal organisms that, unlike machines, have an irreversible end (death), and can therefore attribute an end (purpose) to machines, for example to oxygenate us in intensive care units. Try to imagine an intensive care unit without humans: it would be meaningless, and one would wonder whether it would be worth keeping. Now think of a Web without humans, and you will see the essential role that each of us plays with respect to automation, mainly fed by the profiling of human behaviours and needs operated by Internet platforms.

In short, production, as the mechanical part of humanity, is destined for automation, while consumption, as a properly human fact, cannot be automated in any way, for reasons that are not ethical but ontological. Consumption is not only the cooperation of users in production: it is the ultimate goal of all production. If humankind were to suddenly find itself without needs to be satisfied through consumption, the whole documedia system would implode. Consumption accompanies us until our last day, and even afterwards, if we have made onerous requests for our burial. In this framework, consumption is not the modern surrogate of capital vices, but an organic need inserted in a socio-technical context, which generates all kinds of reinforcement, supplementation and metamorphosis of need that turn it into desire, ambition, luxury, waste, culture. Yet this crucial circumstance will remain hidden until consumption is reconceptualised as labour.

Third sign: mastery and slavery. The third sign also needs to be understood, else humanity will continue to be reduced to the victimhood of voluntary servitude. We no longer need workers (good for them), but we do need humans, including those humans who until not long ago only recognised themselves as producers. The exclusion of workers from production in no way means that humans are no longer able to produce value. Machines always need someone – a living being and most often a human – to give them a purpose. This holds for anything – a watch, a knife, a cheese grater, a computer, a mobile phone. Try to imagine a world without humans, and ask yourself what purpose a mystery novel or a battery charger would have then. Herein lies the enormous and underestimated power of humanity. It is possible that sooner or later there will be a mechanical way to produce and distribute anything, but even a perfect machine would be utterly useless in the absence of humans. Which means, if you think about it, that we are quite literally the masters of steam, the lords of machines, although for some reason – usually related to a wish to absolve ourselves of our own inertia or faults – we are mostly inclined to think of ourselves as slaves to automation.

It is up to us to bring about a conceptual revolution, a turnaround in the way of thinking that has accompanied us through the long but not infinite time when humans were primarily producers. This is less and less the case today, for the reasons I have hinted at here and which I will analyse at length in this book. But precisely because we are living organisms, and in particular organisms that are systematically connected to technological devices, we must understand that without us, without our needs, without our urgencies, without the boredom and anguish deriving from the knowledge that we possess a limited lifespan, automata would be meaningless. In the same way, we must become aware of the fact that, precisely because, fortunately, machines will gradually cover the vast majority of productive needs, humankind will have to rethink its being-in-the-world as mobilisation. And we will have to understand the unprecedented value of that mobilisation, which has effectively taken the place of labour and must therefore be recognised and remunerated. In the light of this, we need a transvaluation of all values and all labours: a transvaluation with no Walhalla, a transformation of a traditional worldview that, although ancient, is not necessarily just.

In other words, the master-servant dialectic must be revived, and to do so we must overturn the misleading assumption that we are masters of nature and slaves to technology. As for our supposed mastery over nature, Covid-19 is a perfect illustration of how easily nature overpowers us – not to mention that, even in normal circumstances, we all eventually die, whether in a hundred years or in a minute. And herein lies the reason for our dominion over technology. This is a point that requires reflection, because what is at stake is our freedom. Now that the gods and the devil are no longer presentable candidates, an even more serious problem than theodicy has arisen. If there is no single principle to which the world can be traced back, for better or worse, whom are we to blame for our misfortunes? The question makes no more sense, from a secular point of view, than the one that inspired theodicy, but, unlike the latter, it has produced a large number of answers, all of which have the defect of distracting us from the analysis of actual reality, and which usually agree in viewing us as slaves to technology. Yet this is a slavery that we have only ourselves to blame for. If someone points a gun at me, I am not a slave to the gun, but to the person holding it. Blaming the Internet for populism is like blaming the radio for Nazism. And talking about the government of algorithms is no different from thinking that the intention to kill Caesar lied in the daggers and not in the conspirators. Humans are not slaves to technology or abstract systems: these are excuses for those who command and those who obey. They can, of course, be slaves to other humans, and their first duty is to emancipate themselves, through political action and philosophical understanding.

Fourth sign: the humanity to come. The fourth sign, lastly, concerns the new world opening up before us and its original characteristics, starting with the centrality of education. This is where we must start. The more humanity progresses, the more sensitive it becomes to injustice and inequality, the more the demand for human rights and the duty to meet these demands are pushed forward. And all this is due, first and foremost, to an essential philosophical reason that is often not given sufficient attention. There is no "human nature" as such: humanity is not something that has been defined once and for all, and consequently the rights (and corresponding duties) of human beings are not set in stone. They are not written in some Neolithic cave, they are not carved in the tablets of the law brought down from Sinai, nor are they displayed in bronze on the walls of Rome, a city rightly considered a symbol of civilisation, but which accepted slavery as a normal condition. Human nature, as well as its rights and duties, is a matter of historical becoming, and curiously enough, those who have condemned ethical relativism as a source of moral leniency or social injustice, undermining the good old values, have failed to consider that those values were not so good – indeed, they were far worse than our own. As there is no end to research, let alone to the history of human education, our current values will undoubtedly appear insufficient and limited to our future descendants, but the latter will still owe us gratitude, whereas we have the human, all too human, right to envy them a little.

The modern world, thanks to automation, has managed to satisfy the needs of an increasing number of human beings and, contrary to what is often claimed, our fundamental problem is not war or hunger. I do not say this to downplay these issues (every death matters, and in fact the only thing that matters is death), but to indicate where the real urgency lies. What we need most in the new world is *education*, understood first and foremost as the ability to produce a humanity that does not feel submissive or lost in the world it has created. So, what stands out as the fundamental necessity for the new world – which will not be paradise with the consequent eternal tedium, but which will certainly be better and fairer than the world we have left behind – is the shift from concern for production to concern for education. Let us not forget: Socrates did not say that a life without production is worthless; he said that a life without research is worthless, and, in productive terms, he did not work a day in his life. That is where we must aim, and if we look at the world without prejudice we will notice that we are closer to that goal than in any historical era before, certainly more so than in Socrates' own.

When Keynes wondered what one should do with one's free time he was thinking of other people – as for him, he knew perfectly well how to occupy it: read, write, attend international conferences, converse, dance and flirt with his Bloomsbury friends. If there is a desirable destiny for humankind, it is that, in a relatively long time, a similar form of life will be granted to every human being. And such a destiny cannot fail to come, for history does not go backwards. Webfare, the digital welfare system that I propose in this book, must pass through education, which will teach us to find new names and new forms, more tolerant and just, for the human needs for security, identity and projection into the future that in the past have been recognised in those old names. And above all, it will teach us how to transform the time given to us by automation into an opportunity for progress. The Web would thus cease to be the machine of discontent, and would become an instrument of emancipation. This, after all, is what it was supposed to be in the initial and naive dreams of its inventors, and this is what it can and must become again.

One very last point before concluding this already overlong introduction. I am sceptical, to say the least, of any criticism that does not come with an alternative, of any deconstruction without reconstruction. What we need is not yet another book about growing inequality, new forms of poverty, the end of labour, the dictatorship of machines and the panopticon state; if anything, we need ideas to avoid all this, because the conditions for doing so are in place today more than in any other time in history. The idea is more or less the following: "You don't like communism? You don't like capitalism? You don't like Webfare? Fine. But then tell me what is wrong with these things, and above all clarify what you *do* think is right. While you're busy thinking about it, let me explain why, in my opinion, we are better off today than we have ever been before, and how we will be better still, if we acknowledge (which you don't) that the path we have been on for a million years is the right one". That is what I will do in the next few pages, if you have the patience to follow me.

Instructions For Use

This work aspires to re-propose a philosophical system. Indeed, the use of systems has fallen into disuse in the last two centuries, for reasons that remain to be clarified and are far from obvious. If, as I believe, from philosophy one expects totality, or at least something like it, then the only alternative to the negative totality of the fragment is the system, or at least the adoption of some form, open and modular as it may be, but a form nonetheless.

The sequence of books that make up this volume suggests that ontology, with its fundamental vehicle, recording, corresponds to book 1; technology, with its fundamental vehicle, iteration, corresponds to book 2; epistemology, with its fundamental vehicle, alteration, corresponds to book 3; and teleology, with its fundamental vehicle, interruption, corresponds to book 4. This structure is repeated in the four chapters of each book, so that all chapters 1 deal with ontology, all chapters 2 with technology, and so on. The same applies to the four paragraphs that make up each chapter, obviously with the modifications resulting from their location.

	Chap. 1 Ontology	Chap. 2 Technology	Chap. 3 Epistemology	Chap. 4 Teleology
Book 1 Recording	1.1	1.2	1.3	1.4
Book 2 Iteration	2.1	2.2	2.3	2.4
Book 3 Alteration	3.1	3.2	3.3	3.4
Book 4 Interruption	4.1	4.2	4.3	4.4

By way of example, 1.3 as well as 2.3, 3.3 and 4. 3 deal with epistemology, except in the first case I talk about knowledge relating to ontology (the solution to the mystery of the commodity form), in the second case about knowledge relating to technology (capital as the vehicle of epistemological progress), in the third about knowledge relating to epistemology (a metaphilosophical reflection on how truth is formed), and in the fourth about knowledge relating to teleology (the struggle for the recognition of mobilisation as a production of value, i.e. the solution to the mystery of labour).

This structure obviously allows for a traditional reading, which is still the one I recommend, if only because it is guided by that tenuous but tenacious thread of the author's intention, and corresponds to a custom established in a great many cultures – although, as we know, at least half the world reads in other ways. According to this tradition, "reading a book" means starting with the first letter at the top left of the first page and ending with the last letter at the bottom right of the final page. This way of reading – which is a regulative ideal that has never been actually implemented fully: one always skips letters due to saccadic movements, if not entire pages or chapters due the dullness of the topic – would entail the following path: in 1 I deal with the documedia revolution, i.e. the ongoing transformation brought about by the Web; in 2 I deal with the anthropological revelation resulting from this revolution (my thesis is in fact that technology does not bring alienation, but a revelation of who we are); in 3 I deal with speculation, i.e. the metaphysical foundations common to both the revolution and the revelation; and in 4 I deal with transvaluation, i.e. the conceptual transformation and political action necessary for a fair redistribution of the social and economic gains generated by the documedia revolution.

This progression, however, is only one of the many possible or complementary uses of the present text, which aims at providing a theoretical matrix to understand the transformation in progress and to explore the future that, hopefully, awaits humanity, or rather, doc-humanity. Indeed, doc-humanity is what we are beginning to turn into, now that the production of goods is reserved for machines and all that remains for us humans to do is generate documents about who we are, what we want or do not want, what we know or think we know. Here are some of these alternative uses.

The first is to read any of the chapters according to whatever interest guides the reader (the documedia revolution for 1, anthropology for 2, speculation for 3 and politics for 4). This would be a non-arbitrary reading, responding to the Hegelian problem of "where to start?" in a system: one starts wherever one wants to and continues coherently. By way of example, those who start from 4, i.e. transvaluation, would immediately come up against the fundamental political proposal of the book, that of a webfare system, i.e. the redistribution of the surplus value produced by the Internet platforms. This would be the practical outcome of recognizing user mobilisation as labour that should be remunerated in terms of welfare, i.e. financial support for consumption, education and invention, which are the specific characteristics of humans as opposed to automata.

Going backwards, the proposal would be substantiated by the metaphysical foundation underlying the capitalisation of human mobilisation and the production of surplus value, namely hysteresis, the fact that acts are transformed into objects, i.e. into actual or potential commodities.

Thus, going back to 2, one would find the anthropological foundation of the mutual dependence between humans and Internet platforms, whereby the latter are the contemporary prototype of the technical complement that constitutes the only defining trait of human nature. At the same time, this supports the fundamental argument for webfare, namely that without human input, technology, and in this case Internet platforms, would be useless and inert, as they would lose their entire *raison d'être*.

At this point, coming to 1, the reader would have the political, metaphysical and anthropological coordinates to understand the documedia revolution in its entirety and depth, without being misled by the surface notions that often condition our understanding of the Web, starting with the idea that the Internet is an infosphere of communications rather than a docusphere that records the world of life, transforming it into profiling, automation and distribution.

Obviously, other access routes are also possible, depending on the readers' interests: those who start from 1 would privilege the ontology of the present, the situation in which we are now (which is why I started from there, following an *ordo naturalis* that immediately and necessarily turns out to be an *ordo artificialis*); starting from 2, on the other hand, one would have access to the structural characteristics of the human being (where "structural" does not mean "natural", since they presuppose a relationship with technology from the outset). Starting from 3, on the other hand, one would find a metaphysical perspective called upon to illuminate the technological, political and anthropological contemporary situation. And, of course, the order to be followed by those who start from 2 and 3 could be progressive (2, 3, 4), inverted (3, 2, 1) or desultory (2, 4, 1).

The second method one could adopt is a thematic reading, which would privilege chapters instead of books. For example, those interested in teleology could read all chapters 4: this path would go from the mystery of labour in 1.4 (a question arises about the purpose of human mobilisation) to the genesis of value in 2.4 (value is what, in a system, indicates a direction), to the explicit reflection on the meaning of teleology in 3.4, up to the revolutionary *telos* of a society freed from labour and emancipated by knowledge in 4.4.

Those who instead favour the issue of technology could start from the examination of the Web as docusphere, that is, as a technological field of document production (1.2), and then broaden the perspective by seeing the docusphere as a particular case of a more general phenomenon, i.e. the human adoption of prostheses (2.2). Still following the leitmotif of technology, in 3.2 readers would find an alternative answer to the conceptualism according to which theoretical under-standing is necessary to grant practical competence. The chapter illustrates the opposite: a competence that does not necessarily come with understanding, and which constitutes the fundamental characteristic of technology. Precisely be-cause there can be competence without understanding, 4.2 illustrates the mecha-nism of surplus value. This derives from the fact that the Web, like any technical device, is not a transparent infosphere (in which the documedia surplus value and the mystery of labour could not take place), but an opaque docusphere generated by competence without understanding, which indeed explains the formation of surplus value.

On the other hand, those who wish to start from ontology should begin with the fundamental character of the Web, namely hysteresis (1.1), i.e. recording and its resources; following with the fundamental character of the human being, namely responsiveness (2.1), i.e. the fact of being an organism systematically connected to mechanisms and technological prostheses. They would then move on to the fundamental character of all natural and social objects, namely emer-gence (3.1), i.e. the fact that the genesis of reality as we know it (or believe we know it) does not consist in a construction driven by principles but precisely in an aleatory bottom-up process made possible by the superabundance of time and space. Finally, these readers would come to the fundamental character of politics (4.1), which consists of deconstruction, i.e. not the critique of the existing state of affairs, but the criticism of the partial or misleading ideologies that prevent us from grasping the emancipatory resources hidden in the existing state of affairs (in our case, the documedia revolution).

Finally, turning to the thematic choice of epistemology, one would start from the fundamental character of the documedia transformation, i.e. the increase in documentation and the consequent solution to the mystery of the commodity form (1.3). One would then move on to capital as a prerequisite not only for the accu-mulation of value, but also for the growth of knowledge (2.3), following with the process of verification as a technological tool (3.3). Finally, one would come to transvaluation, i.e. the political and cognitive change in the understanding of the existing state of affairs resulting from knowledge of the processes (4.3).

The third possible approach is a structural reading, which does not privilege disciplines but the dialectical movements underlying reality, which constitute various articulations of hysteresis as a fundamental metaphysical principle. Thus we have recording, which corresponds to the manifest character of hysteresis (1.1), to the docusphere as the sphere of recording of social acts (1.2), to docu-mentation as the productive consequence of the docusphere (1.3), and to mobili-sation as a vital movement that is transformed into value through recording (1.4).

Then we have iteration, i.e. the technical possibility that arises from recording (what is recorded can be iterated), and which generates the phenomena of responsiveness (2.1, the alliance between organisms endowed with irreversible processes and mechanisms capable of indefinite iteration), prosthesis (2.2, where the prosthesis is an instrument of iteration), capital (2.3, capital being precisely the outcome of iteration and the largest of prostheses) and value (2.4, which arises only within a system of iterations and capitalisations that define its worth, meaning and desirability).

Then there is alteration, i.e. the aleatory process whereby an iteration can deviate from undifferentiated repetition to produce something novel, for better or for worse. The starting point, therefore, is the process of emergence (3.1), which is precisely an area of alteration, i.e. the genesis of something qualitatively new. One then moves on to competence (3.2), i.e. the technical instrument of both iteration and alteration; following with verification (3.3), i.e. the reflective understanding of alteration. Finally, one comes to narration (3.4), i.e. the ability to recognise, in the succession of alterations, a historical becoming that carries meaning.

Finally, we have interruption, namely the suspension of the process (paradigmatically, the death of organisms as an irreversible event), which is what makes the process interesting, since it is the urgency of a finite life beset by need that gives value to the system I have outlined. Here one would start with deconstruction (4.1, the contestation of the usual ways of thinking as a fundamental critical concern), moving on to analysis (4.2, the anatomy of surplus value). Then one comes to action (4.3, the transvaluation whereby every form of value production must be considered labour) and redemption (4.4, webfare as the restitution to the mobilised of the surplus value they have produced). All these actions are marked by interruption because they can only make sense to mortals. Indeed, political action and irrevocable decision-making would be inconceivable for immortal beings, including those which most manifestly populate our world, namely machines.

In addition to the ones outlined so far, of course many other readings remain open. I will simply point them out in broad strokes, but I will not dwell on them, because I am convinced that these possibilities, offered by a combinatory table like the one above, are where readers may find conclusions and paths that I have not thought of, and which for this very reason are more likely to apply to a new world that, by definition, cannot be anticipated.

The first is a diagonal reading, for example the succession that from 1.1 leads to 4.4, not following the pre-established order but proceeding from hysteresis (1.1) to prosthesis (2.2), to verification (3.3), to webfare (4.4). Obviously, one could start from the bottom, that is, from 4.4, or from the bottom left corner, following the diagonal that from deconstruction leads to mobilisation, or vice versa. Of course the diagonal reading can be partial, involving only some boxes, i.e., for example, privileging the connections between competence (3.2) and capital (2.3). Also, one could choose not to start from the corners of the square but from the second or third ones (for example with the triplet of docusphere, 1.2; capital, 2.3; and narration, 3.4).

The second is a modular reading, following paths that readers can shape according to their own interests or needs. One could move like a knight in chess and go from prosthesis (2.2) to transvaluation (4.3), mobilisation (1.4), deconstruction (4.1) or narration (3.4). Or one could follow correlations between non-contiguous boxes, as in the case of answering a specific question. For example: what is the right narrative to unmask the injustice of surplus value without falling into the trap of a negative philosophy of history? Here my choice would be to start from 3.4. (narration) and come to 4.2 (surplus value) via 4.1 (deconstruction of false beliefs) and above all 2.3 (examination of the true nature of capital).

Finally, one could simply engage in the close reading of a single box. One does not have to drink the whole barrel to determine whether the wine is good and, above all, one can have the very legitimate urge to reach a conclusion, for instance by linking the general theme of the book to one's own expertise. Those interested in political economy might focus, I suppose, on 4.2, and those who want to argue for a different conception of labour on 4.3; an economist interested in exposing the naiveté of a philosopher might examine 2.3, and an analytical philosopher wishing to point out the conceptual shortcomings of a continental philosopher might avoid wasting time reading the whole book (not that they would do that, anyway), and might want to only look at 3.3.

That said, of course, there is always the sovereign option available to every reader, which is not to read at all: I will never know, dear *Hypocrite lecteur, – mon semblable, – mon frère!*

1. Revolution: What Is the Web?

We are not slaves to the Web, but to our laziness, or more precisely to our sloth, which prevents us from seeing that in the battle between humans and the Internet it is humans who necessarily have the upper hand, as long as they want to. In fact, there is the evidence of millennia in which humans lived without the Web – not to mention the well-founded, albeit infrequent, suspicion that the Web, without humans, would be perfectly pointless: the Web needs us, our lives, our curiosity, our haste, our consumption. We have the upper hand and we don't know it. How can this be? Perhaps we have not yet understood what the Web is. For decades, in order to capture an essence that is probably still undefined, "the Web" has been called a number of names that are always different, partial, esoteric: the virtual world, collective intelligence, internet, big data, artificial intelligence ... In short, we are composing a collective poem made up of books, essays, articles, debates, and posts singing the thousand names of Vishnu, whose essence, however, re-mains unknown. The question of the thousand names of Vishnu is not just a theoretical problem. It is obvious that when faced with something whose essence is unknown, one cannot but have inadequate answers. The expression "wrestling with the fog" is probably the one that best suits the many reactions to this hidden god. While interpretations conflict, Vishnu, as he should, continues his course undisturbed. So what is the Web, in a nutshell?

Let us start from the one point on which all interpretations converge, namely that we are facing a revolution. However, this agreement is short-lived, as it is broken as soon as it comes to defining the nature of this revolution. Everyone agrees that it is not only a technological revolution, but also a political, econom-ic and social one[1]; and there is broad and almost miraculous consensus that this revolution is the fourth. The point is, though, that there is no agreement on the three that came before it, which is obviously no small matter. For some, it is the fourth industrial revolution.[2] For others, it is a far more radical revolution, affect-

---

[1] M. Bunz, *The Silent Revolution. How Digitalization Transforms Knowledge, Work, Jour-nalism and Politics without Making Too Much Noise*, Palgrave Macmillan, London-New York 2013; J. Lanier, *Who Owns the Future?*, Simon & Schuster, New York 2013.

[2] M. Zuazua, *The Fourth Industrial Revolution Will Change Production Forever. Here's How*, in "World Economic Forum", 18 January 2019. In this framework, the first revolution

ing first and foremost the way we think and see the world: it can be compared to
those triggered by Copernicus, Darwin, and Freud,[3] and amounts to another
wound to our narcissism: Copernicus removed the Earth from the centre of the
universe, Darwin demonstrated that we descend from apes, Freud revealed that
consciousness is merely the tip of an unconscious iceberg, and Turing showed
that machines can think much better than we can.[4]

would be based on water, the second on electricity, the third on information technology and
electronics, and the current one on a fusion of technologies that blurs the distinction between
physical, digital and biological. However, this analysis has two limitations. The first is that it
considers the current transformation as a variant of the industrial revolution, whereas it actually shakes its very foundations, starting from the relationship between production and consumption. The second is that it confines the event we are dealing with to a narrow chronology,
whereas it potentially has the same scope as humanity's transition from the status of hunt-
er-gatherers to that of breeders and farmers.

   [3] L. Floridi, *The Fourth Revolution. How the Infosphere is Reshaping Human Reality*, Oxford University Press, Oxford 2014.

   [4] It seems to me that this genealogy amounts to a confused and ambitious heraldry, for two
reasons. Firstly, it derives from Freud's narcissistic assertion, as he not only considered the
unconscious to be his discovery (which is obviously false), but above all, he presented it as a
wound to the ego, whereas, as everyone can see, it is exactly the opposite: it is indeed unbridled
narcissism and limitless egotism, where one talks about one's dreams over and over and one's
gaffes, slips, actions and omissions are viewed as full of meaning. This, however, is a problem
for Freud and his alleged revolution. As for Turing's revolution, it is simply untrue that we had
to wait millennia to realise that machines can think better than humans. Our remote ancestor
who first carved a bone to track the phases of the moon had understood, well before Turing, that
the human mind is much weaker and more unreliable than external memory. Those who – in a
collective process that is difficult to locate in space and time – gave rise to writing, calculating
devices, and calendars reached the same conclusion. The abacus is certainly not a sophisticated
instrument, yet it allows arithmetic calculations far beyond the reach of a normal human mind.
And the same goes for pen and paper: Euler used to say that his entire mathematics was concentrated in the pencil with which he made calculations. To those who object that if you give a
pencil to a monkey, the latter won't get much out of it, I reply that the same would happen if
you gave the monkey a Turing machine. Moreover, if the revolution really consisted in the
discovery – and gradual acceptance – that we are social animals endowed with language, it
would be a problem, because it would not be a revolution. "Today, we are slowly accepting the
idea that we are not standalone and unique entities, but rather informationally embodied organisms (*inforgs*), mutually connected and embedded in an informational environment, the *infosphere*, which we share with both natural and artificial agents similar to us in many respects.
Turing has changed our philosophical anthropology as much as Copernicus, Darwin and
Freud." (L. Floridi, *The Logic of Information: A Theory of Philosophy as Conceptual Design*,
Oxford University Press, Oxford 2019, p. 210). Did we have to wait for Turing? It seems to me
that Aristotle, along with a plethora of others, philosophers and non-philosophers alike, had
already said this when referring to Homer: "man is by nature a political animal, and a man that
is by nature and not merely by fortune citiless is either low in the scale of humanity or above it
(like the 'clanless, lawless, hearthless' man reviled by Homer, for one by nature unsocial is also
'a lover of war') inasmuch as he is solitary, like an isolated piece at draughts. And why man is

For me, instead, this revolution depends on the exponential increase in recording[5] that justifies the concept of "doc-humanity". Although more than one in two human beings does not yet have a mobile phone, there are 23 billion connected devices, more than three times the world's population. This connection produces more documents every day than all the factories in the world, and drives automation and production: a huge amount of acts, contacts, transactions and traces encoded in 2.5 quintillion bytes. Hence the answer to the question: what is the Web? The Web is the largest recording system that humankind has ever developed,[6] which explains the scale of the changes it has brought about.

But why is recording so important? And, more importantly, why is there so much of it now? As is often the case, the current revolution had a modest and accidental origin, and lies in a difference between analogue and digital that may easily go unnoticed. In the analogue world, first there is a message and then, if at all, there is recording. More often than not, recording does not take place – think of the billions of hours of radio and television viewing that have left no trace except a faint memory in the audience,[7] to say nothing of the trillions of words tossed about in history and prehistory. In the digital world, recording precedes communication,[8] and more generally every interaction with the network leaves a trace of itself, contributing to an enormous growth of documents – especially the documents we call "big data". The latter are in fact metadata or meta-documents, traces of our activity that we unintentionally leave behind, recording the place and date of creation of the document, who has seen the document, the reactions

---

a political animal in a greater measure than any bee or any gregarious animal is clear" (Aristotle, *Politics*, 1253a, 2–8). I would say that, if anything, the present confirms, with unprecedented evidence what we have known for a long time – much like a Viennese doctor found the behaviour of some of his patients to encompass the same tragic and deplorable phenomena already well known to the ancient Greeks. What is true of Oedipus is true of the Web: philosophers, both ancient and modern, teach that man is an animal with language and a social animal, and smartphones and social networks are the best proof of this.

[5] M. Ferraris, *L'esplosione della registrazione*, in *Filosofia del digitale*, ed. by L. Taddio and G. Giacomini, Mimesis, Milan-Udine 2020.

[6] Something that cannot be measured, which makes you think. This was already impossible at the end of the 20th century: R. Albert, H. Jeong, A.L. Barabási, *Diameter of the World-Wide Web*, in "Nature", 401, 6749, 1999, pp. 130–131.

[7] Consider that until the 1970s, the archives of television stations were systematically wiped – the BBC did this until 1978, and even now it has no obligation to keep records, according to http://en.wikipedia.org/wiki/Wiping [11/03/2022].

[8] B. Bachimont, *Between Formats and Data: When Communication Becomes Recording, in Towards a Philosophy of Digital Media,* ed. by A. Romele and E. Terrone, Palgrave MacMillan, Basingstoke 2018; cf. also B. Bachimont, *Patrimoine et numérique. Technique et politique de la mémoire*, Institut National de l'Audiovisuel, Bry-sur-Marne 2017. See also the works of Bernard Stiegler, e.g. *La société automatique*, Fayard, Paris 2015.

it has caused, etc. In a Jules Verne-like dream,[9] everything can become a document.[10] Moreover, thanks to digitalisation, the status of manipulable signs is not only given to numbers and letters, but involves sounds, images, and behaviour, completing the functional homology between the mechanical part of the mind and artificial intelligence. The fact that everything can be documented, and that this documentability is one – i.e. it follows a uniform standard – has changed the world with the same violence of a war in peacetime.

1.1. Hysteresis

So far I have described the technical aspect, which is also a keystone supporting the whole transformation. But the metaphysical question remains: why is recording so important? Why does it change things so much? In order to answer this question fully, I suggest talking not so much of a boom in recording, but rather of a boom in hysteresis, from *hysteron* ("after"). The term, which is widespread in various fields of natural sciences and technology, and which has had some sporadic application in the social sciences,[11] generally indicates a phenomenon of delay whereby a previous event affects the present, and must therefore be taken into account. Quite simply, it is the straw that breaks the camel's back, although not quantitatively greater than the previous ones, but capitalising on

---

[9] In 1895, two Belgian jurists decided to collect universal knowledge and catalogue it with a system that is still in use today, which they called Universal Decimal Classification, establishing the standard and format of bibliographic records, as well as the chests of drawers that collect them. They were Paul Otlet, son of a major industrialist, and Henri-Marie La Fontaine, who won the Nobel Prize for Peace in 1913. Their practice was guided by a theory, which in 1934 resulted in Otlet's *Traité de documentation*, based on the principle that everything can be documented. Otlet dreamt of the day when everyone would be able to access this immense archive from home: books, catalogues and documents would be accessible via television, which was taking its first steps in the 1930s. In this vision, in effect, the Mundaneum becomes a Panopticon. What Otlet did not foresee in his prophecy is that the generation of documents would be automatic thanks to the Web, and that it would be the users themselves who would produce documents through their searches and activities.

[10] S. Briet, *Qu'est-ce que la documentation?*, Édit, Paris 1951. In addition to Briet, a fundamental text for the modern definition of documents is, indeed, P. Otlet, *Traité de documentation. Le livre sur le livre. Théorie et pratique*, Mundaneum, Bruxelles 1934. See also R. Pagès, *Transformations documentaires et milieu culturel*, in "Review of Documentation", XV, 3, 1948; M.K. Buckland, *What is a "document"?*, in "Journal of the American Society for Information Science" (1986–1998), 48, 9, September 1998; R. Day, *The Modern Invention of Information: Discourse, History, and Power*, Southern Illinois University Press, Edwardsville 2001.

[11] I.D. Mayergoyz, G. Bertotti (ed. by), *The Science of Hysteresis*, 3 voll., Academic Press, Cambridge (MA) 2005.

their accumulation. What I propose, however, is a speculative generalisation, intended to show the whole range of meanings and, above all, of processes that are hidden in recording and in the resulting delay effect. As is often the case, what is closest and in plain sight ends up being hidden, or more precisely, goes unnoticed. The Web has had the merit of explicating the implicit and making the obvious manifest.

Hysteresis is in fact the premise of the most varied everyday practises and commonsensical concepts: from the idea that children inherit genetic traits (and sometimes a fortune) from their parents, to the power of exercise in the achievement of bodily or spiritual goals, to the principle that every promise is a debt and that those who know more are more likely to learn further. This applies to every field, but is particularly evident in the social world, which cannot exist without hysteresis – without documents, memories, traditions, language.

Suppose that while reading a word, you forgot the words that came before it. What you would read then could be neither true nor false ("false" in itself, taken in isolation, is neither true nor false). If, moreover, one considers that every language is a coded system, which receives meaning from the position its components occupy in the whole, "false", without a network, would not even make sense. Instead, in a network I not only know what "false" means (that which is opposed to true), but I also know that it can be translated as *falso, faux, falsch...* Likewise, I know that "but" does not mean "tub" even though it is made up of the same letters in a different order, etc. Take this principle to economics, kinship, law, morality, etc. and you will find it unchanged. This too is fairly obvious, but it conceals a principle that, again, is so obvious as to be imperceptible, namely the fact that without recording, the network would not hold, whether it be a physical network (requiring the cohesion of atoms, molecules, and higher components) or a natural, artificial, cultural, social, neural network.

One may ask why I adopt the proud or presumptuous (or at any rate unusual) name of "hysteresis" instead of speaking, quite simply, of "recording". My answer, besides calling attention to a ubiquitous and omnipotent phenomenon that yet goes unnoticed, consists in a paraphrase of what Spinoza said about the body[12]: no one has yet established what hysteresis is capable of in itself. In other words, experience has not yet taught any human what the laws of technology and society alone allow hysteresis to do and not to do, without the intervention of Consciousness, Ideology, or Intentionality. Underestimating its metaphysical scope has led us to appeal to magical entities, i.e. benign or evil beings to which we have superstitiously attributed power over our lives and that we can very well

---

[12] B. Spinoza, *Ethics*, book III, proposition 2, note.

do without. Perhaps these deities do not have such power on their own, but only as the hypostasis of a fundamental function, namely recording.

To stick to this paraphrase, no one knows the machine of hysteresis so well as to be able to explain all its functions, not to speak of the many attitudes observed in artificial intelligence that far exceed human sagacity, or the many actions that we perform in the social world without knowing what we are doing. These examples clearly show that hysteresis, by its very nature, can do a number of things that the mind itself marvels at. As far as the body is concerned, the mystery is still largely unsolved. As for hysteresis, the documedia revolution is allowing us to see things more clearly, showing that recording does much more than we commonly believe. To indicate this "something more", I have resorted to the rather exotic name of hysteresis. I will now analyse its four main functions: recording, iteration, alteration, interruption.

1.1.1. Recording: the network before the network

Let us begin with recording, i.e. the seemingly trivial element. In the *Statesman*,[13] Plato goes into a complicated description of the weaving of wool in order to determine the nature of what he calls *symplokè*, an interweaving or net. And even the most patient reader wonders why he goes to such lengths on matters that are, frankly rather uninteresting. However, little by little, the issue becomes clearer. Plato already spoke of this network in the *Sophist* as the ability to carry out cohesive, unconnected, sensible reasoning, and at the same time to connect being and non-being, thus giving life to the divine.[14] This network is the core of the art of leadership and any political technology that manages to resolve contradictions or smooth out differences. In other words, the network came long before the Web, or, conversely, the Web has come to fill a deep need in humanity. Let us reflect on this for a moment.

Today, when we speak of "the net", we immediately think of the Web, whereas a few decades ago we would have thought of something else – of football nets or fishing nets, of safety nets or, more appropriately, of what the Web is today, namely knowledge and society understood as a network that unites people. If we think about it, however, this latter conception of the Web suggests that there has always been such a thing, or at least that there is no humanity without a Web, or at least that a humanity without a Web is not destined to go far. In fact, the origin of humanity is not to be sought in some heaven-sent endowments that would have made us sociable – we are not too sociable even now – collaborative, and

---

[13] Plato, *Statesman*, 278b.
[14] Plato, *Sophist*, 259c–260b.

intelligent. It should rather be sought in the creation of networks that, by allowing transmission and exchange, have enabled the evolutionary leap that distinguishes us from non-human animals otherwise so similar to us: language, institutions, technology, economics, politics, history. All this cannot exist without a network, i.e. without a code, without an archive and memory, without the possibility of transmission from one individual to another and from one generation to another. That is why, as soon as it came into being, the Web appeared as the most natural thing in the world. And that is why, looking into today's network, which acts as a magnifying glass compared to those that preceded it, we can understand our origins with astonishing clarity and evidence.

Now, what holds the Web together? Recording. In fact, Plato also spoke about this: in *Timaeus*,[15] he theorised about the *chora*, which literally means "space", "place", "location" (it is still a very common name for Greek villages), but which in a metaphysical sense indicates an "intermediate kind", and consists in a great capacity for hysteresis. The demiurge builds the world with ideas, but these ideas need to have an extension in space and a position in time, and the *chora* meets this requirement. Since the appearance of the Web, first in computers, then in telephones and watches, and now in the environment, it is as if the *chora* has become, if not visible, at least conceivable – thus overcoming Plato's difficulty, that of a "third kind" that is neither sensible nor intelligible. The Web, and the recording it brings with it, is just this: a space in which the past is repeated by matter and remembered by memory, with such power that it transforms it into something qualitatively different. The commutation and profound solidarity of space and time are revealed in the circus whereby time cannot but become spatialized, just as a voice, which is time, becomes space as soon as it is written.[16] The fact that there is no being without time therefore means that there is no being without space and that space and time are two functions that derive from an original delay. This makes possible both the perception of the moment and memory, expectation, and the experience of space, which in turn is permanence in time.[17]

---

[15] Plato, *Timaeus*, 49a 6–51a 5.

[16] J. Derrida, *Ousia and Gramme: Note on a Note from Being and Time*, in *Margins of Philosophy*, trans. by Alan Bass, University of Chicago Press, Chicago 1982.

[17] The role of the imagination in the evolution of the human species has been rightly stressed (Y.N. Harari, *Sapiens: A Brief History of Humankind*, Random House, New York 2014). But why speak of imagination rather than hysteresis? What if the transfiguring power lies precisely in hysteresis, with a cycle and a scope that are not limited to humanity, but invest the entire genesis of the universe? Something happens, for example my fingers press some keys or the sun shines on the drying rack, and some time later, when my fingers have stopped typing and the sun has set, what I have written remains and the clothes are dry. In other words, the effects have outlived their causes. We find this normal, and if need be, we take an interest in the search for causes, and perhaps, if we are in the mood for metaphysics, we say that the real glue of the

I will discuss these metaphysical implications at length in 3. For the moment, let us concentrate on an examination that is directly functional to understanding what the Web is.

1.1.2. Iteration: from living labour to dead labour

Let us leave the physical world and turn to the sublunar world, without neglecting an obvious and inescapable fact: if something is preserved, it can be reused, and there is no need to start all over again. This is the principle underlying the processes of acculturation and capitalisation, and it is clearly what drives humans to spend time and energy building technological devices. The effort required to build a shovel is repaid by the resulting enhancement of our shovelling. This obviously does not only apply to shovels – we have always known this, but now we see it better than ever. Our very remote ancestor who chipped a flint to make a scraper did so with an eye to a long iteration of performance that would alleviate his toil. Likewise, our smartphone, which keeps track of our most frequent

---

universe is causality. That is not true: the real glue is postponement, perpetuation and transformation, which explains why memory loss is such an atrocious fate, why a society without documents amounts to so little, and why a quantitative increase in hysteresis, which is among other things what makes a hard disk work, has changed the world, and above all has revealed its true essence.

From an ontological point of view, hysteresis is therefore particularly important, because it allows the macroscopic level of cosmology and the mesoscopic level of anthropology to be united in a single principle that conforms to everyday experience, and thus to what physicists and philosophers call the "anthropic principle"; and it even applies to the microscopic level of subatomic particles, which contrast with many of our mesoscopic experiences, but which cannot do without hysteresis, the force that allows them to interact, and which physicists identify in what, using a technological metaphor, they call "gluons", the post-its of being, the UHU of the universe. However, these difficulties have an advantage, and put us on the right track. As we have just seen, Plato defines space as *chora*, a third genus, an intermediate kind between form and matter, and describes it as a capacity to hold and at the same time to take the forms of everything, thus also being able to iterate and alter. Plato identifies it with various metaphors – including, of course, that of writing – which converge to designate the *chora* as an invisible and amorphous medium, capable of keeping track of everything as absolute hysteresis. Here it is important to note the fundamental difference between the *chora* and the *archai* of the Presocratics and the numbers of the Pythagoreans. The latter were in fact physical principles (water, for example), or purely intellectual ones, such as numbers, called upon to explain the harmony of the world. Whereas the *chora* is precisely an intermediate principle, a "third genus" that stands between the sensible and the intelligible, whose only property is that of keeping track of the forms of ideas, transforming them into sensible entities (*Timaeus*, 48e–49a). This willingness to collect impressions means that the *chora* is never something actual, *always* remaining only in potency, in order to receive the forms that come from the ideas (*Timaeus*, 50b–51b). On closer inspection, Plato's description of space is also a possible solution to the aporia of time in Aristotle. If the latter did not consider it, it is because, as we know, he viewed relation as the

contacts, does not behave differently: it gives us, the distant great-grandchildren of that flint chipper, a chance to iterate and reduce effort. The iteration that results from recording (what is recorded can be repeated) therefore allows for capitalisation, it makes use of what has been recorded. Without going too far, let's think of the "copy and paste" process in document formation[18]: less effort is made, less time is spent and, at least on the machine's side, fewer mistakes are made.

Turing's secret[19] lies here after all: why struggle to understand what can be repeated without too much effort? Why waste time learning mathematics when

<hr />

weakest of the categories (because it "is in the lowest degree a substance and a real thing"; *Metaphysics*, 1088a 22–24 and 30–35). This very fact, which is apparently a limitation, nevertheless allows us to account for the status of space and time, which have the strange condition of being and not being at the same time, because a fully-fledged being or non-being would lead to insoluble aporias. It is necessary to find an intermediate dimension, which Aristotle defines as "relation", and which could be more appropriately indicated as "trace", because a trace has the characteristic of being, at the same time, present, insofar as it is a trace, and absent, referring to something not present, just as ashes are traces of wood that no longer exists. Here again, the writing metaphor seems quite adequate to account for this intermediate element between being and non-being, and this is demonstrated by the frequency with which it appears on the philosophical scene when dealing with these matters. As Kant rightly observes (I. Kant, *Critique of Pure Reason* [1781, A / 1787, 350 B]), if our thought, in representing a spatial line or a temporal course, were to lose the previous representations (points in space or instants in time), not only we could not construct that representation and the thoughts that accompany it, but the first and fundamental representation of space and time could not take place either. Therefore, a retention is necessary, but this refers to the representation of time and space, and not to space and time as such, so that "representation" here has the same ontological weakness as "trace" and "relation". And this weakness, far from being a flaw, makes retention an excellent candidate for solving the difficulties considered so far. The notion of "space-time" that has been imposed in the physics of the 20th century seems to unify and resolve the paradoxes by placing hysteresis, i.e. the possibility of keeping track, memory as matter and matter as memory, at the origin of the universe. Memory is born together with the universe. Time is not external to matter, it is a dimension *of* matter. As Mathematical Physics professor Lamberto Rondoni put it: "If matter had been *in* time, when the Big Bang happened, being unable to interact with anything, the universe would not have evolved. It would have been like an isolated atom, no matter how big or small. Instead, matter, space and time are one. There is no space outside the universe, just as there is no time before or after the existence of the universe" (personal communication).

[18] M. Carpo, *The Second Digital Turn. Design beyond Intelligence*, The MIT Press, Cambridge (MA) 2017.

[19] Cf. especially D.C. Dennett, *From Bacteria to Bach and Back*, Penguin, New York 2017. By the same author, see also *Can Machines Think?*, in *How We Know: Nobel Conference XX*, ed. by M.G. Shafto, Harper & Row, San Francisco 1985; *The Intentional Stance*, MIT Press, Cambridge (MA) 1987; *Darwin's Dangerous Idea,* Penguin, New York 1995; *The Evolution of Culture*, in "The Monist", 84, 3, 2001, pp. 305–324; *Who's on First? Heterophenomenology Explained*, in "Journal of Consciousness Studies", 10, 9–10, 2003, pp. 19–30; *Freedom Evolves*, Penguin, New York 2003; *From Typo to Thinko: When Evolution Graduated to Semantic Norms*, in *Evolution and Culture*, ed. by S. Levinson and P. Jaisson, The MIT Press,

you can simply iterate processes? The work of a copyist or an accountant is much more onerous, tedious and fallible than that of reproducing a digital document or using an electronic calculator. This reduction of effort is clearly analogous to the automation of processes introduced by steam and its successors in the industrial revolution, except that it can be exercised in much more complex and sophisticated domains, and requires much less material. Indeed, today we need much less material, and therefore also less energy, to repeat (to iterate) compared to analogue days. Should anyone object that the Web involves a high consumption of electrical energy, think of how much electrical, mechanical, organic energy and raw materials were needed to build and manage the Library of Congress in Washington with its 164 million volumes.

This capitalisation is already present in human and animal experience[20]: this produces facilitations whose result is the prevalence of the most frequent, and therefore statistically most probable, experience – a process identical to the auto-suggest function in our smartphones, or the training of search engines. This process, which transfers reinforcement learning to machines,[21] underlies dictation, self-driving vehicles, training for complex performances, and the perfecting of machine translation. On the one hand, this explains why automata are now able to carry out such complicated tasks, especially translation: we now have an archive of unprecedented dimensions, which is constantly being supplemented by human contributions. On the other hand, the advantages of this archive are also evident for human agents, who have a larger number of signs to manipulate. And nothing is thrown away. Even mistakes become useful as we instruct the machine *not* to repeat them, whereas an amnesic machine would be condemned to relapse.

Iteration, or at least iterability, since not everything that is recorded is iterated, is a distinctive element of the documedia revolution. And that is why, between the minimalism of the industrial revolution and the maximalism of the narcissistic wound in the heart of humanity, I believe that neither is the most correct genealogy of the fourth revolution. In my view, the latter is not a change in the way

Cambridge (MA) 2006, pp. 133–145; *My Body Has a Mind of Its Own*, in *Distributed Cognition and the Will: Individual Volition and Social Context*, ed. by D. Ross, D. Spurrett, H. Kincaid and G.L. Stephens, The MIT Press, Cambridge (MA) 2007, pp. 93–100; *Competition in the Brain*, in *What Have You Changed Your Mind About?*, ed. by J. Brockman, HarperCollins, New York 2008, pp. 37–42.

    [20] It is described with reference to the human psyche in S. Freud, *Project for a Scientific Psychology (1895)* in *The Standard Edition of the Complete Psychological Works of Sigmund Freud, Volume I (1886–1899): Pre-Psycho-Analytic Publications and Unpublished Drafts*, Stanford University Press, Stanford 1950, pp. 281–391.

    [21] H.S. Seung, *Learning in Spiking Neural Networks by Reinforcement of Stochastic Synaptic Transmission*, in "Neuron", 40, 6, 2003, pp. 1063–1073.

of thinking, but in recording and transmitting thoughts[22] and, I would add, actions. Rather than an ideological transformation in worldviews, we are witnessing a fourth *technological* transformation[23] involving the means of hysteresis and communication, following the transition from orality to writing, the one from scrolls to books, and the invention of movable typewriting. At a first approximation, this is where we should start: the transformation of the technologies of hysteresis. Indeed, we should not limit writing to the simple sphere of the alphabet,[24] but rather reveal its essence – the *recording of acts*.

1.1.3. Alteration: from praxis to poiesis

This brings us to a third element, in which we can see another aspect of hysteresis, namely its capacity to transform acts into objects. The transformation made possible by the Web does not consist so much in connecting already existing documents, but primarily in generating documents from acts that previously left no trace. Hysteresis, as well as iteration, can thus give rise to alteration, to emergence as the production of the qualitatively new. Of these alterations, the most notable is undoubtedly the creation of objects, and specifically social objects, i.e., documents that reify forms of human life: most of my pre-Internet acts, with the exception of the books I wrote, a couple of documents, and a few photographs, have disappeared. Today, in principle, everything is archived, although not by me. The moment they are connected to recording devices, acts lose their traditional character of tracelessness and produce durable objects, which as such can be capitalised (both in terms of primary accumulation and as reuse in terms of automation, profiling and monetisation). This was unthinkable in non-archived life.

Compared to this, transforming water into wine is a minor miracle, something futile and not very good for your health. A venerable tradition[25] holds that life is *praxis* and not *poiesis*, meaning that it does not produce objects and has no ends outside itself. But as soon as human behaviour can be recorded, *praxis* indeed becomes *poiesis*. In other words, it produces documents and value.[26] To be clear:

---

[22] In agreement with G. Roncaglia, *La quarta rivoluzione. Sei lezioni sul futuro del libro*, Laterza, Rome-Bari 2010.

[23] Which is intrinsically more powerful than any ideology: L. Althusser, *On Ideology* (1970), Verso, London 2007.

[24] Falling into a common error, one of whose most prominent and recent manifestations can be found in E.A. Havelock, *The Muse Learns to Write. Reflections on Orality and Literacy from Antiquity to the Present*, Yale University Press, New Haven and London 1986.

[25] Aristotle, *Nicomachean Ethics*, 1131a 10–1132b 9.

[26] Hence the problematic nature of Jonathan Crary's argument (J. Crary, *24/7: Late Capitalism and the Ends of Sleep*, Verso, London 2013), according to which sleep constitutes an "ultimate challenge" to capitalism, due to its unproductive nature. The emblem of this challenge

think of the singular reports that your smartphone presents to you, unsolicited. Events that would have previously vanished are turned into calculable objects, and what we previously considered a waste of time becomes an activity. There are no longer any "halls of lost steps" in the age of pedometers, and if zapping in front of the TV is a waste of time, surfing on a smartphone is not, at least for the Internet platform we use: the difference lies in the fact that zapping left no trace, surfing does. Just as Monsieur Jourdain discovers that he has been speaking prose all his life without knowing it, we find that we are mobilised and produce documents even in our sleep, as long as we have a smartwatch.

This keeping track generates big documents, which matter much more than money, because they are not only numerous, but of greater value. This is the reason why search engines are so successful, as it is not so much about providing information to users through advertising revenues, but rather about capitalising on the information gathered from the users' online activities.[27] To make an example, if the telephone connects two people who merely transfer information between each other, the Internet records actions, behaviours, communications; it transforms them, by fixing them, into objects, i.e. into documents.[28] In so doing, it generates an archive that is available in a small part to the user, who has a record of some of their actions, and mostly to the platform, which can compare the documents produced by one user with those produced by many, many other users. An internet platform has machines with sufficient computing power to find correlations and create profiles, and can therefore capitalise on these documents by accumulating, analysing and selling them.

This is why we are faced with a technological progress akin in scale to the passage from orality to writing. But with the documedia revolution, recordable practices are not only deliberate and intentional ones, but also unintentional ones, often tracked unbeknownst to the actors themselves, as is the case with biometric data. The fact that this happens automatically opens up unimaginable fields of

---

would be the attempt to reduce the night hours by means of the Znamya satellite. Sounds like something out of Jules Verne! Znamya was a Russian project that was abandoned twenty years ago because it failed: obviously, rather than depriving us of sleep, it is more convenient to make our sleep productive: and this is exactly what is happening with the collection of biometric documents through smartwatches.

[27] A. Casilli, *En attendant les robots – Enquête sur le travail du clic*, SEUIL, Paris 2019, which describes how Google represents itself: "Today, the Mountain View giant claims on the homepage [of reCaptcha] that its use serves to 'create value' for the company and boasts about the ability to 'exploit' human beings by transforming them into vectors of its information flow (*human bandwidth*)", p. 151.

[28] It would not be the first time that an increase in documents changes social reality; cf. M.T. Clanchy, *From Memory to Written Record: England 1066–1307*, Blackwell, Oxford 1993. But the impact, this time, was multiplied exponentially.

capitalisation. No sane person would have thought of recording how often humanity coughs – no pharaoh could have forced his scribes to keep track of this. But the fact that billions of human coughs can now be recorded has made it possible to devise (though, as far as I understand, not implement) the algorithm that recognises the specific coughs of Covid-19 patients, and to design a self-diagnostic app.

1.1.4. Interruption: the master and servant dialectic

Now, coughing brings us to the fourth dimension of hysteresis: interruption. Sooner or later, everything comes to an end, at least in our experience: empires and cats, human beings and carbon 14, and certainly sooner or later life on the planet, despite all our efforts first to damage it, then to save it. Our actions will not wipe out the planet – it will end neither with us nor with the life forms that will take our place. Indeed, as far as we can predict by analogy with other events known to us, it will eventually crash into the sun. But let us shift to an anthropic scale from the cosmic one. At the anthropic level, technology is certainly a great arsenal of means and tools, and is therefore "endless" in the sense of an almost unlimited availability of resources. It is also "endless" in the sense that devices, at least those that are not subject to programmed obsolescence, can last much longer than any organism. The chain which, in Constantinople, prevented access to the Golden Horn, as well as the walls of Theodosius, lasted, impassive, as generations of inhabitants, emperors, and strategists were born and died.

But what technology has above all else is an *end* in a third sense, that of purpose. Every technical device has a declared purpose: a chain is made to prevent access from the sea, walls are made to prevent access from the land, a knife is made to cut, a watch to tell the time, a telephone to talk and a smartphone to do almost anything. By having an end, the alleged master of the world (technology) reveals a weakness, its subordination, its inferiority. And in relation to what? To an animal, the human being, who, weaker than other animals, more dependent, more unstable, more unhappy or at least more dissatisfied, has had to acquire supplements to make up for its shortcomings. The most singular thing, when you think about it, is that this animal, like every other living organism, has no ends outside itself – in other words, it has no meaning, it only possesses that shadow of a purpose called "internal purpose": to grow, feed itself, reproduce (not necessarily) and die (necessarily). Devoid of an end in itself, this animal has given ends to the instruments it created, and then, according to a perverse and recurrent mechanism, has become convinced that it is their slave.

Consequently, today the absolute master would be the Internet. But we cannot be content with so little. No act would exist without an actor: without humans, technology would have no meaning or purpose. That is why I began by pointing out that we are only slaves to technology, or to capital, or to society, *if we want to be*; above all, in the case of technology, this only holds for those who *believe* they are slaves, and perhaps want to be, because it is convenient to give up control, and it gives much relief to blame others for one's shortcomings.[29] But the fact remains that voluntary slavery is a free act. Mind you, there is nothing Promethean about this! We are the lords of technology precisely because we are not the lords of nature, but rather disadvantaged animals in need of supplements. The planet has no need of humanity and, strictly speaking, not even of life, precisely because its fate is determined by sidereal destinies that do not concern us. In any case, when this fate is fulfilled, we will have been gone for millions of years. In other words, we can very well think of a world without humanity, but not of a humanity without a world. And with that, the line is immediately drawn, with a peremptoriness that admits no reply.

Humans do not have to set limits to their progress, not because they have inherited lordship over creation and must moderate their exploitation for some moral reason. Quite simply, humanity is defined by very onerous limits: our lifespan, the environmental conditions favourable to us, the balance between needs and resources. It is enough to go beyond those limits for the tale of humanity to be over in an instant. Therefore, rather than *setting* limits, we must recognise the limits we *do* have and be very careful not to cross them. On the other hand, these very limits and the interruption that lies in them define our specific attitude. The human animal, like every other animal, is subject to the urgency of metabolism and the inevitability of death, the awareness of which forms the backbone of our world. The fact that there are hospitals and pension schemes, that cheques cannot be cashed after a thousand years, that we do everything we can to prolong our life ... All this only goes to show how an insurmountable limit hangs over each of us, and manifests itself not only in some *meditatio mortis* (that, we rather spare ourselves), but in every single act of our lives.

From this point of view, in accordance with a principle that I will illustrate at length in 2, one must conceive of the interaction between humankind and the Web as a relationship between souls and automata. The latter provide souls with increasingly efficient mechanisms of recording, iteration and alteration. But souls provide the material for automata, which, as I said, are intended for recording human life forms and what is relevant to them (atmospheric, ecological, economic data and so on). In return, souls give automata their sense. In this relationship, therefore, the soul is in no way subordinate to the automaton, because the

---

[29] I. Kant, *Answering the Question: What Is Enlightenment?* (1784), Penguin, New York 2013.

latter would be pointless without a soul. And the soul, in turn, is driven by what the automaton does not have: an end, in the sense of death. A soul is an organism, subject to those irreversible processes that give meaning and purpose to the history of each of us and to humanity as a whole. Meaning is direction, not rotation, and direction comes from interruption, from the fact that sooner or later hysteresis has a crisis, which consists in the fact that iteration meets the radical alteration of death. If humans are autobiographical animals, it is because they know that there will be a time when all that is left of them are traces. Conversely, to imagine that an automaton, which as such is a very powerful archive, could be concerned with archiving itself is as ludicrous as the idea of automata concerned with setting up a pension system.

Now, right here, far from human subordination to technology, to the system, or to capital, we see the foundation of human *mastery* over technology. I use this antiquated and unpleasant term because, referring to what I said in the Prologue, I am recalling the famous figure from the *Phenomenology of the Spirit*. As we know, following the Haitian slave revolt of 1802, Hegel described the confrontation between the lord, the feudal master who has no fear of death because he is educated in the craft of arms, and the slave, who fears death, and therefore the lord. However, the slave eventually learns to dominate the land and emancipates himself from the lord and his feudal sword, which has become useless and anachronistic in a bourgeois world. *De nobis fabula narratur*: the worker, who controls the machines, is destined to have the upper hand over the capitalist, who merely owns them and does not know how to use them. The master/slave dialectic, as a constitutive figure of revolutionary movements and in general of the progressive ideals of modernity, was based on the assumption that the slave would prevail. Today, exactly the opposite is the case. It is postulated that the master, usually some highly intelligent capital using some very powerful technology, always wins, more or less like the house in casinos, and the only task of reflection seems to be to shed a few tears about it. Alternatively, one may choose not to cry at all, acting as an *esprit fort* and looking at the inevitable with the same resigned strength with which the Athenians watched the Spartans parade by at the end of the Peloponnesian War.

From the master/slave dialectic we have thus moved on to another figure, the unhappy consciousness. However – this is the crucial point – at a time when increasing automation means that machines no longer need workers, not even in a control capacity,[30] this does not mean in any way that machines are taking over.

---

[30] As Marx argued in the Fragment on Machines, one imagines that it can only manage machines. See K. Marx, Fragment on Machines, in *Grundrisse, Foundations of the Critique of Political Economy*, "The Chapter on Capital", Penguin, New York, 1973.

The machine empire is a simple dystopia that can be demolished by a simple consideration. Artificial intelligence may appear to be a last-minute discovery, but in fact humankind has become human *through* technology, which is artificial intelligence. Our ancestor who carved a bone to mark the phases of the moon for predictive purposes was equipping himself with artificial intelligence. Other forms of artificial intelligence include abacuses, books, pens, notebooks, and all the ways in which we have tried to overcome the limitations of natural intelligence. One may conclude that the Golem, the Replicant – in short, an artificial intelligence that strangely enough has the same wishes and flaws as natural intelligence – will eventually come to power. Literature is not the only field that has been very imaginative on the subject: philosophy has also got involved and, what is worse, so have experts in artificial intelligence. We all fear that some Superintelligence will corner us or put us in the attic, if not turn us into slaves – which may even be convenient, because then we would no longer be responsible for anything, we would simply be doing the Superintelligence's bidding. Now, if we take the description of Napoleon in Manzoni's *Cinque Maggio*, we can see the picture of a Superintelligence that goes right and left, commands, calculates, faster than anyone else, unwavering, swift, and so on. Well, if we were to put a computer in the place of a human agent, that which arouses admiration would generate fear.

The point, however, is that such a thing is simply impossible. If we are not to succumb to these superstitions, we must understand that machines exist only in view of human needs, hopefully allowing for the advent of a preferable human condition – provided that humans are able to assert themselves, first and foremost, through self-awareness. Now, as I said, this awareness must come through a different consideration of consumption, and of the death it seeks to postpone. Too often confused with bulimia and mental weakness, consumption, if properly understood, is indeed the human condition. The ability to produce without fail, and with a degree of precision and patience inconceivable for a human, is within the reach of every machine. But *consuming is only possible for humans*, who in fact exaggerate, spend too much, get fat, obsess over goods and kill themselves over lies, which would be inconceivable both for a machine and for a non-human animal. It is vital we understand this, because our happiness depends on it. To do so, however, we must acquire a different understanding of the Web.

1.2. A Copernican Revolution

Let us start from the most important point, a technical detail fraught with many consequences: the Web brought about a Copernican revolution whereby communication *follows* recording. It may seem like nothing, but it subverted the hierarchical order between thought, speech and writing that pervades common sense and is sanctioned by the authority of Aristotle[31]: the assumption that we have ideas that can be expressed in words and subsequently imprinted on a medium, be it the interlocutor's mind or mechanical and non-organic surfaces, such as wax tablets and the like. Have we truly understood this reversal? Have we weighed its consequences? I would say not.

As I was saying, there is much talk of a "revolution", but one may have the impression that the revolution is only apparent, much like the Kantian one. Now, the latter, to all intents and purposes, was a Ptolemaic counter-revolution that put man, as the I think, at the centre of the universe. In our case, it may seem that today's revolution amounts to extending and transforming the sphere of information, which has emerged from newspapers, televisions and libraries, into an all-encompassing horizon. This broader information field has morphed the world in which we live, generating an info-sphere,[32] i.e. a purely quantitative growth of a phenomenon characteristic of the 20th century, namely the mass media age.

Well, let me be clear from the outset. The Web certainly is an infosphere, but this is only the smallest part of it; the infosphere rests on a *docusphere*, i.e. on documents that record human actions without necessarily carrying information, and the latter rests on a *biosphere*, i.e. on the world of life, which – this is the real revolution that the Web has brought to the world – is now in principle always documentable. Life comes out of the realm of silence and oblivion to which, for lack of adequate technical tools, it has been confined since the beginning of the world, leaving us only testimonies that are either very deliberate and rare (poems, sacred texts, solemn contracts, pyramids and triumphal arches) or accidental and not very expressive (chipped plates, arrowheads, wine amphorae). However, this does not mean that the Internet is information *per se*, because in that case we would be confusing the Web with Wikipedia. We must look for for an alternative framework, one that is, in my view, more faithful to the grain of reality, and the first step is to move from a Ptolemaic Web, which sees itself as information and communication, to a Copernican Web, which sees itself instead as hysteresis and capitalisation.

---

[31] Aristotle, *On Interpretation*, 16a 3–4: "Spoken words are the symbols of mental experience and written words are the symbols of spoken words".

[32] The term is due to A. Toffler, *The Third Wave*, William Morrow, New York 1980, and is found again, nine years later, in the science fiction novel *Hyperion* by Dan Simmons. See now L. Floridi, *Infosfera. Etica e filosofia nell'età dell'informazione*, Giappichelli, Turin 2009.

1.2.1. From the Ptolemaic Web to the Copernican Web

The identification between Web and infosphere, based on a Ptolemaic assumption, claims that the Internet is made up of information and thoughts. But is it really thoughts and information that we exchange? And above all, has the revolution enhanced the hysteresis of thoughts and information, or has it rather multiplied the number of acts and behaviours (significant or not, informative or not, conscious or not), that can be recorded? If we admit, as seems reasonable, that a thought is such – in the philosophical and not the psychological sense – only if it is true,[33] then even information is only such if it is true. If information is false it is not information, any more than a fake diamond is a diamond. Now, to speak of the "infosphere" involves an essential ambiguity regarding the concept of information.

It is clear that in classical information theory,[34] "information" is only the computation of the possible states of something: it is very close to hysteresis, which is the presupposition of computation. Unfortunately, however, a confusion arises between information in the ordinary sense of the term and in the technical-informatic sense of it. Of course there is a need to find a strong and clear term for everyone, but this clarity comes at a very high cost. Because it is one thing to talk about information available to everyone, quite another to say that the book of the Web is written in characters accessible only to a few humans aided by very powerful automata. In this sense, there is no shortage of information: any futile or insignificant behaviour I engage in online – whether it be watching videos out of boredom, posting randomly on Twitter, or "liking" things on Facebook – is information for the few who know how to use it to profile me. This is true, but it generates a huge gap between those who can only read information in plain text and those who can *interpret* documents: a difference akin to than that between illiteracy and culture, or between alchemy and chemistry. "This much is known: For every rational line or forthright statement there are leagues of senseless cacophony, verbal nonsense, and incoherency."[35] The infosphere real-

---

[33] G. Frege, *Der Gedanke. Eine Logische Untersuchung* (1918–1919), in "Beiträge zur Philosophie des deutschen Idealismus", I, pp. 58–77.

[34] C.E. Shannon, W. Weaver, *The Mathematical Theory of Communication*, University of Illinois Press, Urbana and Chicago 1949.

[35] J.L. Borges, "The Library of Babel" in Id., *Collected Fictions*, Penguin, New York 1998, p. 114. Thus, there is an essential difference between the infosphere and the theories of absolute knowledge, which, surprising as it may seem, favours the latter, as they prove to be more realistic in their idealism. Aristotle's *noesis noeseos*, the single intellect of the Averroists, Hegel's Absolute Knowledge and, more recently, Marx's General Intellect, Pierre Teilhard de Chardin's noosphere or Frege's or Popper's "third realm" of thought all describe an infosphere. However, except for Marx's General Intellect, in all the other examples we find the description

ises not the General Intellect, but the library of Babel, on the basis of four main characteristics.

The first is virality. Compared to the traditional media world, the documedia world is characterised by a dizzying growth of interconnections that exponentially multiply the number of sources, as we go from a few radio and television channels and newspapers to billions of social network users. The quantitative becomes qualitative: when half of humankind is on the Internet, the world becomes something else, even though everything on the Web has remote antecedents. Indeed, documediality relies on structures, such as writing, that have existed for thousands of years. Through it, it has become very easy to reproduce and globally disseminate all sorts of hysteresis, from images to sounds, from thoughtful writings to the pangs of the soul. The process is fuelled by a myriad of actors with no special skills or infrastructure and by a stunning increase in the speed of transmission, which has no common measure compared to the past.[36]

A second character is persistence. Documents are much more accessible now than ever before: yesterday's newspaper was once the symbol of the ephemeral; now documents (which are not only, or even mainly, news, but *tranches de vie* and comments, blessings and curses, likes and dislikes, prayers and threats, and, above all, photos that, on their own, do not say much) float on the Web without generally notifying the date of the event to which they refer, determining an eternal return of the post, whereby an event seems to repeat itself because of its documedia occurrences. Of course, a technological breakthrough could well wipe it out like a footprint in the sand, but for the moment it is there, always at hand.

A third character is mystification. It is very easy to create fictitious identities, just as it is easy to cut and paste to obtain plagiarised copies which, precisely because of the simplicity of the operation required, are not perceived as such, and above all enjoy free citizenship in the infosphere. Previously, instead, anonymity

---

of a state of law, an ideal condition of thought, and not an actual state of affairs. The underlying assumption is that the individual and limited thoughts we have depend on a universally valid general domain: $2 + 2$ equals 4, not because I think so, but because it is so. In turn, encyclopaedists try to provide a worldly version of this intellect in the form of positive knowledge, but they are well aware that, as Borges said, even the most modest of encyclopaedias is more cultured than the most learned of its compilers. Yet this is just a figure of speech: the compiler knows something, however little or much; the encyclopaedia knows nothing.

[36] For example, on 17 July 1821, Alessandro Manzoni read the news of Napoleon's death in the "Gazzetta di Milano" of the day before, i.e. more than two months after the event. Radio and television cut down such time delays, but they are still synchronous communications: if we are not in front of the screen when the news is broadcast, and of course if we are not browsing the Web, the information falls on deaf ears.

would automatically discredit a piece of news and creating a false identity meant facing practical difficulties and legal sanctions.

The fourth is fragmentation. Broadcasting, whereby a source reaches a large number of recipients, is fragmented into a multiplicity of sources generating much smaller communities of reception and discussion. This results, also through the filter of algorithms, in dedicated sources that say what users and the narrow-caster want to hear, contributing to the production of alternative truths and reso-nance chambers. Moreover, the challenge posed by open, voluntary and free-flowing information forces traditional information channels, already disad-vantaged by the dubious equation between "alternative source" and "truthful source", to a hyper-production that often dispenses with source control. This opacity constitutes the conversational correlate of ideological fragmentation and psychological atomisation.

If this is the infosphere, Candide would say, just imagine what the disinfos-phere is like. Born as the dream of an academy of scholars, the Web has turned into an arena where the four famous "conversational maxims"[37] are systematical-ly violated[38] – a violation that leads to post-truth. The first maxim is that of qual-ity and reads: "Be genuine, and provide truthful information to the best of your knowledge". John says that Jack spies on him, but it is not true. A simpleton would say that he is demented or a liar, or both; but a man of the world would argue that John is expressing an alternative truth. The term "alternative truth" is the tribute that vice pays to virtue, but it is also a formally radical chic construct, which raises the suspicion that the truth is fascist and dogmatic, and claims to emancipate while deceiving.

The man of the world might very well have learned this trick in some good university in which liberal and naive professors, like Rath in *The Blue Angel*, preached the farewell to the truth in the name of justice[39]: solidarity is more im-portant than objectivity, democracy is more important than truth,[40] and especially the given is nothing but a myth.[41] There are at least two weaknesses to this ideal-

---

[37] H. P. Grice, *Logic and Conversation*, in: *Syntax and Semantics*, Vol. 3, Speech Acts, ed. by P. Cole and J. L. Morgan, Academic Press, New York 1975.

[38] M. Ferraris, *Postverità e altri enigmi*, il Mulino, Bologna 2017.

[39] G. Vattimo, *A Farewell to Truth*, Columbia University Press, New York 2011.

[40] R. Rorty, *Solidarity or Objectivity?* (1990), in Id., *Philosophical Papers*, vol. I, *Objectiv-ity, Relativism, and Truth*, Cambridge University Press, Cambridge 1991, pp. 21–34.

[41] What underlies the critique of the myth of the given is the idea that, in the impossibility of accessing things in themselves, ontology is reabsorbed by epistemology. Like every idea, good or bad, it has a strong and a weak version – though in this case, being the two faces of a bad idea, they are both wrong. The strong version is represented by continental philosophy, in the line of thought that leads from transcendental idealism to the radical philosophies of the second half of the 19th century, and from there to postmodernism. Here we are dealing with

istic defence of lies in democracy, or, if you will, two precious lessons that can be drawn from post-truth. The first is that the audience addressed by philosophers is already trained to worship the truth, but must be sensitised to respecting solidarity and otherness. The second is that, after offering an involuntary ideological assist to populists and depriving intellectuals of their only weapon (the pride, if not the courage, of the truth), postmodernists did not consider that a democracy without truth is not a democracy, and that if solidarity prevails over objectivity, this produces an uncontrollable drift (after all, the mafia or amoral familialism are notable examples of the prevalence of solidarity over objectivity).

---

idealism: since there is no being independent of knowledge, the rationality or irrationality of the world is the result of the constructive activity of our conceptual schemes and perceptual apparatuses. The weak version is represented by analytic philosophy, in the line of thought that leads from positivism to logical empiricism, and from there to cognitivism. Here, however, we are dealing with scientism: philosophy has exhausted its cognitive role, which is now entirely fulfilled by the natural sciences. In relation to the latter, philosophy has a purely reconstructive function, i.e. it is called upon to reconstruct the logic of scientific research. Both part of the transcendental fallacy, these two versions of philosophy embrace the basic assumption of transcendentalism, according to which the order we find in the world depends on the order that reason gives to the world, or at least to the part of it that is willing to be influenced. It is the basis of Kantian philosophy and even more so of the rationalisms that preceded it from Descartes onwards, and like these it is a rationalisation of intelligent design. Transcendentalists, and above all Kant, do not commit themselves to the divine creation of the universe, but they presuppose the creationist horizon when they maintain that the order of phenomena depends on the minds that observe them. In its classical formulation (W. Sellars, *Empiricism and the Philosophy of Mind*, 1956), the critique of the myth of the given sounds like this: if I want to be able to use data as the basis for a theory, then these data are not independent of the theory. If, on the other hand, I demand that data be independent of theory, then I can never use a single piece of data to confirm or disprove a theory. As a result, if I want an ontology to serve an epistemology, then the epistemology must construct the ontology. Well, the argument does not hold, because an ontology constructed from an epistemology is an epistemology and not an ontology, and an epistemology that knows an epistemology is not an epistemology either. By claiming that the given or datum is a myth, we implicitly introduce an even more insidious myth: if you want an ontology to serve an epistemology, the epistemology must inform the ontology; but an ontology informed by an epistemology is an epistemology – a strange knowledge to which, curiously enough, there is no corresponding being. But an epistemology informed by an epistemology is not even an epistemology: it is reflection or introspection. Put in these terms it seems absurd, but just think of those who claim that "there is no subject and object, there is only the relationship between subject and object". Well, these philosophers are saying, with the worst argument in the world, that oysters do not exist, and in fact *they* do not exist either, but there is only such a thing as "eating an oyster". Enjoy your meal – but it remains unclear why, shortly afterwards, they, and not – for example – the oyster, will be asked to pay the bill (on the worst argument in the world, cf. D. Stove, *Idealism: a Victorian Horror Story (Part Two)*, in Id., *The Plato Cult and Other Philosophical Follies*, Blackwell, Oxford 1991, pp. 135–178. I have simplified the subject for reasons of space. An excellent exposition

The second maxim, about quantity, goes: "Do not be reticent or redundant."
Aware of the fact that the best reticence is redundancy, post-truth engages in the
industrial production of chit-chat or rather, to use a vulgar expression, bullshit.[42]
In terms of quantity, post-truth is the child of documediality: there is a ceaseless

---

of this argument can also be found in J. Franklin, *Stove's Discovery of the Worst Argument in
the World*, in *Philosophy*, 77, 2002, pp. 615–662). If we reflect on the metaphysical implications of the fallacy, we realise that it involves a blind and perhaps unconscious belief in the
existence of a spirit independent of matter, capable of producing representations and, through
them, things. Thus, if the transcendental fallacy were valid, not only would Newton's laws not
have been true before Newton discovered them, but also the objects to which these laws refer
would have a highly problematic existence. This is what Heidegger claims in *Being and Time*
(Wiley-Blackwell, Hoboken (NJ) 1978, p. 270): "Before Newton's Laws were discovered they
were not 'true'; it does not follow that they were false [...] To say that before Newton his laws
were neither true nor false, cannot signify that before him there were no such entities as have
been uncovered and pointed out by those laws. Through Newton the laws became true; and
with them , entities became accessible in themselves to Dasein. Once entities have been uncovered, they show themselves precisely as entities which beforehand already were." More correctly, one should say that before Newton's laws were revealed, these laws were unknown, but
this is a truism that is out of place in a book of philosophy. I imagine a very natural objection:
how do we prove that the objects of knowledge exist independently of the subjects of knowledge? My answer is very simple: how do we prove that the objects of knowledge do not exist
independently of the subjects of knowledge? We have endless proofs of the contrary, all of
which are based on the fact that to be is to be recorded. In a particularly conspicuous way, social
objects, such as debts, credits, marriages, and wars, exist because they are recorded acts, in
people's heads or better still (it is safer) in documents. Recording is therefore the great hidden
power that makes the social world possible – just think of the little consequence of a wedding
between amnesiacs, without documents or witnesses. This recording has indeed an unprecedented and underestimated power: this is shown by the fact that an explosion of recording like
the one produced by the Web has completely revolutionised our world, even more so than the
industrial revolution. My conclusion, then, is that "nothing exists outside the text" is an obviously false and in itself hyperbolic thesis (there are so many things outside the text, even though
the text is expanding as fast as automation). Instead, to claim that "nothing *social* exists outside
the text" is a solid truism that is not given sufficient attention. Inversely, what can become text,
or more precisely can be recorded, like our heartbeats, our purchases, our contacts, becomes
social, can be bought and sold, accumulated and exchanged. Let's line things up. Physical objects, for example the Tyrannosaurus Rex or the table in front of me, are fine without my concepts. Social objects, such as the Tyrannosaurus Rex in a cartoon or the table in front of me if I
am a judge, also exist without concepts, as long as they are recorded, in the cartoon and in the
records of the clerk's office that record the judge's decisions. Both in the case of natural and
social objects, concepts and knowledge are very useful, and having spent my whole life studying I would be the last to deny it. But they are not ontologically constitutive. The table also
exists for those who are unaware of its existence, and one can be condemned for excellent
reasons even if one does not understand them or are hard to find ("Someone must have slandered Josef K."). Recording is ontologically constitutive, at least as far as social objects are
concerned.

[42] H. Frankfurt, *On Bullshit*, Princeton University Press, Princeton 2005.

production of documents on the Web, and each receiver can in turn become their transmitter and the re-transmitter (the bullshit reaches its critical mass thanks to its re-tweets, the forwarding that heralds virality). Is this production systematic and intentional, as claimed by the Marxist doctrine of ideology, according to which those who control the means of production control the ideas? No: behind bullshit there is no great puppeteer, no intelligent and strategic capital. What we inadequately call "capital" is a documedia system, that is, the union between the constitutive power of documents and the spreading power of the media. Hence a second teaching of post-truth: let's try to explain what is happening with something other than conspiracy theories. In particular, we should rather notice the (very accidental and not very intelligent) convergence between a technological organisation and a natural human weakness: we might understand something more about the world we live in.

The maxim of relation is: "Be pertinent". But pertinence is a rare, burdensome and obnoxious quality, whereas a hoax is mediagenic and prone to going viral.[43] It is gossip, heir of the fairy-tale, the fantastic, and the futurist words-in-freedom. But once again, postmodernism contributed too, by claiming that the world depends on our language and our conceptual schemes. Which, if said at a university seminar, might make you smile or think, but outside the classroom can justify the idea that things are the docile subordinates of words: if you say that there are weapons of mass destruction in Iraq, then there are weapons of mass destruction in Iraq, and if you say, on 1 May 2003, that the war in Iraq is over, then it's over. These hoaxes are much more elaborate than the claim that a given restaurant serves human flesh, but at the same time they manifest the human lordship over language that philosophers and non-philosophers were so passionate about in the 20th century, and of which now (the third teaching of post-truth) we recognize the vanity.

Finally, the fourth and final maxim is: "Avoid ambiguity" and fashionable nonsense.[44] However, people like nonsense – this is an unquestionable truth. It is neither true nor post-true that humans naturally seek knowledge, as Aristotle claimed: rather, they hate the potential consequences of their *lack* of knowledge, and they love being right, which is a very different thing. In short, although the truth sooner or later comes out, the search for truth can hardly be carried out with bare hands and no cultural training. I will return extensively to the factual char-

---

[43] U. Eco, *Fake and Forgeries* (1987), in "Versus", 46, thematic issue *Fake, Identity and the Real Thing* (now in Id., *The Limits of Interpretation*, Indiana University Press, Bloomington 1994); J. Derrida, *History of the Lie: Prolegomena*, in Id., *Without Alibi*, Stanford University Press, Stanford 2002; M. Doueihi, J. Domenicucci (ed. by), *La confiance à l'ère numérique*, Éditions Rue d'Ulm – Berger Levrault, Paris 2018.

[44] J. Bricmont, A. Sokal, *Fashionable Nonsense*, Picador, New York 1998.

acter of truth in 3, in a more specifically speculative context. For now, II will move from the analysis of superstructure to that of structure, that is, from post-truth to documediality.

From what has been said so far, it is clear that post-truth is the inevitable consequence of the quantitative growth of exchanges on the Internet.[45] Understanding the nature of post-truth helps us understand the documedia revolution just as Balzac's and Dickens' novels tell us a lot about the industrial revolution. According to the Oxford Dictionary, post-truth is an unproven claim that relies on emotion rather than reason. However, this definition is insufficient, because in this case even serious religious faith, if associated with doubt, would be a form of post-truth, which it is not. Those who defend a post-truth are convinced that they are rationally and factually right, and not merely manifesting a hope or belief. More precisely, we are dealing with a privatisation of belief. Before, humans gathered in vast communities that shared a belief in paradise for the virtuous or the hardworking, in the inherent inferiority of the enemy of the month, in the inherent superiority of their god, and in the existence of witches and vampires. Now those beliefs have fragmented to some extent, but new totalities of practice and faith have been reconstructed on the Web.

One must therefore beware of the argument that if (1) lies have always existed then (2) post-truth is nothing new. I find it hard to deny that chit chat, nonsense and idle words have been around as long as language itself. Speech was not given to manifest thought, as optimists claim, or to conceal it, as pessimists claim, but to surrogate acts and discharge impulses – therefore, idle talk is the true original function of language. If this is so, bullshit and nonsense must have circulated even in the Altamira cave.[46] Coming to more recent times, many studies have addressed the destabilising effects generated in the Middle Ages by the shift from truth as trust in personal and feudal loyalty to truth as something guaranteed by documents.[47] Nonetheless, the mass media, especially thanks to social networks, are now producing a superfetation of written texts, which are therefore permanent and equally non-biodegradable, at least in the medium term, like plastic bags in the Pacific. Their authors are of the most varied kind: journalists,

---

[45] Cf. G. Maddalena, G. Gili, *Chi ha paura della post-verità? Effetti collaterali di una parabola*, Marietti, Bologna 2017; Ferraris, *Postverità e altri enigmi* cit.; A.M. Lorusso, *Postverità*, Laterza, Bari-Rome 2018.

[46] Lies have always been there, from the Trojan horse to Constantine's donation. See M. Bloch, "Reflections of a Historian on the False News of the War" (1921), translated in *Michigan War Studies Review*, Volume 2013, 20th Century, World War I, 2013-051. What is new is that the bearers of alternative truths are often convinced that they are not only alternative, but also true.

[47] R.F. Green, *A Crisis of Truth*, Pennsylvania University Press, Philadelphia 1999.

philosophers, amateur writers and information professionals. None of this could have taken place without the documedia revolution, which is why, contrary to those who say that there is nothing new under the sun, I assert that post-truth is the characteristic feature of our time for reasons that are technological rather than ideological.

Indeed, the postmodern idea that there are no facts, only interpretations, has found a technical expression in what we so obscurely call "populism", whereas it is simply one of the legitimate if undesirable outcomes of democracy, namely ochlocracy. In fact, the practice of political leaders, who in this respect inherit the detestable tradition of lying for the greater good, manifests a disregard for truth that would have delighted a postmodernist: or, more accurately, would have made them unhappy by persuading them that it was not a good idea to say that truth is merely the expression of the will to power and that it must be replaced by solidarity. The fact that this mass diffusion of the reduction of facts to interpretations is taking place as a result of a technological transformation is of the utmost importance, because it answers an inevitable question: have the ochlocrats read postmodern authors? No. What I mean is that they did not need to read Nietzsche to implement the world of post-truth any more than Henry Ford needed to read Marx to implement the world of industrial capital, or Hearst needed a master's degree in McLuhan studies to understand the importance of the media. All of them were bound by a technical horizon: manufacturing in the case of Ford and Marx, the media in that of Hearst and McLuhan, documediality in the arc that leads from postmodernists to ochlocrats. This is a mode of production that generates facts and factoids through interpretations, simply because it exponentially multiplies the number of documents in circulation.

Once again, what matters is not the superstructure (information and misinformation), but the structure, i.e. documediality, the technical, social and economic system that makes it possible. And this is why, leaving behind the Ptolemaic Web of information, I propose moving on to the Copernican Web of hysteresis. The prevailing hysteresis of acts suggests that between the empyrean of the infosphere and the fertile lowlands of the biosphere there is a docusphere, a realm of inscription of facts, hysteresis of acts and connection of documents that need interpretation much more than they bring knowledge or even collective intelligence. On the Internet, long before exchanging information and knowledge, humans record actions fixed in documents that, in a documedia environment, become ubiquitous and automatic. These documented actions, if the tools are available, can become information, but as such they are a closed book whose code we do not know: they are like Enigma before Turing cracked it. Talking about the "infosphere" and "information" as in the Ptolemaic Web evokes controversial and ancient ideas, according to which in order to act we need to know,[48] and we only really know what we ourselves have done, namely society and not nature.[49] Instead, the Copernican Web offers a completely different image: the competence needed to act does not require conceptual understanding or prior knowledge any more than a theory (and a correct one at that!) is needed to pay for beer at the pub; and, although I have no doubt that society is the work of man, and not of God or nature, I have a well-founded suspicion that I am not that man, so much so that some aspects of society seem to me no less obscure than celestial wonders and mysteries of nature.

Let us reverse the Ptolemaic way of thinking and come to the Copernican Web. To sum up: 1) the Web is above all hysteresis, and not only communication; it functions not as a television, but as an archive; 2) it is action and performativity rather than information, it does not merely accumulate knowledge, but also defines a space in which social acts such as promises, commitments, and orders take place; 3) it is real rather than virtual, i.e. it is not merely an immaterial extension of social reality, but instead constitutes the elective space for the construction of social reality[50]; 4) it is mobilisation rather than emancipation, i.e. it does not provide immediate liberation, as was believed when the Web took its first steps, nor is it simply an instrument of domination, in accordance with the view prevailing today, but rather an apparatus that mobilises, i.e. capitalises on and promotes further actions; 5) it is emergence rather than construction, in the sense that it is not someone's deliberate project, but rather the result of many components that have converged in a non-programmatic way; 6) finally, it is opacity rather than transparency,[51] i.e. it does not clarify itself, but, on the contrary, requires to be clarified, again revealing a close isomorphism with social reality. In the light of this reversal, let us return to the three spheres and put them into perspective.

---

[48] "We do things with information, to paraphrase Austin (1962). It is because we know how to do this and that creatively, interactively and collectively that we can rightly report to know that such and such is the case passively and individually. Propositional knowledge is the glorious conclusion of the informational process, not its humble beginning" (Floridi, *The Logic of Information*, p. 29. Floridi refers to J.L. Austin's book, *How to Do Things with Words*, Oxford University Press, New York 1962).

[49] Ibid., p. 39. "A white box is a system about which the observer knows everything, because the observer is actually its maker."

[50] B. Latour *et al.*, *Enquêtes sur les modes d'existence. Une anthropologie des modernes*, La Découverte, Paris 2012; N. Marres, *Digital Sociology*, Polity Press, Cambridge 2017, pp. 106–115.

[51] Cf. Y. Van Den Eede, *In Between Us: On the Transparency and Opacity of Technological Mediation*, in "Foundations of Science", 16, 2, 2011, pp. 139–159.

1.2.2. Infosphere

Indeed, it is indisputable that information is produced, collected and distributed on the Web.[52] However, if I do not speak Hungarian, all I can say about a book in Hungarian is that it is a book. That's not much information at all. For the majority of non-Hungarian-speaking users, the Web is – in the most of its truly decisive functions, those related to the collection and calculation of behaviour – a systematically truthful Wikipedia[53] in Hungarian, and it is an authentic infosphere only for those who speak Hungarian and know how to distinguish between true and false. That is, to be precise, it is an infosphere only for an omniscient Magyarist – not that such a person, if they existed, would need an infosphere at all. For everyone else – i.e. for the whole world, provided there are no omniscient humans (and therefore also hungarophones) – the infosphere is, strictly speaking, a riddle wrapped in a mystery within an enigma. To be apocalyptic, but not too much, this infosphere is not a transparent pleroma surrounding humankind, but a small circle of platform operators, who do not know much either, except that their business is extremely profitable. Identifying the Web with the infosphere is a form of Platonism, which focuses on the tree – there is certainly information on the Web, just as there are needles in a haystack – and fails to notice the forest. The supporters of the infosphere are aware of this, but at this point there are two possibilities: either we find another name, or we engage in a fight that is doomed from the start, i.e. to moralise the infosphere, to exhort us to parrhesia, to teach us all to become truthful and multilingual, to make intrinsically circular appeals

---

[52] L. Floridi (ed. by), *The Onlife Manifesto. Being Human in a Hyperconnected Era*, Springer International, London 2015; P. Lévy, *Collective Intelligence*, Perseus Books, Cambridge (MA) 1997.

[53] L. Floridi, *Information: A Very Short Introduction*, Oxford University Press, Oxford 2010: "In many respects, we are not standalone entities, but rather interconnected informational organisms or inforgs, sharing with biological agents and engineered artefacts a global environment ultimately made of information, the infosphere. This is the informational environment constituted by all informational processes, services, and entities, thus including informational agents as well as their properties, interactions and mutual relations" (p. 9). It is clear that here "information" is understood in a very broad sense (that of information theory). Which is not obvious, because one wonders whether any contact, viral transmission, or nourishment, can be considered "information" – I, for one, am deeply perplexed. If one then uses "information" in the ordinary sense of the term (transmission of news or concepts), one arrives at formulations that are problematic to say the least (or, to be honest, completely erroneous), such as that of an alleged collective intelligence. And it is worth noting that Floridi also shares this perspective: "an initial account of semantic information as *meaningful data* will be used as yardstick to outline other approaches"; cf. L. Floridi, *Semantic Conceptions of Information*, in *The Stanford Encyclopedia of Philosophy* (Spring 2017 Edition), ed. by E.N. Zalta, https://plato.stanford.edu/archives/spr2017/entries/information-semantic/ [12/03/2022].

to reputation (who is watching over the reputation of reporters?) and, most improbable of all, to know the whole truth: *eritis sicut deus scientes bonum et malum*.[54] This is a grand design, difficult to implement, whereas it is more within reach to recognise the Baconian *idòla* that militate in favour of the infosphere.

The *idòla tribus* characterise us as a species. They are manifested in the idea that knowledge is the greatest human desire. Yet a quick self-examination should expose this view – the description of humans as naturally inclined to knowledge – as somewhat presumptuous, considering how late knowledge appeared in the history of humankind, how little knowledge – not belief, not habit – counts in the lives of people, including professors, and how rarely our access to the Web is dictated by cognitive needs.

The *idòla specus* are the prejudices we all have, especially those – so influential at the beginning, so irrelevant at the end – that drive the inventors of a technical device. As I said, the inventors of the Web had in mind some sort of academy of scholars, whereas their creature has become a bazaar, an arena, a red-light cinema. Focusing on the infosphere means missing the whole Web, only paying attention to a minor part of it: that of explicit and recognised communications. In doing so, the problem is narrowed down to the ethics of communication, perpetuating trends that were already underway in the 20th century. This also means talking about communication in general instead of the Web, whose specificity is lost.

The *idòla fori* derive, as we know, from inaccuracies in language. In the case of the Web, these inaccuracies are genuine errors, and manifest themselves in the form of paleonymy – in simple terms, in keeping names that no longer fit the things they refer to. The acronym ICT, which designates Web technologies, is eloquent in this respect: it stands for "Information and Communication Technologies", but the technologies to which it refers, today, are about information and communication only in a derivative form. Here one risks falling into the same misunderstanding by which the Italian telephone network, when I was a child, was called "SIP": this acronym was supposed to mean, inexplicably, *Società italiana per l'esercizio telefonico* ("Italian Society for Telephone Use"),[55] but was actually the result of the reconversion, in 1964, of the Società Idroelettrica Piemontese (Hydroelectric Company of Piedmont), active since 1918 and *then* specialised in telephony.

The *idòla theatri* derive from the philosophical doctrines of the past, and, because of their authority, I believe they played a predominant role in the choice of

---

[54] "You will be like God, knowing good and evil", Gen 3:5, NIV.

[55] Hence one of my first unanswered questions: why not call it SIET, in line with the actual initials?

such a problematic term. The infosphere has a respectable pedigree, because ever since Plato, philosophers have been fond of imagining a world of ideas or a general intellect, either as a region of spiritual being or as a possible outcome of social evolution. In this context, the infosphere is a re-proposition of the noosphere,[56] a "sphere of reflexion, of conscious invention, of the conscious unity of souls ".[57] But, as we have seen, it is anything but that; rather, it is the exact opposite. This theoretical misunderstanding has two practical and political consequences: one is the idea that the Web is a surveillance machine, a bit like Bentham's Panopticon; the other is that it is a wondrous machine destined to take the working class to paradise, by turning workers into scientists.

So, the first dystopian misunderstanding is that the Web is the expression of surveillance capitalism.[58] The expression "surveillance capitalism" if taken seriously, is a self-contradiction like "institutional revolutionary party" – and the fact that such a party exists does not alleviate the contradiction. We are dealing with an oxymoron here: either there is capitalism, which does not want to surveil but to make profit, or there is surveillance, and then there is no capitalism, because surveillance is a cost. It's obvious that if I build a spy satellite, it's for surveillance, and that surveillance is the essence of the art of war, since it ensures the priceless commodity of information. Nevertheless, to claim that these instruments, which respond to a need as old as humankind itself, can be transformed into a system of universal control is to assume that there may be some interest beyond the commercial one, such as to "control" humanity, as if there were nothing better to do. The fact that the West of commercial platforms is now facing enormous difficulties in the fight against the pandemic is the clearest proof of the

---

[56] P. Teilhard de Chardin, *Hominization*, in Id., *The Vision of the Past*, Harper & Row, New York 1923.

[57] Ibid., p. 71.

[58] *Locus classicus*: S. Zuboff, *The Age of Surveillance Capitalism: The Fight for a Human Future at the New Frontier of Power*, Profile Books, London 2019. See also P.J. Rey, *Alienation, Exploitation, and Social Media*, in "American Behavioural Scientist", 56, 4, 2012, pp. 399–420; N. Carr, *The Shallows: What the Internet Is Doing to Our Brains*, Norton, New York 2010; A. Romele, F. Gallino, C. Emmenegger, D. Gorgone, *Panopticism Is not Enough: Social Media as Technologies of Voluntary Servitude*, in "Surveillance & Society", 15, 2, 2017; R. Laudani, *Disobbedienza*, il Mulino, Bologna 2010; O. Leistert, *Resistance against Cyber-Surveillance within Social Movements and how Surveillance Adapts*, in "Surveillance & Society", 9, 4, 2012, pp. 441–456; R. Gori, M.J. Del Volgo, *L'idéologie de l'évaluation: un nouveau dispositif de servitude volontaire*, in "Nouvelle revue de psychosociologie", 2, 8, 2009, pp. 11–26; G. Chamayou, *Petits conseils aux enseignants-chercheurs qui voudront réussir leur évaluation*, in "Revue du Mouvement Anti-utilitariste dans les Sciences Sociales", 33 (*L'Université en crise. Mort ou résurrection?*), 2009, pp. 146–165; M. Andrejevic, K. Gates, *Big Data Surveillance. Introduction*, in "Surveillance & Society", 12, 2, 2014, pp. 185–196.

fact that states and health organisations have no control or governance over behaviour.

Despite the fact that our presumption nurtures the idea, or more accurately,
the mirage and the miracle, that someone out there would do anything to steal
our souls, the truth is that everyone is simply trying to sell us something. And,
unless I am mistaken, situations where a soul has a price are only found in literature – Faust and Mephistopheles, for example, though Mephistopheles was already rich in his own right, and was interested in the soul of Faust, which, incidentally, was no ordinary soul. Obviously, capital, assuming such an indetermi-
nate entity can be identified, is not interested in our ideas, but in our money: we
do not matter to it as individuals to control, but only as a profitable mass. The
limit of surveillance capitalism is that, with a mechanical transposition of the
microphysics of power,[59] it treats capital as a totalitarian state, and humanity as a
symposium of ideologues.

Both of these circumstances are far from the truth and hamper effective political action, creating phantoms and shadows. If one tries to buy a Kalashnikov
online (in this case, on the "dark" Web), the last thing the platform wants to do is
report them. The next time, it will simply suggest Molotov cocktails ("people
who bought this item were also interested in this"). Indeed, you can buy *Das
Capital* and *Mein Kampf* on the same platform, with the only consequence that if
you purchase the former, they will recommend *The Manifesto*, and if you buy the
latter, *The Protocols of the Elders of Zion*. To think that computers are interested
in our beliefs and consciences and want to manipulate them is an idea worthy of
President Schreber, the patient so dear to Freud[60] who was convinced that God
was concerned with the regularity of his defecation. Such an inquisition would
require a lot of patience and enthusiasm, and I doubt that the computer, or the
computer programmers for that matter, have any. The accumulation of documents is betting on our future behaviour, though not in terms of ideological orientations, but consumer habits. And why should it be otherwise? Unless it has
changed its mind in the meantime (a rather far-fetched suggestion), capitalism is
all about *laissez faire* and *laissez aller*, since its principle is "get rich" and its
only interest is profit.

"Surveillance capitalism" is therefore an empty and strange expression, more
or less like "cervical sciatica". The same argument can be applied to the government of algorithms.[61] If a government is algorithmic, it is not a government,

---

[59] M. Foucault, *Discipline and Punish*, Allen Lane, London 1977.

[60] S. Freud, "Psychoanalytic notes on an autobiographical account of a case of paranoia
(Dementia Paranoides)", in the *Penguin Freud Library*, Volume 9, Case Histories II, Penguin,
New York 1979, pp. 131–226.

[61] That today there is algorithmic governmentality (T. Berns, A. Rouvroy, *Gouvernemental-*

since an algorithm has no objectives; if it is really a government, then the algorithm follows the humans who engineered it just as the troops follow the general. It is problematic to postulate some collective intentionality,[62] since – as we shall see in 3.4.3 – intentionality, representations, and the intentions that derive from them all depend on corporeality, and are therefore individual. Consequently, it is all the more doubtful that these purposes should be available to an algorithm. Thus, blaming algorithms, say, for an accident involving a self-driving car or a racist bias is ultimately like blaming an accident on the car rather than the driver,[63] and the racism of the Ku Klux Klan on the bizarre costume worn by its adherents.

There is also a theoretically relevant point, which has to do with the nature of surveillance. Panopticon theorists assumed that there might be a genuine interest

---

ité algorithmique et perspective d'émancipation. Le disparate comme condition d'individuation par la relation?, in "Réseaux", 177, 1, 2013, pp. 163–196; A. Rouvroy, *Face à la gouvernementalité algorithmique, repenser le sujet de droit comme puissance*, unpublished paper) and that decisions on the stock exchange are made by computers is just one of the endless examples of the primacy of documentality over intentionality.

[62] J.R. Searle, *The Construction of Social Reality*, Free Press, New York 1995; Id., *Making the Social World: The Structure of Human Civilization*, Oxford University Press, Oxford 2010. Before Searle (or in conjunction with him) the notion of "collective" or "shared intentionality" was used by other philosophers and psychologists from the late 1980s onwards: M. Gilbert, *On Social Facts*, Routledge, New York 1989; Id., *Group Membership and Political Obligation*, in "The Monist", 76, 1993, pp. 119–131; Id., *Walking Together: A Paradigmatic Social Phenomenon*, in "Midwest Studies in Philosophy", 15, 1990, pp. 1–14; P. Pettit, *The Common Mind: An Essay on Psychology, Society, and Politics*, Oxford University Press, Oxford 1996; R. Tuomela, *The Importance of Us*, Stanford University Press, Stanford 1995; Id., *The Philosophy of Social Practices*, Cambridge University Press, Cambridge 2002; M. Bratman, *Faces of Intention*, Cambridge University Press, Cambridge 1999; A. Thomasson, *Fiction and Metaphysics*, Cambridge University Press, Cambridge 1998. Resorting to collective intentionality is, trivially, a form of anthropocentrism. For a critique of anthropocentric social ontology cf. B. Epstein, *The Ant Trap: Rebuilding the Foundations of the Social Sciences*, Oxford University Press, Oxford 2015.

[63] On a more strictly economic level, it is often observed that the algorithm punishes the poor, who are recognised as such through profiling on consumption and behaviour and cannot access credit. Now, on the one hand, banks are often interested in providing loans to insolvent people in order to take away their homes, so algorithms should favour the granting of loans. On the other hand, if they were virtuous banks, there is no reason why they should expose themselves to the risk of losing their customers' money. However, it is the precise duty of the state, or rather of states, to support those who lack resources. Instead of accessing credit that they will not be able to repay, the poor should receive subsidies to which they are duly entitled, and these should not be paid for by tax-payers (perhaps reimbursed by the state, if the banks go bust by dint of allowing bad loans), but by the surplus value accumulated by platforms, which should support those truly in need, not those who simply pretend they are poor (the profiling is indeed reliable in this sense).

in keeping one or more people under observation. In the case of Bentham, who had invented a model prison, the interest was obvious. In the case of Foucault, who thought of a French-style absolute state, it may yet work. But in the case of Internet platforms we have a *metabasis eis allo genos*: there is nothing in common between a centralist state and the decentralised and rhizomatic documedia capital. From this point of view, the fact that in Europe and the US, unlike in China, platforms are private and respond to commercial interests presents a huge opportunity, that of a taxable production of value that does not involve an ideological inquisition against the users. Moreover, there are good reasons to doubt that the concept of "surveillance" makes any sense at all if it is not combined with the concept of "punishment": in fact, the seminal book on the subject was entitled *Surveiller et punir*. You cannot write a book on surveillance without punishment because then even a birdwatcher would be a Benthamian bird warden; a student of fish archaeology would be a Benthamian bone and jewel controller; and an astronomer would be a cosmic Benthamian supercontroller. Internet platforms are not a Big Brother punishing those who stray from the path; on the contrary, they promote dissidents, because they reveal market niches that need to be filled.

Above all, one must not lose the sense of history. A small community is a more effective control system than any computer device, and the "public opinion" of which Locke spoke did not refer to the tabloids or the Web, but to malicious judgement and scrutiny among neighbours.[64] It is lamented that police departments use predictive software, as if they did not previously rely on more crude tools like gender or skin colour (if we consider that reporting goes hand in hand with the demand for universal access to data, the solution is to give equal cognitive opportunities to organised crime). We complain that there are studies on population density, as if censuses had never existed; or about wealth, as if the Tax Police were a new institution. We have exact maps of cafes, churches, schools (what's wrong with that? Would it be better to wander around aimlessly?); "and even the phases of the moon" are reported with precision[65] – as if Palaeolithic people (the first to note them down in order to predict them) were acting on behalf of some kind of platform.

On this side of the Palaeolithic, the 110 linear kilometres of files on the citizens of the DDR, collected at a time when platforms were not even remotely

---

[64] J. Habermas, *The Structural Transformation of the Public Sphere: An Inquiry into a Category of Bourgeois Society*, MIT Press, Cambridge (MA) 1989.

[65] M. Chammah, M. Hansen, *Policing the Future. In the Aftermath of Ferguson, St. Louis Cops Embrace Crime-predicting Software*, in "The Verge", 2016, https://www.theverge.com/2016/2/3/10895804/st-louis-police-hunchlab-predictive-policing-marshall-project [15/11/2021].

known, show that surveillance came long before the Web – and that the Web only adds the possibility of transforming this collection of documents into a resource, whereas the Stasi files are rather a huge waste of resources, justified only by a high and (in its own way) noble interest in the ideological convictions of those under surveillance. In the documedia age, the only surveillance capitalism we have known so far is not capitalism, but a deep, powerful and technically advanced communism: the Chinese state. China has nationalised its Internet platforms, realising the two fundamental aspirations of communism: social welfare *and* the control of ideas through the reduction of individual freedom. I suspect, however, that a book entitled Surveillance Communism would cause much less of a stir.

What is certain is this. Commercial platforms have no use for what we think, for the little or big secrets of our lives. If Locke, instead of writing his *Essay on Human Understanding*, had composed – as de Maistre maliciously suggested – an essay on his own understanding, the theoretical interest of the exercise would have been close to zero. The ideal that usually drives the political and social reflection on the Web, however, is to safeguard humans who cherish discretion very much in theory, and not at all in practice. Another ideal is transparency, in a sphere that was once considered to be the realm of transparency *par excellence*, and now turns out to be rather opaque. These two objectives cannot be achieved in a simple way – e.g. by demanding free and universal access to documents – both because a conflict would arise between protection and transparency, and because the docusphere is intelligible but not intelligent. The problems of the infosphere thus include trust, piracy, the digital divide, the surveillance society, privacy,[66] freedom of expression, artificial partners and cyberwar.[67]

These are all genuine problems, but in all likelihood they cannot be solved by drawing up a charter of principles. The fundamental issue lies in two facts: firstly, the structural rootedness of the transparent infosphere in an opaque docusphere and, secondly, the dependence of the docusphere on the biosphere, and therefore the centrality of the contribution of humans as bearers not of knowledge or ideas, but of needs, urgencies and goals. The solution, in other words, is to enhance the role of humans as the fundamental engine of the process, and not to demand (again) some kind of etiquette for information – such things have already been written, and, above all, have no better chance of being applied than President Wilson's 14 points after the American contingent withdrew from the European war scene. Having said that, I am in no way opposed to considering the problems

---

[66] The notion of "information" is no less ambiguous and problematic; see J.J. Thomson, *The Right to Privacy*, in "Philosophy and Public Affairs", 4, 1975, pp. 295–314.

[67] Floridi, *The Logic of Information*, p. 50.

of privacy and information ethics as serious and relevant. But their relevance for documediality is no greater than their relevance for the industrial economy. The importance of the docusphere is quite different, because that is where documediality produces a new economy and a new surplus value whose very existence was unknown until a few years ago. If someone had suggested, in 1919 or 1929, "It's the economy, stupid!", perhaps we could have prevented the economic crisis in Germany from producing Nazism and all that followed.

Of course, the increasing quantity of leisure time – it is challenging to write a hate speech while standing by the assembly line – and human unsociable sociality have their share of responsibility. However, what counts most is the enormous amount of unpaid and unrecognised labour that is provided through mobilisation. The exchange between users and platforms contains a particularly well-hidden surplus value. A very concrete consequence is this: on its own, the protection of privacy is a distraction. Remember old station notices, where one train could hide another? Well, the concern for privacy masks much more substantial processes, starting with the fact that the Web is not interested in us individually or in our ideas, but rather in the workforce of humanity as a whole. Not to mention that privacy is obviously the least of the problems for the abundant half of the world that posts content on social networks, and for the quasi-totality of the world that automatically consents to the use of cookies, as they urgently need this or that service. It is not a question of bourgeois discretion, of decorum, of minding one's own business with due reserve, but of producing value. Neglecting this fact has a cost that is not only cognitive, because it replaces the political and economic problem with an ethical drift. Now, I'm not sure how useful ethics is here, since the Web, like the road to hell, is paved with good intentions, and the first to list their good intentions are the platform owners. As we are looking for the nearest emergency room, they make sure to tell us that our privacy is the most important thing in the world for them, and that they are ready to respect it, provided we give up the service they offer.

After the dysphoric misunderstanding of the infosphere, surveillance capitalism, we come to the euphoric misunderstanding, i.e. cognitive capitalism.[68] The

---

[68] M. Hardt, A. Negri, *Empire*, Harvard University Press, Cambridge (MA) 2000, ch. III, § 4, p. 409: "The multitude not only uses machines to produce, but also becomes increasingly machinic itself, as the means of production are increasingly integrated into the minds and bodies of the multitude. In this context, reappropriation means having free access to and control over knowledge, information, communication, and affects because these are some of the primary means of biopolitical production. Just because these productive machines have been integrated into the multitude does not mean that the multitude has control over them. Rather, it makes more vicious and injurious their alienation. The right to reappropriation is really the multitude's right to self-control and autonomous self-production".

latter shares with the surveillance dystopia the assumption that power is first and foremost knowledge. However, it claims that – just as in the Hegelian master-slave dialectic – work on the Web constitutes knowledge and emancipation as such. Now, there have been theorisations about a collective intelligence that would develop spontaneously on the Web,[69] as well as messianic visions according to which our age would be the age of transparency.[70] Going even further, the theorists of cognitive capitalism[71] re-propose, as I said in the Prologue, the relationship between lordship and servitude, but in the classical form, that according to which the servant defeats the master by taking over the production processes. In other words, they fail to consider that although Hegel's dialectic is still relevant, lordship does not consist in the domination of production processes, now perfectly automated, but in the circular yet profoundly unjust economy of need. In other words, capital does not need workers to produce value: it is happy to fire them all, and produces more and more value using not their intelligence, but their consumption and mobilisation.

So, utopia is very well and good, and revolutionary hope (the hope for a bloodless revolution like the one enunciated in the oldest systematic programme of German idealism)[72] takes the place of the victimhood of surveillance capitalism. However, it rests on the false assumption that the Web is exclusively an infosphere. Unfortunately, access and labour are not the same thing. Saying that they are generates a noble illusion, similar to the idea that, upon entering the library of Babel, we will immediately come across the right book. Like any utopia, it is as old as human beings, and in the case of documediality, as old as the advent of digitization. Universal access to platforms was in fact the hope with which Lyotard ended *The Postmodern Condition* over four decades ago.[73] But since the search for truth is fuelled by strong practical interests, the libertarianism of document access would be a valuable tool not only in the hands of jealous partners (fair enough) but also in the hands of criminal gangs, mafias, and terrorist organisations, which are probably more motivated and equipped to read the data than ordinary and hopefully disinterested citizens. As for the latter, instead, are we

[69] P. Levy, *Collective Intelligence*, Basic Books, New York 1997.

[70] G. Vattimo, *The Transparent Society*, Johns Hopkins University Press, Baltimore 1992.

[71] P. Virno, *General Intellect, Exodus, Multitude*, in "Fine Art Critical Studies", 1, June 2002; Hardt, Negri, *Empire* cit.; Idd., *Assembly*, Oxford University Press, New York 2017; C. Vercellone (ed. by), *Capitalismo cognitivo*, Manifestolibri, Rome 2006; C. Marazzi, *Capitale e linguaggio. Dalla New Economy all'economia di guerra*, DeriveApprodi, Rome 2002.

[72] Anon. The Oldest Systematic Programme of German Idealism in F. Beiser (ed. by), *The Early Political Writings of the German Romantics*, Cambridge Texts in the History of Political Thought, Cambridge University Press, Cambridge 1996, pp. 1–6.

[73] J.F. Lyotard, *The Postmodern Condition* (1979), Manchester University Press, Manchester 1984.

sure they have any interest in accessing data, considering that the vast majority of it would be incomprehensible without appropriate calculation tools?

Virtue and knowledge do not arise spontaneously, but require effort, education, and sometimes even predisposition. It is true that, as Socrates noted, a slave, if well trained, can demonstrate the Pythagorean theorem: but that would require training and, above all, interest. As for the former, we all know the pain of learning, and the world teems with deserted libraries. As for the latter, one cannot but agree with Proust and Simenon. Proust reminds us that the search for truth is primarily driven by interest, and that there is no inquirer more demanding and tireless than the jealous lover, ready to spy on the slightest gestures of the beloved with a degree of attention that they would hardly squander in a laboratory or library. Simenon, along with Conan Doyle, Poe and many others, teaches us that one of the few truths the world is interested in is "whodunit". There is no reason to think that hypothetical open-access platforms would be overcrowded, not only because access would require more than literacy, but especially because the motives that might induce someone to access documents rarely have anything to do with pure knowledge. Plato saw this clearly: it is not enough to know *grammata* to be literate. Documents are proof of this: only at a very high level of interpretation does knowledge come from data. The mobiliser, i.e. the platform, possesses real documents, because they reflect the actual behaviour of the mobilised; of course, algorithms could be created to confuse the results, but quantitatively speaking most of the documents would still remain true. On the contrary, the mobilised person is bound to sail across a sea of post-truth.

There are very few people who know how document collection works, and in order to be able to analyse data, they need instruments with enormous computing power. The rest of us – i.e. the whole of humankind, as well as the platform operators themselves in their non-professional lives – are simply free lenders of labour that we are often unaware of providing, and whose significance we ignore. This is not due to the malice of capital, but to the naivety of those who have established an undue equation between the growth of competence and the growth of knowledge. We know very well that the relationship between these two things, if there is one, is inversely proportional: the more technological competence grows, i.e. the more things can be done automatically, the more epistemological knowledge decreases, since it is not necessary to know how something works in order to obtain results.

The cognitive illusion thus carries a political price: if capital really did give us knowledge, our mobilisation would be rewarded, but this is not the case. There is therefore a very real risk of transforming – against the will of those who theorised it – the call for collective intelligence into an imaginary reward for real and unrecognised labour: the production of documents that are exploitable information for some (platforms) and not for others (users), who thus become unpaid workers in a job which they are not even aware of doing. This is the crucial point of the ongoing revolution, which is not understood, yet as annoying as a phantom limb. We work and we don't know we are doing it – that's all, if we want to boil it down. So, why limit ourselves to the infosphere when we can design a more comprehensive concept describing a system of inclusions, where the infosphere is the tiny subset of the docusphere, and the docusphere in turn is the result of a biosphere that does not think but acts? All things considered, this is another narcissistic wound, which does not consist in saying that machines think better than we do (this has always been the case), but rather in recognising that we think much less than we think. In this sense, *ego cogito, ergo sum* is not the description of a normal condition for humankind, but the thought experiment that a man of genius carried out once in his life, after doing many other things including being a soldier, tearing apart cats in the cellar and mourning a dead child.

1.2.3. Docusphere

In 1967, when personal computers, not to mention the Web, were still completely unheard of, Jacques Derrida opened his *Grammatology* with a prophetic claim loaded with consequences.[74] Perhaps we were indeed moving towards "the end of the book" understood as an organic and complete totality, like a Hegelian encyclopaedia. But what Derrida was surely right about – in an age when McLuhan talked about the end of writing[75] – was the unprecedented growth in hysteresis. This boom of hysteresis was a boom of recording, of the archive, of the external and mechanical memory called to compensate for the insufficiencies of organic memory. The problem is that Derrida's prophetic intuition was accompanied by a hyperbolic phrase: "il n'y a pas de hors-texte".[76] As if the world could be re-

---

[74] J. Derrida, *De la grammatologie*, Les Éditions de minuit, Paris 1967, chap. 1: "La fin du livre et le commencement de l'écriture".

[75] M. McLuhan, *The Gutenberg Galaxy: The Making of Typographic Man*, University of Toronto Press Toronto 1962.

[76] "Il n'y a pas de hors-texte" (Derrida, *De la grammatologie* cit., p. 227), or less cryptically, as Derrida says shortly afterwards, "Il n'y a rien hors du texte" (ibid., p. 233). The ontological significance of this phrase was reiterated the same year in *La Voix et le Phénomène*: "S'il n'est rien en-dehors du texte, c'est que le texte n'a pas de bord" (J. Derrida, *La Voix et le Phénomène*, PUF, Paris 1967, p. 373). The expression did not fail to arouse controversy, allowing Michel Foucault to score a point without too much difficulty. He qualified Derrida's philosophy as "a historically well-determined little pedagogy [...] which teaches the student that there is nothing outside the text" (M. Foucault, *History of Madness*, Routledge, London 2006, p. xxiv, 573). J.R. Searle, another philosopher who is not exactly an admirer of Derrida, had it easy on the same point. Derrida's defences in the course of these polemics have been uncon-

duced to a text. Now, to claim that nothing, in general, exists outside the text is, in the best case, a matter of confusing epistemology, what we know and the words we use, for example "red" or "rough" or "sad", with ontology, what there is: a red rose, sandpaper, a bulldog's snout. In the worst case, instead, making such a claim amounts to throwing in a catchphrase – a gesture that can be perfectly understood on the part of a great philosopher who was still young and full of hope, at a time when everything seemed possible.

Instead, to argue that there is nothing *social* outside the text is to recognise the constitutive role of hysteresis in the emergence of social reality. Indeed, there is nothing social outside the network of inscriptions that invade our lives, populating them, enriching them and giving them a meaning that they do not inherently possess beyond their organic survival – an internal purpose that is only a shadow of a purpose. Inscriptions construct social reality, either by enacting our intention, such as when we make a promise; or by counteracting it, such as when we have to keep a promise and no longer feel like it; or – even more interestingly in my opinion – by arousing it. An essential difference thus opens up between action and intention, with action playing the leading role. In fact, action is an ontological notion: it refers to something that exists, and which acquires relevance insofar as it is recorded. Intention, on the other hand, is an epistemological notion; rather than being the motive for action, it appears as its explanation: X performed action Y for motive Z. That this is the case seems to be proven by the accuracy with which algorithms recognise our tastes based on our purchasing habits. We did not know we had certain intentions and inclinations, and we discover them through our behaviour, just as we often only discover our intentions when we communicate them to others, or put them down in writing. Therein begins the mystery of mobilisation that I will focus on in 1.4.

---

vincing, and have consisted either in reducing the thesis to a banality, i.e. the obvious assertion that nothing exists outside of a context and everything requires interpretation (J. Derrida, *Limited Inc*, Galilée, Paris 1990, pp. 252–253: "'Il n'y a pas de hors-contexte'; déconstruire, c'est prendre en compte cette 'structure a priori' dont l'analyse n'est jamais politiquement neutre") or in relaunching the hyperbole, complicating the thesis even more. Cf. J. Derrida, *Limited Inc.*, Northwestern University Press, Evanston (IL) 1988, p. 148: "I wanted to recall that the concept of text I propose is limited neither to the graphic, nor to the book, nor even to discourse, and even less to the semantic, representational, symbolic, ideal, or ideological sphere. What I call 'text' implies all the structures called 'real,' 'economic,' 'historical,' socio-institutional, in short: all possible referents. Another way of recalling once again that 'there is nothing outside the text.' That does not mean that all referents are suspended, denied, or enclosed in a book, as people have claimed, or have been naive enough to believe and to have accused me of believing. But it does mean that every referent, all reality has the structure of a differential trace, and that one cannot refer to this 'real' except in an interpretive experience. The latter neither yields meaning nor assumes it except in a movement of differential referring. That's all."

This is why the infosphere is the shallow, and notably murky, water[77] of a much larger sea, which I define "docusphere", and which consists of the domain of documentality[78]: the capitalisation of acts carried out on the Web. These acts, however, like any document, need to be interpreted and understood – something that is not within everyone's reach, indeed almost no one's reach, considering the novelty of the phenomenon and the complexity of the tools required for this interpretation. These instruments are almost exclusively available, at the moment, to platforms, certainly not to users, and to a large extent not even to liberal states and traditional industries. On the one hand, this explains the enormous competitive advantage of platforms, but on the other hand, we must not overlook an element that will play a decisive role in the rest of my argument: for the docusphere to be produced, hysteresis is not enough, it also takes acts to feed it. These belong to what I call the "biosphere", the world of life in its human – and not natural – form: of course, atmospheric events or animal behaviour can be recorded, too, but they would still be of interest only to humans. If, like me, you have time to spare, you may have found yourself thinking, say, on a cab ride to a restaurant,

---

[77] An early critique of the limits of the infosphere can be found in Day, *The Modern Invention of Information* cit.

[78] M. Ferraris, *Documentality. Why it is Necessary to Leave Traces*, Fordham University Press, New York 2012. The role of documents in the construction of social reality is the focus of many recent studies: B. Kafka, *The Demon of Writing: Powers and Failures of Paperwork*, Zone Books, New York 2012; M. Bunz, *The Silent Revolution*, cit; L. Gitelman, *Paper Knowledge: Toward a Media History of Documents*, Duke University Press, Durham 2014. For Marx it is public debt (i.e. a record) that gives money the capacity to "reproduce" (K. Marx, *Capital*, Volume One, Penguin Classics edition, New York 1976, p. 711), i.e. to lay the foundations for the creation of mercantile capital as a precursor of industrial capital. On money as a debt witnessed before the whole community, see G. Simmel, *The Philosophy of Money* [1900], Routledge, London 2004. This thesis has been taken up and updated by N. Dodd, T*he Social Life of Money*, Princeton University Press, Princeton 2014. See also H. De Soto, *The Mystery of Capital: Why Capitalism Triumphs in the West and Fails Everywhere Else*, Basic Books, New York 2000; C. Herrenschmidt, *Les trois écritures. Langue, nombre, code*, Gallimard, Paris 2007; T. Piketty, *Capital in the Twenty-First Century*, Harvard University Press, Cambridge (MA) 2014. More recently the role of documents has also been recognised by Searle, *Making the Social World* cit. Documentality explains the resources of planning, which constitutes the secret of military effectiveness, much better than collective intentionality. More broadly, it is the foundation of social action, as demonstrated by the importance of bureaucracy in the formation and management of power; see esp. A. Armando, G. Durbiano, *Teoria del progetto architettonico. Dai disegni agli effetti*, Carocci, Rome 2017. Cf. also S.A. Shapiro, *Massively Shared Agency*, in *Rational and Social Agency: Essays on the Philosophy of Michael Bratman*, ed. by M. Vargas and G. Yaffe, Oxford University Press, New York 2014; A. Gangemi, *Norms and Plans as Unification Criteria for Social Collectives*, in "Journal of Autonomous Agents and Multi-Agent Systems", 16, 3, 2008, pp. 70–112; B. Smith, *Diagrams, Documents, and the Meshing of Plans*, in *How To Do Things With Pictures: Skill, Practice, Performance. Visual Learning*, ed. by A. Benedek and K. Nyíri, Lang, Frankfurt a. M. 2013, pp. 165–179.

that the car wouldn't go anywhere without the motivation and urgency of the organisms inside it. And if the machine in question is the Web, what this act produces are documents. Feelings, ideas and opinions aside, there is something minimal and very powerful that is recorded in the docusphere: the actions connected to those opinions, ideas or feelings – actions that, more importantly, are often performed by chance or out of boredom, yet are filling an unprecedented archive with documents, i.e. recorded acts.

The reference to the archive and documents seems to me of primary importance. In fact, there is an intrinsic obscurity in the notion of "data", as it includes, in public discourse, both what is expressly conceived as a communication (such as what is published on social networks) and reactions to this communication, such as likes. Most of all, this notion also encompasses so-called metadata, which we produce unintentionally, and which are by far the most interesting asset for the purposes of profiling, which consists in translating the biosphere into the docusphere. It is therefore necessary to swap the generic definition of "data" for the more exact definition of "document". "Document" translates the Latin *documentum*, from *doceo*, and means "that which shows or represents a fact". This description fits the three spheres in which we usually speak of documents: the historical one, where "document" designates everything that appears relevant for the reconstruction of the past; the informative one, where the term includes everything that conveys information; and the juridical one, where it indicates what has legal value. Intuitively, it is this third sense, which is also the oldest and most traditional, that appears to be the most specific. The other two values, the historical and the informative, derive – historians say – from the juridical. Here "legal" is to be understood in an extensive sense, inherent in the overall process of inscribing all that is socially relevant, from economics to religion.

Documents are first and foremost a guarantee of rights, not in some hyperuranium, but in a historical world of which we must keep account and track, otherwise we would not understand where we are and where we are coming from.[79] If this is the prevailing meaning in the definition of "document" – whether it be public or private; dispositional, such as a law; or testimonial, such as a passport, a driving licence, or a diploma – then it is appropriate to integrate and clarify the description of the document as "representation of a fact" with that of the document as "capitalisation of an act". It is in fact an essential feature of social reality that it is necessary to keep track of the acts that constitute it, generating the docusphere.

---

[79] V. Crescenzi, *La rappresentazione dell'evento giuridico. Origini e struttura della funzione documentaria*, Carocci, Rome 2005.

Documents are divided, in the Web as in any other domain, into two spheres: strong documents, as records of acts, and weak documents, as records of facts.[80] Strong documents are the traditional components of social reality. It is in fact in documentation that the formation of social objects takes place, according to the constitutive rule Object = Recorded Act (O = RA): the social object is the result of a social act that takes place between at least two people, or between a person and a delegated machine, and which has the characteristic of being recorded.[81] With the documedia turn, the sphere of intentional documentation is not reduced to canonised acts, such as weddings, bets, and the like, but is archived by making relationships, opinions, and professional experiences usable. If strong documents are what is usually called "data", weak documents correspond to metadata, i.e. the capitalisation of facts that goes hand in hand with the production of acts, in a way that is neither conscious nor intentional, and which generates a much richer capital of documents than the already abundant one produced by strong documents. So, hysteresis allows the capitalisation of otherwise volatile documents. I will analyse in detail the metaphysical implications of this circumstance in section 3, and I will draw its political and economic consequences in section 4.

The docusphere, of course, came long before the documedia revolution, as it is both the necessary and the sufficient condition of social reality. Demonstrating that documentality is the necessary condition of social reality, i.e. that there can be no effective social acts without technological hysteresis, is a fairly intuitive issue that I have dealt with at length elsewhere,[82] so I will limit myself to a few simple references. Social reality is constituted by social objects that are the result of social acts and that respond to the law O = RA. In this framework, documentality is the *conditio sine qua non* for the constitution of social reality, since an act of which there is no trace – be it a baptism, a wedding, a debt, a promise, an honorary title, a rite, a myth, a symbol or a linguistic or semiotic system – would not contribute in any way to the constitution and development of social reality. A more complex issue, however, is demonstrating that documentality is also the *sufficient* condition of social reality, i.e., that it is capable of producing intentionality. Again, this is a theme I have addressed elsewhere,[83] but I will develop it in

---

[80] Ferraris, *Documentality* cit., pp. 247 ff. (for the distinction between strong and weak documents, see especially p. 249).

[81] And not simply a speech act, as such evanescent, as posited by Austin, *How to Do Things with Words* cit. The first intuition of the necessity of recording is found in J. Derrida, "Signature, Event, Context" (1971), in Id., *Limited Inc.* cit. A discussion of this topic can be found in A. Burkhardt (ed. by), *Speech Acts, Meaning and Intentions. Critical Approaches to the Philosophy of J.R. Searle*, De Gruyter, Berlin-New York 1990.

[82] Ferraris, *Documentality* cit.

[83] M. Ferraris, *J.R. Searle, Il denaro e i suoi inganni*, Einaudi, Turin 2018.

a new form in 2.4. For the time being, I would like to draw attention to a circumstance that is crucial for the continuation of my argument, which ties in with what
I said above about the Web being emergence[84] rather than construction.

Approaching the Web as a docusphere means, in my view, attaching new historical depth to the documedia revolution, as well as greater metaphysical clarity.
Answering the question "who created the first documents?" is no different from
answering questions like "who invented chess?" or "who came up with jokes?".
If social reality were constructed, it should be as simple as pointing out the inventor of Juno, Indian castes, Indo-European institutions, or the wheel. But this
is not the case, and not only for empirical reasons linked to the difficulty of going
back so far in time (an issue not found in jokes), but because things simply did
not happen that way. There was never a time when someone, or a group of
friends, to overcome the boredom of an afternoon in Isfahan, decided to invent
chess. First there were battles, then games imitating them, and finally the setting
of rules, the end result of which was the game of chess. The same goes for gods,
forms of government, castes, and of course, going back a long way, languages.

What this constructionist genesis overlooks is precisely its obscurity, that is to
say, the givenness of myth, the positivities we encounter in the world, the values
we inherit and share long before we understand them, the languages we speak,
the gods we worship, be they Juno, Shiva, or Sapiens. The world is the product
of an ontological superabundance[85] that in addition to time, space and matter has

---

[84] M. Ferraris, *Positive Realism*, John Hunt Publishing, London 2015. Emergentism classics: J.S. Mill, *A System of Logic, Ratiocinative and Inductive*, Harper, New York 1875; S. Alexander, *Space, Time and Deity*, Macmillan, London 1920; C. Lloyd Morgan, *Emergent Evolution*, Williams & Norgate, London 1923; C.D. Broad, *The Mind and Its Place in Nature*, Routledge-Kegan Paul, London 1923. Later discussions: T. Nagel, "Panpsychism", in *Nagel's
Mortal Questions,* Cambridge University Press, Cambridge 1979, pp. 181–195; *Consciousness
and Content*, in "Proceedings of the British Academy", 74, 1988, pp. 225–245; J. Van Cleve,
Emergence or Panpsychism: Magic or Mind-dust?, in "Philosophical Perspectives", 4, 1990,
pp. 215–226; B. McLaughlin, *The Rise and Fall of British Emergentism*, in *Emergence or Reduction?*, ed. by A. Beckerman, H. Flohr and J. Kim, De Gruyter, Berlin 1992; J.R. Searle, *The
Rediscovery of the Mind*, MIT Press, Cambridge (MA) 1992; T. Horgan, *From Supervenience
to Superdupervenience: Meeting the Demands of a Material World*, in "Mind", 102, 408, 1993,
pp. 555–586; D.M. Armstrong, *Emergence and Logical Atomism*, in *A World of States of Affairs*, Cambridge University Press, Cambridge 1997, pp. 152–153; T. Crane, *The Significance
of Emergence*, in *Physicalism and Its Discontents*, ed. by B. Loewer and G. Gillett, Cambridge
University Press, Cambridge 2001. At the origin of all this, though, I would place Leibniz:
G.W. Leibniz, *A New System of the Nature and the Communication of Substances, as well as
the Union Between the Soul and the Body* in Loemker L.E. (ed. by) *Philosophical Papers and
Letters. The New Synthese Historical Library (Texts and Studies in the History of Philosophy),*
vol 2. Springer, Dordrecht 1989.

[85] F.J. Dyson, *Infinite in All Directions*, Harper & Row, New York 1988.

required a single ingredient – hysteresis, the possibility of recording and all that this entails, up to the genesis of conscious beings like you and me. Constructionism grasps a fair point, namely the relevance of action, but makes it depend on a concept or a contract, i.e. it cultivates the idea that everything, from society to values, is produced by a variant of the social contract: we sit around a table and agree all together, plan B being that the overbearing and solipsistic create their own values and rules in the privacy of their studies.

This is why the technological enhancement of hysteresis has changed our lives so radically. If we look at what it is – and not what Tim Berners-Lee, who should be considered more of an onomaturgist than an inventor, thought it would be – the Web is a great form of collective emergence.[86] Compared to the remote emergencies of the past, the Web has the enormous advantage of having developed right before our eyes. Therefore, there is no need for conjecture: what the Web has become, surprising those who designed it, occupies a short and perfectly controllable historical period, of which I, like so many others, have been an eyewitness. The conditions of this emergence were, in this case, not an overabundance of time, but rather the pervasiveness of the medium, the number of users, and the speed of capitalisation of an immense archive. All this in a context in which documedia machines, being endowed with memory, were able to exploit all the potentialities of the biosphere, translating acts into documents according to the law $O = RA$.

Now let us broaden our perspective. Exceptionalism is always a dubious counsellor, and there are good reasons to think that the emergence of the Web is not the only case in nature and history.[87] There are much older ones. For example, the emergence of the pharaonic civilisation[88] shares some stages with the emergence of humankind in general, such as the full passage from *homo erectus* to *homo sapiens*, which occurred a hundred thousand years ago. Others, on the other hand, appear to be exclusive to Egypt: a culture that gravitates around the Nile, between 45,000 and 20,000 years ago; the transition to microlith, between 15,000 and 10,000 years ago, with the first traces of harvesting crops; and finally, 5,500 years ago, the shift to agriculture, the birth of writing, and the emergence of truth, and with it the pharaonic society – the development of the funerary culture, the origins of pyramids and the animal totems that prelude to the Egyptian pantheon. Each of these conditions must have coincided with the emergence of a certain number of meanings, which grew as they multiplied through society and writing.

---

[86] For more on this, please refer to my *Emergenza*, Einaudi, Turin 2016.

[87] S. Johnson, *Emergence: The Connected Lives of Ants, Brains, Cities, and Software*, Scribner, New York 2001.

[88] As explained in the first chapter of N. Grimal, *A History of Ancient Egypt*, Wiley, Hoboken (NJ) 1988.

We can thus imagine that the most sophisticated emergencies took place with the sharing of some previous meanings, derived from previous life forms (e.g. dominance) and produced new ones (e.g. pharaoh or vizier), linked to more articulated life forms. In these emergencies, there is a movement from documentality to society, intentionality, truth and responsibility.

And now let us take the pyramid.[89] It would be naive to think that some Egyptian architect, in order to please a pharaoh with a desire for immortality, came up with the idea of a pyramid. The way in which pyramids are built, which points to the proximity of spirit and nature under the sign of technology, recalls rather the origin of termite mounds and hexagonal bee cells.[90] Initially, the pharaoh was laid to rest in a simple hypogeal mortuary chamber. However, as the goods collected in the chamber attracted thieves, the access was protected with a pile of stones, and the piles resembled pyramids for obvious reasons of statics. Gradually, the focus shifted from the underground chamber to the mound, in much the same way as a fetishist's attention shifts from the beloved's body to objects that recall it. Finally, the pyramid, having become large, autonomous and recognisable, incorporated the death chamber. And only then did the functional building acquire a symbolic meaning, as the soul's ascent to heaven, as the sun's ray projecting itself onto the Earth, as the tower of Babel coming to confuse languages, and so on. The process is systematic: inscription (hypogeum), iteration (cumulus) and alteration (sign, symbol); at the origin of all this there is a teleology (death) that sheds light on archaeology (who we are and where we come from).

Thinking that there are forty centuries of history behind us when we look at the alleged uniqueness and novelty of the Web gives us (to say the least) some historical perspective, and most importantly prevents us from embracing a Pentecostal and magical approach when we look at the present and its supposed unprecedentedness. What do I mean by "Pentecostal"? As many of us were taught as children, Pentecost refers to the descent of the Holy Spirit fifty days after Christ's ascension to heaven. As you can imagine, that was not a common event, and one of its most remarkable effects was to turn the disciples into polyglots, so that they could become apostles and preachers. All this may appear very ancient and seems to require a lot of faith, yet the Pentecostal attitude is one of the most widespread among people who are supposed to be completely alien to it. The Pentecostal image is based on the assumption of a short and finite time and space, and dreams of a spirit that descends from above, giving life to the dead and the

---

[89] I elaborated on this interpretation of pyramids in my *Piramide e coscienza*, in M. Ferraris, *La filosofia e lo spirito vivente*, Laterza, Rome-Bari 1991, pp. 97–120.

[90] Marx, *Capital* cit., book I, section III, chap. V; G. Ciccotti, M. Cini, M. de Maria, G. Jona-Lasinio, *L'ape e l'architetto. Paradigmi scientifici e materialismo storico*, Feltrinelli, Milan 1976.

inert, animating that which is without reason and without movement. It is the foreign soul in this world, which ends up in the body as in its tomb. It is Plato, if we want to name names.

The emergent image I propose in this book follows the opposite path. To name names again, it is Aristotle, or the Big Bang. A sublime superabundance of time and space has left the world with all sorts of disorder, nonsense and dissipation: the economy, let us not forget, is only worthwhile for those who have little time, and nature has plenty, if only the notion of "time" has hardly any meaning outside of an anthropic horizon.[91] Remember what Darwin said? "What a book a devil's chaplain might write on the clumsy, wasteful, blundering, low, and horribly cruel works of nature!".[92] But in the end, here we are. This also applies to society. By contrast, the vast majority of philosophers who deal with society maintain that it rests on some "collective intentionality",[93] which would make us capable of social interaction and of sharing principles and values. The great-great-grandparents of these philosophers spoke of the general will and the social contract, but little has changed. Compared to the social contract, even the idea of language as a divine gift seems much more reasonable, describing an emergence that takes place independently of consciousness. It is also astonishing that those who would rightly reject the idea of a collective unconscious as a Romantic superstition accept the idea of a collective consciousness, and those who would scornfully reject both adhere unconditionally to the idea of collective intentionality, which is the sum of the two.

Now, why should we be sceptical of Pentecost but believe that society and technology are a construction (deriving from a concept, as in the infosphere) rather than an emergence (which can – but need not – produce a concept, as in the docusphere)? Conceptually, there is no difference, and besides, both are subject to strong empirical refutations: no one has ever become polyglot overnight, just as no society has ever been omniscient, as it obviously should be if collective intelligence were more than an empty expression. This also applies to collective intentionality. Whatever collective intentionality wants, or does not want, and provided such a unitary feeling exists, when a state prints too many banknotes people do not become richer, as they should if the value of money depended on

---

[91] Hysteresis does not only explain emergence, but, when technically possible, makes it evident and easier to reconstruct. See D. Roy, *The Birth of a Word*, TED talk 2011, available at: http://www.ted.com/talks/deb_roy_the_birth_of_a_word [02/02/2022].

[92] F. Darwin, (ed. by), *More Letters of Charles Darwin,* 2 vols., D. Appleton & Co., New York 1908, 1:94.

[93] I have developed the critique of collective intentionality in various places. See in particular *Where Are You? An Ontology of the Cell Phone*, Fordham University Press, New York 2014 and *Anima e iPad*, Guanda, Parma 2011, pp. 85–115.

collective intentionality, but poorer, as money loses its value. Why is it that, although everyone agrees that 1,000 marks are a lot of money, they are suddenly worthless, and a stamp like 100,000 or 1,000,000 has to be superimposed on them? Rather than a collective intentionality, it would appear to be a collectively masochistic intentionality. Finally, if it were collective intentionality that determines the value of money, phenomena such as financial crises would be inexplicable. Just like natural phenomena, they cannot be controlled, with one significant difference: namely that the disappearance of collective memory and documents would put an end – albeit dramatically, decreeing the end of civilisation – to any financial crisis, whereas it would not stop the rain, nor would it undermine the law of gravity.

Let us think about this, every time we look at the Web. Apart from exceptional circumstances, social reality is not actively constructed, but passively endured. This same mechanism underlies our ordinary assumption that money is intrinsically valuable and that the traffic police are entitled, if the conditions so dictate, to take away a person's driving licence. Maybe, at some point, in certain cases and in certain people, a realisation takes place. However, this is a possibility that may never occur, and it is neither the norm nor the prerequisite for being in the social world. Likewise, it would be absurd to claim that revolution is the norm. What is a revolution? If it is the product of good intentions, it is not a revolution. Marx's great intuition was precisely this: to see revolution not as the outcome of human thoughts, of the ideas one professes or the books one has read, but rather as the result of immanent necessities, not unlike the chemical reactions that determine the destiny of the quartet in *The Elective Affinities*. The point is that the greatest revolutions, the ones that leave their mark, are *technological* revolutions – which is not surprising when one considers that the human being is a creature specifically defined by its relationship with technology. And the transvaluation to which philosophical reflection is called consists precisely in a reinterpretation and redefinition of our value system in the light of the transformations – which are first and foremost revelations and emergencies – brought about by technology much more than by ideology.

Now, if there is no construction but rather emergence, the point becomes precisely to determine how emergent processes develop from hysteresis. On the anthropic scale, as we shall see in 2, the human being – as a human, and not simply as an organism – is connected to a system of automata, that is, of supplements, which in the social world are called documents. Documents are what shapes the consciousness of humans and thus makes them capable of acting as humans – not as generic souls, as is the case with many non-human animals. We should not forget, however, that even in the case of non-human animals we have abundant evidence of complex forms of sociality, which it would be difficult to attribute to some naturally infused benevolence.[94] Moreover, if social reality were indeed constructed,[95] planned economies would have proved successful, which was not the case, at least until the documedia revolution. I thank you for putting up with this digression, because it will make everything clearer in the remaining pages of this volume. Above all, it will clarify a key concept to which I attach the utmost importance: that of mobilisation, to which we shall come shortly.

1.2.4. Biosphere

Are we all embodied and embalmed compendiums in the docusphere? No. Much more classically, we are whitewashed sepulchres. To claim that the human is a document like any other, as some have done,[96] is to confuse the map with the territory; rather, humans, in their race through history and towards death, are the origin of any process of social documentation, that is, they are the soul that feeds the automaton. The sphere from which the docusphere draws its nourishment is life, behaviours linked to our metabolic needs, intertwined with the social structures that specifically characterise the human life form. This life has nothing virtual,[97] liquid,[98] or solipsistic[99] about it, contrary to what was implied by the claim that television kills reality.[100] Therefore, the most correct way in which we could translate "onlife"[101] today is not so much life down here or up on the cloud, but rather: On Life, as two words, *De vita*, *Perì tou biou*. Even more important than the docusphere, in order to understand the essence of the documedia revolution and the role of us humans within it, is the biosphere: the sphere of vital impulses that give meaning and matter to the docusphere. These drives are what

---

[94] It has also been argued that the tendency to cooperation is typically human (cf. M. Tomasello, *The Cultural Origins of Human Cognition*, Harvard University Press, Cambridge (MA) 1999; Id., *A Natural History of Human Thinking*, Harvard University Press, Cambridge (MA) 2014). I think this is an anthropocentric perspective, since the most distinguished examples of cooperation come from insect superorganisms.

[95] P.L. Berger, T. Luckmann, *The Social Construction of Reality: A Treatise in the Sociology of Knowledge*, Anchor Books, Garden City (NY) 1966.

[96] O. Ertzscheid, *L'homme, un document comme les autres*, in "Hermès", 53, 1, 2009, pp. 33–40.

[97] Responding to a climate typical of the second half of the 20th century, which is well represented in the posthumous writings collected in G. Châtelet, *L'enchantement du virtuel*, Éditions Rue d'Ulm, Paris 2010.

[98] Z. Bauman, *Liquid Modernity*, Polity, Cambridge 2000.

[99] S. Turkle, *Alone Together: Why We Expect More From Technology And Less From Each Other*, Basic Books, New York 2011.

[100] J. Baudrillard, *The Perfect Crime*, Verso, London 1996.

[101] Floridi (ed. by), *The Onlife Manifesto* cit.

gives meaning to the automaton, making humans the masters (albeit submissive, listless and self-victimised) of technology.

Indeed, what do documents record? It would be a mistake to assume that the object of recording is thought. This has never been true: a vinyl record of a symphony does not record a thought. What is recorded are the acts of a soul. It is no coincidence that Aristotle defined the soul, in its three dimensions – vegetative, animal and intellectual – as the entelechia,[102] the "form of a natural body that has life in its potential"[103] or, in perhaps slightly clearer words, the internal purpose of a living body. As such, it is the centre of needs and of the resources to satisfy them: it is the principle of the nutritive, sensory, rational and motor faculties.[104] Reciprocally, having an organism is the condition for having a soul as a property emerging from the organism, which, under certain conditions, can give rise not only to the vegetative and animal strata, but also to the intellectual one. This has an important consequence that does not concern the quality of what is done – both my smartphone and my mind can perform calculations or translate from one language to another – but their meaning. I can do maths or translate words to meet practical needs, or to pass the time. Instead, my smartphone acts because it is programmed to do so: it could also do nothing, and it would be the same for it, since an automaton has no internal purpose, only an external one – i.e. it is made by souls to meet their needs.

That is why we will never see an automaton with withdrawal symptoms. Automata are not bored, afraid, wishful, or power hungry. This apparently virtuous ataraxia derives only from the fact that an automaton feels the pressure of time and need far less than an organism does. And this is simply because the organism's only purpose is itself, i.e., its own survival. So too, if a watch runs out of power, it does not die: it is made to be rechargeable. Switching on and off are not unique events, they are a serial succession. The same applies to cars, traffic lights and hairdryers. This means that clocks and Christmas trees have no hurry, no boredom, no anxiety, and no internal purpose. They are made to mark the time or to indicate the place for presents, so they have an entirely manifest purpose as their *raison d'être*. However, this purpose does not lie within them, but in the humans who designed them: to realise this, one need only ask oneself what purpose or meaning a Christmas tree might have if for some reason it survived the disappearance of all humanity.

This circumstance has immense consequences for the revolution that we can and must make, giving up the victimisation that ascribes to hidden forces what

---

[102] Aristotle, *De anima*, 412b 9–1.
[103] Ibid., 412a 20–21.
[104] Ibid., 413b 10ff.

actually depends on us, and on us alone, for better or for worse. Let us think of the current situation in which I am writing these lines and which I hope will have improved by the time you read them. "A virus is haunting Europe – the vector is capitalism"[105] – that is a headline that was often read in the first months of the pandemic, which expresses in the best possible way the superstitious attitude behind victimhood. Some have even stated, in all seriousness, that "digital corporations are spreading this insidious virus and at the same time weakening our immune system in order to make us even more dependent and addicted, so that they can control us better."[106] The result was that criticism and analysis gave way to negationism. Now, far be it from me to prevent anyone from cultivating, in their own internal forum, their own professed convictions – I certainly do not agree with the Socratic principle that the philosopher wishes to die. However, in this respect, I am convinced that our judgement of Socrates would be very different if, instead of drinking hemlock, he had sought death by becoming infected with the plague, at the risk of taking others with him into the long dreamless sleep that he considered most desirable for himself.

I would suggest an alternative line of reasoning. I suppose I am not the only one who, at the appearance of the virus, thought of Odradek, the undefined being in Kafka's *Cares of a Family Man*.[107] The two are closely related, and by reflecting on Odradek we can understand something about the virus even if we are not virologists, epidemiologists or scientists, but only potential hosts of this most unpleasant little creature. It is true that, unlike the virus, Odradek is able to move on his own and does not need our legs; he wanders around the house, making the call "stay at home" useless. Above all, he does not kill anyone, yet the father is worried, and asks himself two only seemingly contradictory questions. The first is whether Odradek can die and, therefore, whether he is alive. The second is whether he will outlive him, and whether he will outlive his children's children. I do not know about Odradek, but the virus certainly cannot die, simply because it is not alive. We don't know whether viruses are pre-life forms, or lives that have been degraded for some reason, but they certainly need the hospitality of a living being to live. Now, "anything that dies has had some kind of aim in life, some kind of activity, which has worn out", whereas the virus has never had any purpose whatsoever, nor any activity, and when we call it an "invisible enemy"

---

[105] B. Montague, *A Virus is Haunting Europe – The Vector Is Capitalism*, in "The Ecologist", 18 March 2020.

[106] B.C. Han, *Noi, schiavi felici della pandemia digitale*, in "La Repubblica", 31 October 2020.

[107] F. Kafka, "The Cares of a Family Man" (1917) in *Franz Kafka: The Complete Stories*, Schocken, New York, 1971, pp. 427–29.

we give in to anthropomorphism – understandable, of course, but anthropomorphism nonetheless.

This is the point we need to think about carefully, because it holds the key to understanding today's problems. What is it that holds together such diverse entities as the virus, the Web, and webfare, i.e. the welfare system that I propose to draw from the Web? Let us start with the obvious. The virus, without humans, would be perfectly useless. It needs our hands, our feet, and the needs, desires and urgencies that drive us to meet other humans. If patient zero had been a sloth, we could be sure that the pandemic would not have happened, indeed that we would never even have spoken of Covid. In its destructive power, however, the virus has accelerated processes that were already under way, involving the transfer of a huge number of activities onto the Web. That the virus would do nothing if it did not have humans to carry it seems quite obvious. But the Web would also be perfectly useless without humans, and this fact is not given enough thought, as people prefer to complain about the tyranny of machines and the dictatorial designs of platforms. Now, machines do not tyrannise anyone. The tyrants, if anything, are the humans behind them, and platforms would not make sense without humankind. It is precisely this principle, which is systematically underestimated and yet crucial, that lies at the heart of webfare: the virus gives nothing back, the Web gives something back, in terms of services and information, but it should give much more because it owes us much more, quite simply because we can think of humans without the Web, but not the other way around.

The Web is nothing without hands tapping and clicking, and above all without bodies feeling desires, a temporal urgency, a physiological pressure. That is why, as I said, those who see the Web as collective intelligence and fear that one day it will take over are wrong. It will do no such thing, because to strive towards something you have to be alive, and neither the virus nor the Web are. Imagine a world without humans, but with the Web: it would be a world just like ours, where houses serve as shelters for animals or are swallowed up by vegetation like the Mayan pyramids, but where mobile phones, screens, and other devices remain inert, because they are useless to sloths, beavers, elephants, and of course viruses. We must therefore avoid the anthropomorphism whereby, when we speak of "artificial intelligence", we picture a kind of reproduction of the human intellect including characteristics such as finality, the will to power, desire and hope. And of course, the fantasy of enframing[108] is also a relief from human responsibilities, because it makes us the passive executors of orders coming from

---

[108] M. Heidegger, *The Question Concerning Technology and Other Essays*, Garland Publishing, New York & London 1977, p. 20.

elsewhere.[109] On closer inspection, we are still in the grip of Asimov's laws,[110] which unanimously and fallaciously assume that automata can have human intentions. This anthropomorphism was understandable at the time, but not today. A gorilla perhaps could take over, and a disused technical tool may become a totem, but not without the incense offered to it by submissive humans. An automaton can never do that. Documediality does not make decisions, but executes decisions that have already been made. It is a fairy-tale naivety to imagine the growth of automation as an enhancement of the will to power of machines, which will take the place of humans. What machines need, on pain of paralysis – and this is their only kinship with vampirism – is not human intelligence, but organic life, and in particular its mobilisation.

Between the virus and the Web, however, there is a fundamental difference. The former gives nothing back to humans, it merely takes their lives away when things go wrong both for the virus itself, because it stops spreading, and for us. The second, on the other hand, gives back. There is a common feeling that there is something amiss with this restitution, but it is dissipated by great distractions (the Panopticon, the protection of privacy, the exploitation of micro-work), and great imaginary inquisitors, such as capital and the dominance of technology. Let us leave distractors and inquisitors aside and imagine this apocalyptic picture: the ecological catastrophe makes all human souls disappear, but all the automatons remain, billions of smartphones, smartwatches, and tablets as far as the eye can see, ATMs, self-published and heteropublished novels, leopard-print sofas, and Mickey Mouse watches. All this for what? For whom? Not for beavers (suppose they are the life form that has benefited most from the disappearance of humans): it is hard to imagine a beaver watching a pornographic film on leopard-print couch. At that point, a hypothetical Survivor God would realise that this highly imperfect being – the human – was the purpose of the whole system.

Now, humans could come to that conclusion even without the ecological catastrophe – indeed, while trying to prevent it and hopefully succeeding. Far from kenosis, it would be an apotheosis. I personally rely on it, but this does not mean

---

[109] This fatalistic and self-absolving view can be found in F. Kittler, *Gramophone Film Typewriter* (1986), Stanford University Press, Stanford 1999 as well as in K. Kelly, *What Technology Wants*, Penguin, New York 2010 and a plethora of other books. This form of self-absolution must be overcome. And yet, on 29 October 2019, "La Repubblica" published an interview with Jaron Lanier who first says (rightly) that we produce value with our mobilisation on the Web, and therefore we must be paid, but then goes on to claim that a "mass manipulation of consciences" has taken place, as if there were consciences in themselves, not manipulated by culture, religion, political ideologies. And as if our consciences could be of interest to anyone. Cf. J. Lanier, *Tanti auguri, internet. Ma ora paga per i nostri dati*, in "La Repubblica", 29 October 2019.

[110] I., Asimov, *I, Robot*, Doubleday & Company, Inc., Garden City, New York 1950.

– needless to say – that I aspire to bliss, simply because humans are never happy, and that is why they stole fire from the gods, believing they would become as happy as them and not understanding that the problem is not fire, but human nature. We must therefore lay the groundwork for understanding that humans, when they consume information, produce documents that are not necessarily accessible to them, and are much more valuable than the information and services they receive as users. In the exchange between users and Internet platforms, therefore, a production of value takes place to the exclusive advantage of the latter, which I call "documedia surplus value" and which I analyse in 4.2. This surplus value, if reasonably taxed, could be the basis of webfare, the Web-based welfare system that I propose in 4.4. Of course, one may wonder whether the exchange is really unfair, and whether we should rather speak of mutual benefit, especially if we refer to the infosphere: after all, who would have thought of having free libraries and music archives? In the same way, it is legitimate to ask whether the lowering of prices allowed by the automation and fine-tuning of distribution that platforms derive from the profiling of the docusphere does not already constitute compensation for the production of value generated by users.

However, even if this is true, it remains that the infosphere and docusphere would be worthless if they were not animated by the biosphere, by the needs that have always mobilised humans and which now determine the necessity and purpose of the Web. Once all this is acknowledged, not only is a Webfare system conceivable and legitimate, but we also find the right place for the three spheres: the biosphere is the fundamental life-world that gives purpose to the Web as well as to any other technical device; the docusphere generates value through consumption, and is the sphere in which human souls meet automatons; and the infosphere is a zone for responsible communication and correct information which, far from constituting the essence of the Web – but rather, as we have seen, a very minor part of it – represents its *ought*, the development that humanity must follow in the form of education and rational progress. In this transformation, two great turning points must be recognised: the solution of the mystery of the commodity-form and the rise of the mystery of labour.

1.3. Solving the Mystery of the Commodity Form

Let us recall Marx's description of the mystery of commodities. There is no mystery in use value: if I build myself a table, it is clear that I am the one who made it, and that the wood that goes into it is sensory material. Things change, however, when we move on to commodities, i.e. exchange value. Here, a trans-substantiation takes place that makes the table something "sensibly supersensible".[111] This mystery does not derive either from a transformation of the material, or from the effort that was necessary to produce it, which remains the same in use value as in exchange value. The transubstantiation depends exclusively on the commodity-form assumed by the table (or any other product), and consists in a reification whereby the social characteristics that have contributed to production have been transformed into the natural characteristics of the product. And, conversely, the natural characteristics of the product are transformed into social ones. Thus, long before the Internet of things, a relationship between objects concealed a relationship between people, which is all the more difficult to recognise because it is broken down into many different activities due to the characteristics of industrial production. Now, once we realise that the infosphere, which is the contemporary equivalent of the mystery of the commodity, is merely the tip of a docusphere and a biosphere, we will find the solution to the mystery of the commodity.

As documents that record the forms of human life, it is quite clear that commodities manifest a relationship between people, instead of concealing it, as they did in classical industrial production. The commodity, as a document, is a social object, and this explains the extension of the docusphere, which however always presupposes a biosphere, constituting the human capital I will discuss in 2.3. Two things follow. First, the commodity's transformation into a document is not a deviation, a spiritualisation or a cognitive metamorphosis, but simply the revelation of its essence. Second, machines and documents are always representatives of a human biosphere, which motivates the production of documents and determines their ultimate purpose: in nature, in a herd of non-human animals, but also in a hypothetical tribe of computers, we can be sure that there would be no stock exchange.

There is nothing sensibly supersensible about a process that starts and ends in the biosphere, i.e., human needs and the mobilisation resulting from them. This biosphere is capitalised in a docusphere, the structure of which the infosphere is the super-structure. It is characterised by two fundamental processes: the weak transformation whereby commodities are increasingly documented, and the strong transformation whereby documents become the fundamental commodity in the documedia economy. If industrial production was the production of consumer goods, here it is mobilisation, the biosphere, which generates documents (the docusphere), which in turn enable automation and thus production. Life is now the great container taking upon itself a double purpose: that of providing

---

[111] "The form of wood, for instance, is altered if a table is made out of it. Nevertheless the table continues to be wood, an ordinary, sensuous thing. But as soon as it emerges as a commodity, it changes into a thing which transcends sensuousness." Marx, *Capital* cit., p. 163.

documents and that of constituting the ultimate goal of the whole system, which can be automated as much as one likes, but remains subordinate to human needs and consumption. This characteristic is reinforced in the documedia age now coming to the fore. That is what the enhancement of memory is for: the construction of a great virtual workshop in which algorithms, trained by our behaviour, will work, eventually achieving perfect automation.

1.3.1. Information

Here, too, let us start from the surface: the infosphere. Originally anonymous and amnesic – apart from the producer's signature, which we already find on Babylonian bricks – traditional goods are now hyper-documented and, above all, can be read even by robots, thanks to bar codes. In this respect, the Internet fulfils needs that are much older than its current manifestation as the Web. The name of a pharaoh and, even closer to home, "Johannes de Eyck fuit hic" in the *Arnolfini portrait* are in fact documentation that strips the commodity of some of its mystery. This circumstance was overlooked by Marx, even though it increased precisely with the industrial economy and production. Style was the characteristic sign of the artist, the mark of the craftsman in his work, as Descartes wrote when speaking of the idea of infinity, which would be God's signature in us.[112] However, as soon as industry began to mass-produce, the notion of "brand" came forward, marking out not the individual but the species. And there is nothing surprising about the fact that the work of art, in the age of its technological reproduction, finds in the commodity brand the substitute for the artist's signature.[113]

Warhol's *Brillo Box* and *Campbell's Soup cans* are the culmination of a story that goes back at least to 1812: then, in his account of the catastrophe of the Grande Armée, Joseph de Maistre recalled that the Cossacks were selling Breguet watches stolen from the French. In 1838, his brother Xavier, in *Le Philosophe de l'Odenwald*, spoke of a "wonderful gold Breguet watch". A family heirloom? No, a sign of the times, which we even find in Pushkin's *Eugene Onegin*. There was a moment when the brand, the product of industrialisation, entered history and art: Porter beer in *The Magic Mountain*, the Waterman pen in *Acquainted with Grief*, Hemingway calling Ezra Pound "Stetson" because of his wide-brimmed hat, Rossetti shoes in Thomas Bernhard, the use of the adjective "brand-

---

[112] R. Descartes, *Meditations and Other Metaphysical Writings*, trans. Desmond M. Clarke, Penguin, London 1998.

[113] M. Ferraris, *From Fountain to Moleskine*, Brill, Leiden 2019; A. Donati, *Law and Art. Diritto e arte contemporanea*, Giuffrè, Milan 2012. See also G. Ajani, *I diritti dell'arte contemporanea*, Allemandi, Turin 2011.

ed" as a mark of quasi-aesthetic appreciation, as secularised sanctity or gentrified nobility, up to car brands almost everywhere.[114]

The logo perpetuated this trend by making a product immediately recognisable and certifying its authenticity, acquiring the same function – and exposing itself to the same potential counterfeiting – as a signature on a document. We must therefore bless the fact that we are a logo- and icon-driven civilisation, because otherwise countless pieces of information about our past would have disappeared. In this sense, documentation is first and foremost passive, so that document-goods are documented goods: for example, the type of fruit, the degree of ripeness, the length of preservation, the sugar content. This is the cognitive advantage of the hyper-documentation of products, which is a growing phenomenon, whereas in the past the fundamental information came from the price – a very poor summary, as one can imagine.

This process has not stopped evolving. Origin data on food products, nutritional stats, information on fabric composition and washing methods mean that traditionally anonymous products are now full of signatures, dates and data. This is most evident in digital stores, where the interaction, so to speak, between what is in the text (the document) and what is outside the text (a bottle of wine, a cheese, a dress) is heightened to levels of hyper-documentation. Buying food online, receiving all the information on composition, origin, and expiry date through documents, running simulations to find out whether the clothes we order online will fit us, and whether the furniture available in the Web catalogue will suit our home – all this means giving more and more importance to the map over the territory. The territory, i.e. the use value, does not change. But the map, or exchange value, is extended, articulated and expanded.

To take a lesser known example, consider BIM (Building Information Modelling), which offers a documented perspective on buildings by evaluating them not only according to the three spatial dimensions, but also through a fourth dimension (the time needed to realise them), and a fifth dimension (the cost of building and, in perspective, of maintenance).[115] In addition to these spatio-temporal and financial relations, BIM deals with an analysis of light, geographical information, building components: in short, it implements the reduction of substance to a bundle of properties that Locke had carried out in metaphysics. In the latter sphere, the reduction was improper, because it transformed an epistemological character of their analysis into an ontological property of the objects, whereas in the field of social ontology and its objects it is perfectly legitimate.

---

[114] N. Klein, *No Logo: Taking Aim at the Brand Bullies*, Picador, London 1999.

[115] E. Asadi, M. Gameiro da Silva, C. Henggeler Antunes, L. Dias, L. Glicksman, *Multi-objective Optimization for Building Retrofit: A Model Using Genetic Algorithm and Artificial Neural Network and an Application*, in "Energy and Buildings", 81, 2014, pp. 444–456.

Houses are hopefully not documents, but there is no doubt that documents are needed to make satisfactory houses.

1.3.2. Storing

Up to this point, we have seen the continuation and growth of old practises. The novelty introduced by the Web is the automation of the system and the consequent spontaneous expansion of the docusphere, with a first and very central process of capitalisation. Information is the highlighting of pre-existing data. Archiving, on the other hand, is the creation of a capitalisation field as large as the docusphere and capable of collecting an enormous amount of recorded acts from the biosphere that were previously lost like tears in the rain. We are witnessing a shift from the production of commodities through commodities to the production of documents through documents, ensuring ever more complete automation and ever more effective distribution. Or, more precisely, we are witnessing the production of documents through behaviour.[116] In the case of strong documents, the ones we produce on social media, this is an extension of the classic exchange of opinions and daily chitchat, except that it can now go on for longer, as it is not incompatible with work activities that most often consist of document production and, above all, unlike the social exchanges of the past, it leaves traces – it is itself a production of documents. However, the strong documents are only the visible part of a much larger continent: the weak documents, the metadata, the hysteresis of our biosphere-related activities. As is often the case, David beats Goliath: the most valuable document is the smallest. It is not the explicit one accessed by those who surf the Web in search of information or services, but the one (or rather, those) that the user leaves as a trace of their surfing, often without being aware of doing so.

Where does this value come from? Simple: what is archived is capitalised, becomes someone's property, and can be sold. The fundamental characteristic of the documedia revolution is that it has transformed every kind of document, from maps to photographs, from messages to likes, into an actual or potential commodity. This is the greatest and by far most significant transformation: traditional goods are sometimes offered for free or at very low prices, provided that in return the buyers, effectively doing the fundamental work of mobilisation, leave their documents behind. These documents are worth much more than money

---

[116] Cf. G. Mari, "Il lavoro 4.0 come atto linguistico performativo. Per una svolta linguistica nell'analisi delle trasformazioni del lavoro", in *Il lavoro 4.0. La quarta rivoluzione industriale e le trasformazioni delle attività lavorative*, ed. by A. Cipriani, A. Gramolati and G. Mari, Firenze University Press, Firenze 2018, pp. 321–335. More than a linguistic shift, mobilisation is about a textual shift: the recording of acts.

because they do not speak of what consumers have but of what they *are*, of their beliefs, weaknesses and hopes, not as individuals but as part of a virtual class, that of all those in the world who are similar to them in some way. If industrial production was the production of goods for consumption, here it is consumption (biosphere) that generates documents (docusphere), allowing automation and hence production.

As a result, every day that goes by (with a progression that is sometimes slow, sometimes very fast, but always uninterrupted), production requires less work – if any – in the ordinary, formal sense of the term. The aims of document acquisition are to automate production by learning about human behaviour and to perfect distribution by knowing human needs in detail. Indeed, since producers and consumers have always been human, there is nothing better than recording human behaviour in order to automate production – sending humans home, now useless as producers – and to make distribution perfect. Humans are and will always be consumers, except now they can be at home, happily waiting for the delivery of a good that will arrive on time and at a price they could never have hoped for in the past.

This becomes clear as soon as we shift from the Ptolemaic Web to the Copernican Web. The former is eidocentric, the latter ethocentric. Eidocentrism claims that the primary function of the Web is to collect, disseminate and capitalise on the idea (*eidos*) and its variants, not meant literally as forms, but as thoughts, convictions, beliefs, information and general feelings. Ethocentrism, on the other hand, is based on the hypothesis that the aim of the Web is to collect, classify and capitalise on forms of behaviour (*ethos*): machines automate processes because they record and classify our acts, habits and desires, so that what we call "artificial intelligence" is simply the great archive of the human comedy. Now, the eidocentric view neglects an essential circumstance: what AI offers is not intelligence but capitalisation, which in itself has nothing cognitive about it, consisting in the mere accumulation of iterable, alterable, and therefore monetizable documents.

The shift from eidocentrism to ethocentrism is not only theoretical, because it affects the understanding of the political objective of the Web: for eidocentrism the latter is political control, while for ethocentrism it is economic profit. Also, this transition involves two different views of the Web as a whole. In fact, eidocentrism adopts the information → document → action model: information is fixed in documents and produces actions. Ethocentrism, on the other hand, follows the action → document → information model. That is to say: first there are behaviours that are recorded in documents and that can, under certain circumstances and with the appropriate calculation and interpretation tools, be transformed into information, which is generally neither transparent nor in the public domain.[117] This is why documents are so valuable for automation: they are actions that are crystallised like insects in amber, a bit like the Ashanti people,[118] who, upon subjugating a neighbouring territory, would take possession of the "books" of the former, for instance the treaties that the king had signed with other potentates, and demand that those treaties be valid also for them.

1.3.3. Profiling

In this context, profiling should be conceptualised as a phenomenon analogous to the stock market, where it is readily accepted that prices are created by betting on the future and on human behaviour: only, it is much more profitable, because here it is not a matter of guessing the growth of a stock or a sector, but of knowing exactly what human needs are, thus reducing business risk and maximising profit and automation.

The first and most obvious aim of profiling is the rationalisation of distribution. Work is not only production; it is also – and mainly – market, i.e. distribution and exchange, and these can take place at any time, especially if not only the market, but work itself is globalised.[119] We are witnessing a shift from industry to the market[120]: just as computers tend to be able to do everything, technically speaking, so too hysteresis enables the conversion of everything into everything. In this case, commodity-documents produce and distribute (at low prices and without waste, i.e. with the advantages of a fully efficient planned economy),[121] the most diverse products: jogging shoes, coffee pods, election programmes, books, television series, etc. In fact, producers of traditional goods, be it shoes, coffee, or government treaties, buy them at a high price from platforms and try to create their own platforms by tracking consumption. Likewise, concept stores do not sell objects (which are deliberately heterogeneous: beds, suitcases, shirts,

---

[117] See the remarkable book by A. Badia, *The Information Manifold. Why Computers Can't Solve Algorithmic Bias and Fake News*, The MIT Press, Cambridge (MA) 2019, which has the merit of highlighting how information has not only a semantic value, but also a syntactic and pragmatic one, showing how the infosphere is rooted in the docusphere (syntax) and the biosphere (pragmatics).

[118] J., Goody, *The Logic of Writing and the Organization of Society. Studies in Literacy, the Family, Culture and the State*, Cambridge University Press, Cambridge 1986.

[119] M. Graham, *The Rise of the Planetary Labour Market – and What It Means for the Future of Work*, in "The New Statesman", 9 January 2018.

[120] V. Mayer-Schönberger, K. Cukier, *Big Data: A Revolution That Will Transform How We Live, Work, and Think*, Eamon Dolan, Boston-New York 2013; V. Mayer-Schönberger, T. Ramge, *Reinventing Capitalism in the Age of Big Data*, John Murray, London 2018.

[121] L. Phillips, M. Rozworski, *The People's Republic of Walmart: How the World's Biggest Corporations Are Laying the Foundation for Socialism*, Verso, London 2019.

pens, biscuits ...), but rather forms of life, not unlike travel sites. The distinctive feature of a concept store, however, is that it does not even sell concepts, but rather behavioural patterns or, if you prefer, reference narratives: *ethos*, not *eidos*. But, of course, here too, documents are essential: to sell life forms, you need to know them.

The fact that many services are still offered for a fee today is due to the fact that we still rely on a market with classical jobs that can be paid for. But the growth of automation will eventually produce two results: the lowering of the costs of services and goods thanks to automation and, in parallel, the disappearance of most productive jobs. This is why I speak of "capitalisation": documents are not a good in themselves, but are used to sell goods or services. No exchange value, starting from its linguistic meaning, is a good in itself, and certainly not even stock market shares, Chanel No. 5 and *War and Peace* are goods in themselves. However, this value, which is as relational as any other value in the social world, is magnified as the costs of producing goods plummet due to automation, so that the real benefits are derived, as I said, from market knowledge for production and distribution purposes.

We are dealing with an active documentation, where we witness the creation of market-enterprises that allow for the movement of goods. Goods have to move, so either we go to the goods, as was traditionally the case, or the goods come to us, as is the case in the documedia world. But for goods to come to us, documents are needed to ensure their transfer. And this is where the new forms of intermediation come in, each taking the place of previous intermediations. Now the process is much more effective, and capable of self-training. Traditional intermediations were not Internet platforms, but basilicas, courts, travel agencies, parliaments or taxis. Today's intermediations, which for incomprehensible reasons are often interpreted as disintermediations, are archives for document management. This is how goods arrive at our homes. Platforms generate documents, sort them, and at the same time acquire a gigantic flow that enables them to perfect their algorithms and thus obtain surplus value, which is rightly perceived by users as a reduction in prices and an improvement of services, but which nevertheless results in a primary accumulation of documents by the platforms.

1.3.4. Automation

We can begin to see not only why automata are important to us, but also why our mobilisation is so interesting to automata, or more precisely to their owners. Human intelligence can easily be replicated by an automaton. But no automaton will ever be able to replicate life in the illogical multiplicity of its needs. It is on this basis that powerful and effective cognitive enterprises such as translation machines can be built. The machine is not more intelligent than the human, it simply has more patience, memory and speed. But only a human can feel the need to translate from one language to another, or lose patience, or feel tired. Automation thus consists not in machines taking over, abandoning traditional executive tasks and formulating intentions of their own, but, quite the opposite, in the multiplication of tasks that require execution rather than intention: and, often, in the discovery that those tasks we thought required intentions can be fully done by running a programme.[122]

Thus, the more the human life form is profiled, the easier it is to automate its productive processes. To automate means to imitate, in a mechanism, properties that are typical of an organism,[123] it being understood that one can only imitate actions, not the motivations for those actions. The dictation mechanism with

---

[122] Automation means that production consists of running a programme, where any human intervention is equated with text writing, since the same interfaces are used for writing as for production. This takes place in three senses. First, and most obviously, the interface with which commodities are produced, in a highly automated environment, becomes identical to the interface with which documents are produced – the keyboard of a smartphone, the voice commands of a cooking machine, or a dictating device for metaphysical musings. The second sense is that the commodity opens up a twofold documentation process: on the one hand, it is rich in information about itself, its mode of production, expiry dates, conformity to standards, etc., all of which is strictly absent from traditional commodities. We can call this type of documentation "expressive documentation". On the other hand, goods are programmed to generate documents and collect information. When we are asked in a shop if we want the receipt by email, it is in order to create a system of information gathering and capitalisation. We can call this type of documentation "receptive documentation". This means that goods are produced by documents and with documents. And above all it means that, precisely because of automation, humans are now very important as producers not of goods, but of documents generated by their mobilisation. They are also very important because it is their needs that give meaning and time to the production of goods; but I will come back to this later. This is why so many distinctions of the bourgeois world are disappearing, starting with the distinction between blue-collar and white-collar workers. This circumstance, however, is not too sociologically relevant, since the number of people working with 3D printers will in any case be insignificant compared to assembly-line workers. Here, goods reveal their subordination to documents: they are manufactured by machines based on programs that have the same physical interface – a screen or a keyboard – that is used to produce documents.

[123] Cf. A. Leroi-Gourhan, *Gesture and Program* (1964–65), MIT Press, Cambridge (MA) 1993, pp. 250–251, italics mine: "Not until the last twenty years has the mimicry of living matter by artificial matter achieved a reasonably high standard. A century of familiarization with electricity and, later, with electronics was needed before this became possible. The resulting machine represents a synthesis of all previous stages. Mechanical organs of execution, actuated by as many energy sources as efficient articulation may require, are set in motion by a program that, at one stage at least, is recorded on tape. The essential difference lies in the presence of what amounts to a real nervous system through which the central organs transmit commands and monitor their execution." These words were written more than half a century ago, but they shed an unparalleled light on the present.

which I am writing these lines finds its condition of possibility in the vocal recognition that is perfected through training, which is nothing more than exposure to a large quantity of vocal traces. That is why it is so keen to offer us its services – it needs us to expand the archive and improve its performance. Indeed, this is also why a child can be even more petulant than a voice recognition mechanism. As we shall see at length in 3, intelligence, far from being something exclusive and inimitable in humans, is the part of humans that can be mimicked by a mechanism: it only has to record, manipulate, calculate, and repeat. Judging by abstract functioning, there is no way of differentiating human intelligence from artificial intelligence; instead, as we shall see, what makes the difference is that natural intelligence is rooted in a body, which defines its manifestations and aims.

Excluded from the production of traditional goods and many services, human agents are not losing relevance, but are assuming a new and preferable function. At a time when life, in its interaction with the Web, is capable of perfecting automation through profiling, and at a time when it is clearer than ever that the essence of humans is not to replace machines, but to consume what machines produce, a progressive biopolitics should devise forms of redistribution of the surplus value accumulated by machines. Above all, it should be understood that every living act, insofar as it can become data, produces value and should therefore be recognised as labour and rewarded as such, for the greater glory of human life and for the sake of stability, progress and social justice. By optimising resources, automation is able to generate wealth which, if redistributed, would promote mobilisation in all senses. And mobilisation should be conceived as the very essence of human beings, their completeness or at least their necessary and beneficial goal. It will come as no surprise that automation is growing so impetuously, and that labour, in its narrowest sense, is disappearing to a directly proportional degree.

1.4. The Mystery of Labour

Automation progressively reduces explicitly remunerated labour, the labour that produces professional pride but also alienation and burn-out. Perfectly indifferent to such things, automation is the manifest destiny of documediality. The significant feature of the present is thus not the obvious disappearance of traditional jobs due to automation, but the emergence of a mystery of labour as the symmetrical counterpart of the mystery of the commodity in industrial production. This is the great mystery of our age: in the face of the reduction of explicit and paid labour brought about by automation, we have the emergence of uninterrupted, implicit, and unpaid labour – which allows for automation itself.

If the never-fulfilled dream of roaring capitalism was to make children work at night, this dream is now coming true with the added advantage that children do not ask to be paid and are not dying of starvation, whilst their parents provide the means of production. This new type of invisible and unrecognisable labour generates a widespread and contradictory sentiment, according to which, on the one hand, we are witnessing a decrease in labour and, on the other hand, there are growing complaints of an unprecedented exploitation of human labour, which would constitute, so to speak, the extreme phase of capitalism, inclined to make the whole of humanity work every day and at all hours,[124] providing the premise of a new slavery.[125] These complaints are welcome because in human experience there is a contradiction between the end of labour and the feeling of overwork.[126] We have the impression that we work fifteen hours a day[127]: Why are we tired? Of what? How is this possible? Why are jobs disappearing, and yet people are working more and more? Let us try to answer these questions, so as to find an effective remedy to a situation that is unavoidable and, I would add, largely desirable, too, for the excellent reason that a job that can be done by a machine is most likely unworthy of a human being.

1.4.1. Rarefaction

Forty thousand silent and disciplined bosses, neatly combed and booted, precise and confident with umbrellas and coats folded on the arm or laid like cloaks on the shoulders, loose or tight-fitting, never flapping. Forty thousand bosses! How can this be! There must also be the fourth-ranking managers and the older ones out of office. There must also be trainees, aspiring candidates, relatives, neighbours, acquaintances, owners of houses, hotels, restaurants, cafés, managers of bakeries, grocery stores, pharmacies, tailors, teachers, accountants, insurers, renters, agents, bank clerks, as well as ruffians, gamblers, undertakers, shareholders, savers, war heroes and pensioners, journalists, reporters, booksellers, football and boxing technicians, antique dealers, restorers and furniture makers.

---

[124] Crary, *24/7* cit.

[125] B. Stiegler, *L'emploi est mort, vive le travail! Entretiens avec Ariel Kyrou*, Fayard, Paris 2015; R. Ciccarelli, *Forza lavoro. Il lato oscuro della rivoluzione digitale*, DeriveApprodi, Rome 2018; Casilli, *En attendant les robots* cit.

[126] U. Beck, *Risk Society: Towards a New Modernity*, Sage Publications, London 1992; Id., *Schöne neue Arbeitswelt*. Campus Verlag, Frankfurt a.M. 1999; Id., *The Brave New World of Work*, Polity Press, Cambridge 2000.

[127] O. Burkeman, *If we feel busy all the time, it's because of shadow work*, in "Internazionale", 15 October 2018, https://www.internazionale.it/opinione/oliver-burkeman/2018/10/15/shadow-work [17/07/2021].

These were the participants in the forty-thousand-strong march in protest against workers, as described by Volponi.[128] It was 14 October 1980. Now workers have disappeared, but the categories listed as supporters of the anti-union demonstration have also thinned out.

Between then and now, there has been an obvious and totally non-mysterious phenomenon: the rarefaction of traditional, modern, industrial jobs, much as had already happened with agricultural jobs. A great many jobs will change and many others will disappear.[129] The reason for this is all too obvious to be worth mentioning: it is because of automation, and especially the particularly effective automation that comes from the archiving of human acts and forms of life provided by the Web. This transformation, which has been announced for almost a dec-

---

[128] P. Volponi, *Le mosche del capitale*, Einaudi, Turin 1989, p. 320.

[129] On the transformations of work in the 20th century: A. Appadurai, *Globalization*, Duke University Press, Durham-London 2001; H. Arendt, *The Human Condition*, University of Chicago Press, Chicago 1958; P. Askonas, A. Stewart, *Social Inclusion: Possibilities and Tensions*, St. Martin's Press, New York 2000; Bauman, *Liquid Modernity* cit.; Beck, *The Brave New World of Work* cit.; P. Bourdieu, *La précarité est aujourd'hui partout*, in Id., *Contre-feux: propos pour servir à la résistance contre l'invasion néo-liberale*, Liber-Raisons d'Agir, Paris 1998, pp. 95–101; R. Castel, *From Manual Workers to Wage Laborers: Transformation of the Social Question* (1995), Transaction Publishers, New Brunswick (NJ) 2003.; L. Castelvetri, *Il diritto del lavoro delle origini*, Giuffrè, Milan 1994; S.R. Chakravarty, *Inequality, Polarization and Poverty: Advances in Distributional Analysis*, Springer, New York 2009; P. Chirumbolo, *Letteratura e lavoro*, Rubbettino, Soveria Mannelli 2013; M. Colonna, E. Pugliese (ed. by), *Il futuro del lavoro in Europa. Occupazione, diritti civili, diritti sociali*, Edizioni Scientifiche Italiane, Naples 2007; A. Gorz, *Critique of Economic Reason* (1988), Verso, London 2010; Id., *The Immaterial: Knowledge, Value and Capital*, Seagull, London 2010; E. Brynjolfsson, A. McAfee, *The Second Machine Age: Work, Progress, and Prosperity in a Time of Brilliant Technologies*, Norton, New York 2014; A. Condello, T. Toracca, *Lavoro, identità: riflessioni tra letteratura e diritto*, in "Il Ponte", 2, March 2016, pp. 120–126; M. De Vos, *From Labour Market Crisis to Labour Market Reform: Lessons from and for a Divided European Union*, in *Labour Regulation in the 21st Century: In Search of Flexibility and Security*, Cambridge Scholars Publishing, Newcastle 2012, pp. 117–131; L. Gallino, *Il lavoro non è una merce*, Laterza, Rome-Bari 2007; G. Lovink, *Social Media Abyss: Critical Internet Cultures and the Force of Negation*, Polity, Cambridge 2016; R. Munck, *Globalization and Labour. The New "Great Transformation"*, Zed Books, London-New York 2002; J. Rifkin, *The End of Work: The Decline of the Global Labor Force and the Dawn of the Post-Market Era*, Putnam, New York 1995; R. Rowthorn, R. Ramaswamy, *Deindustrialization: Causes and Implications*, in "IMF Working Papers", 42, 1997; R. Sennet, *The Fall of Public Man*, Penguin, New York 1976; Id., *The Corrosion of Character: The Personal Consequences of Work in the New Capitalism*, Norton, New York 1998; B. Stiegler, *L'emploi est mort, vive le travail!* cit; A. Supiot (dir.), *Au-delà de l'emploi*, Flammarion, Paris 2016; T. Toracca, *Labour Between Law and Literature: Historical Similarities and Critical Propositions on the Present*, in "Pólemos", 11, 2, 2017, pp. 361–377; M.A. Toscano (ed. by), *Homo instabilis. Sociologia della precarietà*, Jaca Book, Milan 2007; V. Tranquilli, *Il concetto di lavoro da Aristotele a Calvino*, Ricciardi, Milan-Naples 1979; V. Trevisan, *Works*, Einaudi, Turin 2016.

ade,[130] is now at the heart of every investigation into the future of labour.[131] The World Economic Forum predicts that within five years at the time of writing there will be a 6.4% reduction in the most repetitive jobs, almost entirely offset by the 5.7% growth in new professions. Rarefaction, like precariousness,[132] is a reduction of the workforce, and this is happening in every field. Armies, which in the Second World War were made up of millions of people, are now reduced to limited numbers of mercenaries aided by drones. Factories have suffered the same fate a few decades earlier. Stock exchange sessions, which used to be picturesque and noisy shouting matches, are now replaced by quiet algorithms working round the clock. The place where the human input has not decreased is precisely in the sphere of consumption: stadiums continue to be full, and so do cities of art, beaches, trains, planes and restaurants – not coincidentally, the activities hit the hardest by the Covid crisis.

This rarefaction generally goes hand in hand with the disappearance or reduction of toil. The Runner, a figure whose *raison d'être* lies in the disappearance of physical exertion, is more symbolic of this transformation than the Rider, because the latter will eventually disappear, while the former will become more and more common. As an alternative or complement to Dieters, Runners run in order to burn energy that they do not consume at work, but – this is the crucial aspect – their mobilisation can produce more wealth than that generated by toil, because it allows them to consume goods, industrially produced at a very low price, that would otherwise remain unsold, or whose use would not make sense. What would a pedometer be for in a Fordist factory? With earplugs in your ears you would get your hand crushed in the press. The ongoing transformation is much more pronounced than the industrial revolution: then, peasants became workers and moved to the city, but toil remained. Now, on the other hand, toil has disappeared.

---

[130] C.B. Frey, M.A. Osborne, *The Future of Employment: How Susceptible Are Jobs to Computerisation?*, 17 September 2013, http://www.oxfordmartin.ox.ac.uk/downloads/academic/The_Future_of_Employment.pdf [05/01/2022].

[131] MIT Work of the Future Task Force, *The Work of the Future: Building Better Jobs in an Age of Intelligent Machines*, The MIT Press, Cambridge (MA) 2020; OECD, *OECD Employment Outlook 2020: Worker Security and the COVID-19 Crisis*, OECD Publishing, Paris 2020, https://doi.org/10.1787/1686c758-en [13/02/2022]. Cf. also *The Future of Work*, in "Nature", vol. 550, 19 October 2017 (https://doi.org/10.1038/550315a [13/02/2022]).

[132] The literature on precariousness is very extensive; see, among others: P. Cingolani, *La précarité*, PUF, Paris 2017; Id., *Révolutions précaires. Essai sur l'avenir de l'émancipation*, La Découverte, Paris 2014; Id., *Un travail sans limites? Subordination, tensions, résistances*, ERES, Toulouse 2012; S. Contarini, M. Jansen, S. Ricciardi (ed. by), *Le culture del precariato*, Ombre corte, Verona 2015; S. Contarini, L. Marsi (ed. by), *Precarietà. Per una critica della società della precarietà*, Ombre corte, Verona 2015; Colonna, Pugliese (ed. by), *Il futuro del lavoro in Europa* cit.

The point, of course, is that wages have disappeared as well; it would, however, be a strange way of reasoning to postulate that when toil reappears, wages will reappear too, perhaps with a raise. Should we be worried about the end of toil? I don't see the rational reason why, so long as we know how to keep calories under control. *Homo faber* has a specific history and geography: it did not exist when our ancestors were hunter gatherers, and these conditions were perpetuated in areas marginal to Western culture, for example in the Amazon, or at the top of it, such as Roman senators or English dukes. But it is gradually disappearing, replaced by people who travel, write, handle phones, run with watches that count their steps and heartbeats and store them who knows where. Seen from the perspective of traditional labour, rarefaction is an asymptote whose result would be an outright disappearance of labour and an unconditional assertion of automation. While both asymptotes are entirely plausible, the real issue is not to stop automation, but to conceptualise alternative forms of labour that are not simply the "new jobs" of the old world.

As with any revolution, the human tendency is to lash out at the wrong targets, projecting into the present and future the shadows of past idols and a nostalgia that is rarely sincere. It would be interesting, going back in time, to find a historical period in which humankind declared itself to be perfectly fulfilled: I bet it would not be found, but even if I lost the bet, I would be ready to bet again that it would have been the saddest time in universal history. In the face of every machine, from the lever to the mobile phone, humans have had a surge of resistance as they made changes to their lives; but they soon outgrew it, fortunately for them, as those changes were improvements in the vast majority of cases.[133] However, who would really want to be a typist or an accountant or an assembly-line worker *à la Modern Times*? It is time to shed nostalgia for a past to which no one would be actually willing to return and which, more seriously, more often than not has never been present in the first place. One is reminded of the situation described by Descartes about the husband who mourns his dead wife, but who would not mourn any less if he saw her resurrected.[134]

There are, we read, a lot of underpaid workers in Ethiopia, and others all over the world, of all genders and ages, who act as serfs of automatic intelligence for a pittance, helping computers with recognition tasks they are not yet able to perform. It is true and sad: they are there, but they too will be gone soon. Forty or

---

[133] "Les objets techniques [...] sont ostracisés non parce qu'ils sont techniques, mais parce qu'ils ont apporté des formes nouvelles, hétérogènes par rapport aux structures déjà existantes de l'organisme qu'est la culture". S. Gilbert, "Psychosociologie de la technicité (1960–1961)", in *Sur la technique* (1953–1983), Presses Universitaires de France, Paris, 2014, pp. 25–129.

[134] Descartes, *Passions of the Soul*, in *The Philosophical Writings of Descartes*, Cambridge University Press, Cambridge 1985, pp. 325–404, § 147.

fifty years ago, there were factory workers in Turin. I remember them without regret, first of all because I don't think any of them misses that life and those years. It is vain to follow the 1970s myth of full employment; it is like dreaming of the *ancien régime* in the France of the Restoration – with one decisive difference, namely that someone might have longed for the *ancien régime*, with its certainties and its *douceur de vivre*, while no one would regret the mists of yesteryear, waking up at dawn, the cold, the yellow lights of the trolley bus, and then the repetitiveness, toil and racket of the factory.

Now, the people left doing unwanted jobs are cleaners, who disappear as soon as office work starts and clerks come in (to do jobs that may be relatively uncreative or unfulfilling, but which no one would exchange for those of factory workers or cleaners). Then there are bricklayers, road builders, and the often-mentioned riders who are soon to be replaced by drones. Not only because they are worse off, but also because they are the most difficult to relocate. Of course they monopolise concerns, which is understandable and necessary, but not to the detriment of a broader perspective. Also because the real problem, especially for the impoverished West, is that there are many precarious workers in a world whose fundamental structures have not yet been made precarious. Instead of fighting for the restoration of permanent jobs, it would be advisable to set up strategies to mitigate the risks arising from precariousness.

In fact, it is useless to rely on preconceived optimism, as is often the case.[135] The technological revolution has taken place, but we must not lag behind with the conceptual revolution, which would consist, as I will argue in 4.3, in understanding as labour all production of value, starting with the exclusively human and non-automated property that is consumption. In other words, it is a question of finding labour even where there is no exploitation. We should not be limited to the classical framework of industrial labour, noting instead that the human input in digital labour, often underpaid, is predominant.

Before going any further, let's try to look at rarefaction from another point of view, for example that of the unemployed, i.e., those who have lost their traditional jobs. It is difficult to imagine them sitting on their hands, because at the very least they are holding a mobile phone. Let's not take this lightly and think about it. It is certainly easy and right to observe that the end of labour was claimed by Lafargue at the end of the 19th century,[136] predicted by Keynes in the

---

[135] Brynjolfsson, McAfee, *The Second Machine Age*, cit.

[136] P. Lafargue, *Le Droit à la paresse: réfutation du "Droit au travail" de 1848* (1883), Bureau d'éditions, Paris 1929.

1930s,[137] noted by Rifkin at the end of the 20th century,[138] and yet we are still here working. However, it is no less true that the work we do today bears little resemblance to what Lafargue, Keynes and even Rifkin were thinking of, and that, above all, today the greatest production of value takes place at times and in ways that we ourselves have not yet conceptualised as work.

Work does not occupy a central, monolithic space that defines people's identity: it is dispersed and hidden in the folds of our lives. From this point of view, the mystery of labour has the characteristics of "dream-work" according to Freud: the *condensation* whereby a function that has a ludic purpose is at the same time a production of value; the *displacement* whereby when we are at work we can safely not work, yet work will follow us through life; and the *reversal* whereby consumption, traditionally conceived as the inverse of labour, is today a crucial element in the functioning of the production system. The same can be said of holidays, entertainment, events, and all the apparently non-work spheres in which the essence of mobilisation unfolds as a notion that preserves but overcomes the narrow and traditional notion of labour.

Condensation is a weak transformation, which consolidates trends that have been ongoing for some time. A lot of work is partly done by prosumers (producer-consumers): indeed, this is the case whenever we deal with an automated interface, from the motorway toll booth upwards: printing boarding passes, making reservations, transferring money, assembling furniture, weighing purchased products, passing codes under readers, and so on. Software searches for the flight and generates the ticket, another software makes the payment, and the human agent exists not only as the beneficiary of the service, but also as the active producer of the meaning of that service as well as its diligent executor. When we become our own clerks and shop assistants and assemble our own furniture, we are certainly working unwittingly, and this explains why we may find ourselves tired at the end of a day in which we have not worked in any canonical sense of the term. However, it also means – and this is a much less prominent aspect than the previous one, to which I will return at length in a moment – that we have broken down one of the fundamental characteristics of the capitalist form of life, namely the division of labour and the resulting alienation. The universal platform supersedes not only traditional mediations, but also traditional specialisations.

At this point, through a process of *repression*, workers feel subjectively tired, but they do not know what to blame their tiredness on, because they are not aware that they are working and to some extent they hide it from themselves.

---

[137] J.M. Keynes, "Economic Possibilities for our Grandchildren (1930)," in *Essays in Persuasion*, Harcourt Brace, New York 1932, pp. 358–373.

[138] Rifkin, *The End of Work* cit.

What was once the producer's responsibility is carried out by the consumer, generating shadow labour that is not recognised as such. People who assemble their own furniture at home do not generate value: they simply reduce their costs. Today, however, in addition to these savings, prosumers generate real value with the documents they create through online orders. This is the radical innovation, which also entails a significant reversal of the relationship between company and consumer. There is no longer too much need for advertising to make the company known to the consumer, because the real protagonist is the intermediary, i.e., the platform: have you noticed how difficult it is to find the names of publishing houses on book sites? The platform needs hysteresis to make the consumer aware of the platform itself, and ironically the company may even be buying advertising space on the platform, actually contributing to its own self-destruction.

Hence the most important aspect, the *reversal*, which relates to the general phenomenon whereby consumption, if recorded, generates the automation of the production and distribution of goods: it is the exchange of values whereby in an environment rich in memory, *otium* is much more profitable than *negotium*. This is the true and fundamental metaphysical transformation that calls for a radical rethinking of the human condition. If properly addressed, this rethinking can provide emancipation by imposing a reversal of the power relations between capital and labour, as well as between humans and automata, which currently depend largely on our inability to conceptualise this transformation and draw the necessary conclusions.

1.4.2. Dissemination

We will come back to this, but in the meantime, let us examine the second phenomenon, which is just as obvious: the dissemination of many jobs, generally related to services. This trend is accelerating today, but it has already been underway for at least three decades. Moreover, it does not affect the immutable nature of labour, but simply the form it took in the industrial world. With the euphemisms of "agile" or "smart" work (unequivocal synonyms for "rip-off"), dissemination potentially extends the amount of work required by breaking the unity of time, place and action characteristic of industrial work. Beware! This is still the old work in new guise, the work that will largely disappear insofar as it is automatable and non-creative! The change is only organisational, not qualitative.

The characteristic divide of the bourgeois world and the administered world stigmatised by Adorno, the principle of "work when you work, play when you play", disappears, resulting in a Schillerian continuity between play and work. But, as usual, Adorno mistakenly viewed processes that were taking place in his distant homeland as exclusive to America. "Government offices will work faster and less bureaucratically. It does not leave a good impression when the office closes on the dot after eight hours. The people are not there for the offices, the offices are there for the people. One has to work until the work is done."[139] Who said that? A theorist of smart-working? Not exactly: it was Joseph Goebbels when, on 18 February 1943, he incited fourteen thousand fanatics gathered in Berlin's Sports Palace to total war. "I ask you: Do you want total war? If necessary, do you want a war more total and radical than anything that we can even imagine today?"[140] This has been explained as the masochistic enjoyment of the crowd ready for self-destruction,[141] while more prosaically Goebbels was able to make sense of it by considering his enthusiastic proselytes as idiots.[142] The ensuing decades and the spread of the Internet have shown that dissemination is the destiny of all work.

Madding crowds aside, this is in fact what happens with agile work, which is strictly speaking "total work". In fact, it entails the mutation of traditional jobs, which either disappear (only occasionally making way for new paid jobs) or are radically transformed, utterly disregarding the canonical times and places of work. As we can all see in our own experience, the difference between working time and living time is vanishing.[143] At any moment in our non-working lives we may be called upon to perform work activities, but at the same time, at any time in our working lives we may perform activities that have nothing to do with the immediate task at hand. And the fact that today holidays, entertainment, and celebrations are one of the fundamental sectors of value production remind us of how much water has passed under the bridge compared to the not so distant past,

---

[139] Quoted in R. L. Bytwerk, *Landmark Speeches of National Socialism*, Texas A&M University Press, College Station (TX) 2008 p. 128.

[140] Ibid., p. 136.

[141] S. Žižek, *The Sublime Object of Ideology*, Verso, London 1989.

[142] "This hour of idiocy! If I had said to the people, jump out the fourth floor of Columbushaus, they would have done that too." Quoted in S. Lehrer, *Hitler Sites: A City-by-City Guidebook (Austria, Germany, France, United States)*, McFarland Inc. Publishers, Jefferson 2002, p. 90).

[143] At the computer, and even if we only perform actions related to the work task at hand, we contribute to the production and accumulation of documents, and this makes it impossible to draw a line between work and non-work – a difference which, let us note, has begun to waver with the introduction of the telephone into offices, but which is made possible by writing in general: after all, Kafka wrote his novels in the offices of Bohemian and Moravian insurance companies. In the course of work tasks, one can carry out operations that are part of private life, but which nevertheless share with work tasks the formal character of document production, which takes place through the same channels and devices. Inversely, we can be reached by a work request at any time and in any place; not to mention, of course, the large number of contacts which, especially in management, are halfway between work and life: which category does a business dinner belong to?

when – from the English weekend in the pre-Thatcher era to the taboos of the Shabbat, which are generally easier to circumvent – rest was as opposed to work as night to day. With dissemination, work loses its not so much traditional[144] as typically industrial-age characteristics, and becomes a widespread activity, for-feiting its identity and centrality. Work is now ubiquitous, and in this ubiquity it becomes indistinguishable from life, as it was at the time of our hunter-gatherer ancestors. Work is now an activity.[145]

Let us take a break. This is where the most conceptually relevant aspect of dissemination manifests itself. Let us think for a moment and consider the words with which Marx, in the *German Ideology*, sketches the difference between the alienated condition of the worker in bourgeois society and the finally liberated world of communist society.[146] In bourgeois society, we read, human beings, transformed into prostheses of machines, are forced to repeat the same meaning-less gesture ten hours a day, six or even seven days a week, and for the rest of their lives. But in the communist society promised at the end of the liberation struggles, everyone will be able to do what they like, whenever they like: they will be able to go fishing in the morning, write critical essays in the afternoon, tend to their cattle in the evening. That's for one day, for others they will decide according to their mood. The question at this point is very simple: are we not these liberated people?[147]

Of course, we are talking about the fortunate part of the world, the one with a fully developed capitalist economy, in accordance, after all, with Marx's per-spective. Here – at least in the pre-Covid world – in the morning we hand over the keys to our house to a person who has contacted us on a platform, then we are driven to the airport in a car driven by a unemployed waiter and take a low-cost flight to another city where, in another house rented on a platform, we follow a TV series while eating a pizza brought to us by a delivery person who, after clocking off, will also watch ten episodes. In an uberised world, the same person who may be called upon to be present in the middle of the night may, during working hours, write a Wikipedia entry or, more likely, update their status on social networks. Above all, one's work, if and when it exists, is but an island in a sea of activity: social media today and formal emails tomorrow, an online class

---

[144] This transformation lies at the heart of Richard Sennett's research. See in particular Sennett, *The Corrosion of Character* cit.

[145] A. Accornero, *Era il secolo del lavoro*, il Mulino, Bologna 1997.

[146] F. Engels, K. Marx, *German Ideology* (1845), International Publishers, New York 2004.

[147] P. M. Menger, *Portrait de l'artiste en travailleur. Métamorphoses du capitalisme*, Seuil, Paris 2003. The worker's condition is now the same as the artist's has always been. See also R. Florida, *The Rise of the Creative Class*, Basic Books, London 2002.

in the morning and a critical essay (not necessarily intelligent, needless to say) in the afternoon, finishing up with a talent show at night.

Is this not our life? Are we really alienated in the sense in which young Marx conceived alienation? Of course, "alienation" means many things and not only negative ones. As determined negation (*Entäußerung*) it is for example the effort to get out of oneself that we all make when trying to understand something or someone other than ourselves. But also as *Entfremdung*, as abstract negation, it means many different things: alienation from the products of one's work; alienation from one's activity, i.e. forced and unfulfilling work; alienation from one's own essence, which would only come about in free and creative work; alienation from others in a society that compels us to work and nothing else. Let us examine these kinds of alienation in detail: some of them are disappearing, others are changing completely, and all of them reveal that overcoming alienation, literally understood, is heaven only if seen from hell.

Alienation from the products of one's work. This alienation traditionally concerned any kind of work: even Picasso was alienated when he sold a painting, so it is not a specific character of industrial society. The point is that the focus on immaterial goods typical of the documedia society significantly reduces this alienation, since the paradigm of production appears more akin to the production of ideas (from which one is not alienated even if one gives them to others) than to the production of objects. Alienation from the product of one's work is inevitable, since every producer receives compensation by giving away the fruit of their activity. And the image of the poet as possessed by the gods – alienated when composing and therefore alienated from the product of their own work – reminds us of how much inconsistency and approximation lies behind the appeal to alienation and its positive counterpart: the immune and whole individual.

Alienation from one's activity. As for alienation from one's activity, it is again too vague a notion to be meaningful. On the one hand, it seems obvious that an athlete mechanically repeating an exercise, or a schoolboy practising Greek verb conjugations, are performing an alienating activity. Is there, however, a reason to divert them from what they are doing? Conversely, there are good reasons to condemn assembly-line work, and it is precisely this type of activity that is disappearing and yet some people seem to regret. Not to mention, I repeat, that there are excellent reasons to see "human nature" as a suspicious, sloppy notion that has historically authorised the worst kinds of repression. The claim that by handing over documents, one is selling one's life is questionable for at least three reasons. Firstly, in many cases, the surrender takes place voluntarily, through the publication of content on social media. Secondly, what I do not give up voluntarily, especially since in most cases I am not even aware that I am giving it up, is of no value to me, but only to those who possess calculation capacities that manage to make the document meaningful. Thirdly, if alienation from one's own activity were an evil in itself, what could be more alienating than *La Recherche du Temps Perdu*, a life objectified, dispossessed, put in unknown hands?

Alienation from one's essence. As for the workers' alienation from their essence, the question arises as to who could ever make the exorbitant claim to be in possession of one's essence. Werther certainly did not own his essence, and yet few would be willing to consider his condition as a form of alienation. And if "essence" means the free manifestation of good or bad potential, then we have never been so close to our essence as we are today. Except our essence is apparently not as wonderful as we would hope: instead of poems we write hate speeches, instead of scientific communications we post, when we are lucky, the food we have on the table, instead of Elizabethan dramas we watch their contemporary correspondents, with four or five people in a house pretending to live for real. There is far too much creative work: what we lament is the poor quality of its products, but this depends on our humanity and not on the Machiavellian schemes of some convenient monster summoned to be blamed for faults that are ours alone.

Alienation from others. As for alienation from others in a society that forces us to just work, well, if there is one thing the documedia world does not force us to do, it is indeed work. One really wonders what kind of harmonious paradise was being pursued in describing a condition of organic fusion between workers: more like a sinister caricature, a *Kraft durch Freude* that seems to have only occurred with the seven dwarfs, and probably under the charismatic influence of Snow White. Here it is worth introducing a daring argument, not because it evokes evil in God, but rather because, more seriously, it postulates the imbecility of all humans. Although reckless, this discourse aspires to be fair and serious, as is clear from the anthropological examination conducted in 2.

If the reference to alienation plays a major part in explaining our unhappiness, it is due to a linguistic misunderstanding. People often say "alienated" but are thinking "imbecile", or perhaps just "unhappy" or "dissatisfied". And so it is that "alienation", in common parlance and with an unacceptable translation from medical nosography to popular sociology, indicates fashion victims, karaoke singers, members of the excited society,[148] victims of the cultural industry, and probably also violent fans, the kinds who kill passers-by because they look happy. These alienations, however, have nothing to do with Hegel and Marx. The first thing to do would be to differentiate between the alienated in the sense of the insane and the alienated in the sense of the exploited. But most of the time this is

---

[148] C. Türcke, *Erregte Gesellschaft. Philosophie der Sensation*, C.H. Beck Verlag, München 2002.

not done, causing a fuss in which everyone, to a greater or lesser extent, is alienated, and therefore everyone is more or less a victim, except for a few executioners who are very far away and in any case are not part of our professional circles – we will never meet them at a conference, and they will never review one of our books.

To be harsh but fair, we are imbeciles (a technical term, the meaning and implications of which I will explain in 2.2) who, out of vanity – which is the constant companion of imbecility – believe to be alienated. We think we are fallen aristocrats from imaginary dukedoms, telling ourselves, whether we know it or not, the fairytale of a perfect soul full of merit, foreign to this world, disfigured by technology and society, awaiting he moment when it can return to its heavenly homeland. The misunderstanding here is not only linguistic, but conceptual: it makes no sense to speak of alienation and of being the master of oneself and one's own acts as the ideal of the human race, because – in what only appears to be a paradox – the non-alienated essence of the human race is alienation; the human being is a defective creature that only by transforming itself into something other than itself becomes human and not merely animal.

Consider a trivial phenomenon. A guy is watching a series on his smartphone. His ancestors listened to storytellers, and before that they went to see Aeschylus' tragedies. Is he conditioned by Power and the Market? Or is this due to the fact that we went from killing humans, to sacrificing animals, and finally to performances? In other words, here we are dealing with exquisitely human characteristics. Or let's take another example. A girl is wearing a necklace she bought on the Internet. We look at it and realise that it reminds us of something, namely Punic jewellery we saw in an archaeological museum in Sardinia. The Carthaginians were merchants, it may be argued, but the first *sapiens* also made ornaments, and before them, Neanderthals also had something of that kind – does this mean we have always been alienated? No, it means that we are and have always been structurally insufficient and in need of recognition. Another scene. In a nightclub, young people are dancing and many of them have taken drugs. Is this Power enhanced by technology, for example by the decibels of amplifiers, or the development of chemistry, that alienates them and drives them to behave in this way? Perhaps, but I don't see why we should speak of "alienation" instead of "imitation" or "tradition", since these behaviours can be found in Dionysian rituals, in shamanism, and in Sufi whirling.

Please note that I am in no way claiming that overcoming alienation means reaching heaven. Quite the contrary: we are not in heaven, and to get there, I fear there is still a long way to go. However, since even the longest path begins with a single step, I would suggest defining concepts such as "alienation", which explain everything and nothing, atrophying our ability to look reality in the face so as to be able to transform it. The essential thing here is not to be influenced by the dystopias we invent to make ourselves feel less responsible for our actions, blaming them on computers, smartphones, and perhaps – why not? – cuckoo clocks. And let us not forget that every ideology has its zealots, its madmen, its mythomaniacs and above all its liars.

1.4.3. Mobilisation

Let us continue our examination of the metamorphosis of work. The third phenomenon is one that, so to speak, is less phenomenal than the others – which is a serious problem, since it is precisely this third change that marks the fundamental turning point. When automation renders human action superfluous, humanity, far from falling into inaction, deploys an unprecedented mobilisation. At first glance, it may seem that in this mobilisation – more so than in dissemination – all life is alienated, so as to realise, albeit weakly, the dream of working 24/7. But there is much more, and mobilisation is the future, while dissemination still has one foot in the past. Don't be confused, because confusion between future and past can be fatal: in the early days of firearms there were still halberdiers; soon, however, they disappeared, usually to the sound of rifle fire.

Now, what do the mobilised do, apart from mobilising themselves? Some are teachers or professors, some are cooks or chefs, some are nurses or carers, some are gangsters or mobsters, some are writers or novelists ... And some do nothing, because they are children or old people. Even if all these jobs are filled – some of which, mind you, are forbidden by the law while others are not recognised as jobs – there will always be some who have no occupation. And what will become of them? The answer comes from another question. How many of those who in 2019 were wandering the streets of Venice or crowding Stansted at 5am on a Saturday morning were aware that they were working? Still, they produced value: much more than if they had been working in a mine, where their mining would have been useless as the sources of energy are now different. Those who crowded beaches or shopping malls may have been similar to Fordist workers, but they *wanted* to be doing what they were doing, and they were better off, they could be happy or dream of being happy, however you want to define this obscure term. It would be an unforgivable mistake, and an injustice to their human dignity, not to take this state of mind into account, and to trace it back to a form of false consciousness. This applies at all levels. Managers reading handbooks on how to be a good leader do not necessarily produce more value in their professional activity than they do while surfing the net to reward themselves for their efforts.

Così fan tutte, after all. There is no point in fighting against precariousness and calling for permanent contracts when Web-specific jobs employ very few people, and even the most traditional companies will push automation to the extreme in order to remain competitive. As you may have guessed, my strategy is not so much to mourn the disappearance of the jobs of the past, but to understand what is disappearing with those jobs, and above all to recognise the revelations about the essence of labour that these transformations are offering us. When we were faced with the problem of how to occupy the free time left by machines, we could not have foreseen the invention of a universal typewriter that would prevent us from sitting on our hands and mobilise us for fifteen hours a day. We travel, we consume, and above all we interact at all times with the documentation devices we have equipped ourselves with, which generally look like telephones or watches. However, this mobilisation, which in the true sense has nothing to do with classical work, is the greatest production of value that history has ever known.

Here we come to the heart of the mystery. While we think we are living our life outside work, satisfying our needs, pursuing our desires and expressing our ideas, we are actually taking over the functions of banks, newspapers, advertising and travel agencies. Above all, we are filling unknown archives with highly detailed files on our likes and dislikes, our habits and the exceptions that make us unpredictable for those who do not know us, including ourselves. These archives are not meant for censors, but for salesmen, eager to provide us with what we want, without investigating or inquiring into our lives otherwise, because it is not their business. To be paid for living undoubtedly seems too much to ask, but if our life produces value, why should our activity not be recognised, when the production of goods becomes a sacrifice to an unknown god? After all, not only are there machines, but they are perfectly automated, so that, in principle and in the strictest sense, not a minute's work is needed. If it is true that living is a job, it is just as true that working is tiring, even if not in terms of physical effort or repetitive routine.

This mystery and the question of mobilisation lend themselves to being misunderstood, and I was the first to do so, some years ago, in a book based on a real-life experience I had:

It is the night between Saturday and Sunday, the night that is traditionally consecrated to respite from the world of work. In the dead of night, I wake up. I want to see what time it is and, obviously, I check my smartphone, which tells me that it is three o'clock. But in doing this I notice that someone has sent me an email. I cannot resist my curiosity, or better: anxiety (the mail concerns a matter of work) and so nothing can stop me. I read it, and reply. I am suddenly working on Saturday night, no matter where I am.[149]

---

[149] M. Ferraris, *Mobilitazione totale*, Laterza, Rome-Bari 2015, p. 3.

Why was I doing this? The best and most credible answer is: because of the document I received, which activated my tendency to be mobilised. The idea that this was some kind of order was undoubtedly an exaggeration, since I doubt that anything serious would have happened to me if I had not responded. Behind a self-victimising explanation, my first step into the world of mobilisation concealed two elements which, after a more mature (or at least protracted) reflection, I consider more important.

The first element concerns humanity, which derives pleasure from mobilisation, it seems, certainly more than from knowledge.

All men naturally desire knowledge. An indication of this is our esteem for the senses; for apart from their use we esteem them for their own sake, and most of all the sense of sight. Not only with a view to action, but even when no action is contemplated, we prefer sight, generally speaking, to all the other senses. The reason of this is that of all the senses sight best helps us to know things, and reveals many distinctions.[150]

Are we sure that what Aristotle is talking about is a desire for knowledge that masquerades as pleasure and not a desire for mobilisation? One has to recognise labour in the modern (rarefaction) and postmodern (dissemination) sense of the term as a minimal and subordinate portion of mobilisation. This recognition has significant analogies with the recognition of the biosphere as a dimension above the docusphere and infosphere. Just as the infosphere is merely the superstructure of the docusphere and the biosphere, so modern and postmodern labour is merely the superstructure of dissemination and mobilisation. And again, there is no reason to assume that this is a technical imposition or a strategy to alienate us. It may well be the manifestation of a profound vocation.

In order to recognise mobilisation, it is necessary to leave aside on-demand jobs and micro-jobs, which are at least recognised as such, and which are destined to die out as soon as a more evolved automation means we no longer need them. What should be brought into the open is what is called "online social work", but which is more correctly qualified as the production of value through mobilisation. This is implicit labour that needs to be made explicit, and which is not adequately captured by the term "social work" because there is very little sociality about most of what we do online and, above all, there isn't much of it that the platforms are interested in. This is work that no one formally asks of us or authoritatively imposes on us, and that we most often do out of sheer restlessness. If there was ever any doubt that the "in the beginning was the Act"[151] is much more convincing than the "in the beginning was the Word",[152] the last few

---

[150] Aristotle, *Metaphysics*, 980a 21–29.
[151] J.W. Goethe, *Faust*, Princeton University Press, Princeton 1994, vv. 1224–1237, p. 33.
[152] Jn 1:1.

decades have provided hardly disputable evidence to support this. We could have stayed at home, immobile and connected, yet until the pandemic we had never moved as much; we could entrust everything or almost everything to the machines, and instead the machines thrive by capitalising on our every act; we could have fulfilled what some incurable optimists[153] consider to be our essential vocation, namely contemplation, and yet we fret to no end, making it difficult to disprove the pessimists claiming that "people love darkness instead of light".[154]

Mobilisation therefore meets a basic human need, above all as an impulse release, as is very clear in children. Indeed, sections of the population that are not mobilised by work or war engage in dances, processions, tournaments and rodeos. All this is usually a waste of time, but documendiality has turned it into capital. Just as you can transform land into capital through property contracts,[155] so you can transform mobilisation into capital through the documentation of acts. This is because mobilisation is anything but an abstract movement. It is not a click on a screen, not least because screens and clicks will soon no longer be necessary for mobilisation. Instead, it is the prominence of the body and its needs, of life in all its aspects and in its total context: as we have seen, we can chat at any time, but in the middle of our chitchats, at any time, we may receive a work-related request, for those who have a job, and therefore associate the regime of mobilisation with that of dissemination. This dissemination occurs between age groups (even children and pensioners are mobilised); between specialisations (the value of the information produced by the mobilisation of a theoretical physicist is equal to that of a football fan, who could also be a theoretical physicist, since it is not about ideas, but about behaviour); and between social classes, all equal in front of a smartphone.

Let us reflect for a moment on this condition without rushing too hastily into the loopholes of alienation and technical imposition. We are condemned to be free, but, just like Sartre, we rarely ask ourselves why, because we are, just like him, perpetually busy writing, he at the Café de Flore, we on the Internet. And in addition to writing, we leave traces of all our activities, so much so that every day we produce documents that have little to envy, by extension of course, to the *Critique of Dialectical Reason*. This mobilisation, which depends on the needs, interests, strengths and weaknesses of humans, is the truly great labour of the future, which will not be a part-time job, but a full-time occupation, and which must be recognised as labour, in accordance with the proposal I make in 4.3. One can work without toil, one can produce value without feeling alienated. This was

---

[153] Aristotle, *Nicomachean Ethics*, 1177a 1–11.

[154] Jn 3:19.

[155] De Soto, *The Mystery of Capital* cit.

traditionally a character of the liberal professions, the *eureka* in the bathtub. Thus, if a philosopher has a good idea at a child's birthday party, they write it down and reserve the right to work on it as soon as they get home. The example taken from a rare profession should not mislead us, because with mobilisation, the production of value, mainly through behaviour, becomes an ordinary phenomenon. In fact, if we conceptualise labour as the production of value, and if we admit that the main effect of the documedia revolution has been to transform every act into a document, and therefore into potential value, then the project of a fifteen-hour work week would not make more sense than the project of a week in which, by decree, one is obliged to live for no more than fifteen hours.[156]

This mobilisation is not simply a *vita activa*[157] as opposed to a contemplative life; it is a mixture of action, understanding and misunderstanding. The new technologies extend human-machine interaction to every moment of life, with a veritable division of labour where the machine takes on all the dead work, and delegates to humans the remaining living work, which is the conferring of meaning, first and foremost as a direction of travel. This is a new element, though hidden and often condemned as consumerism, which skips traditional intermediaries: agreements between users to offer services, creation of alternative sources of information, use of the Web as a form of professional self-promotion. By exploiting the resources of hysteresis, platforms capitalise on human acts, transforming mobilisation into value and harvesting the product. And – with the process of documedia surplus value, which I will examine in 4.2 – they do not need to pay the mobilised because in return they offer them services (information, intermediation, entertainment) nor, above all, do they expose themselves to trade union claims, because mobilisation is not recognised as work by anyone. Should we criticise them? Of course, but not in the same way as one would criticise a 19th-century factory owner, because platforms are no more factories than we are assembly-line workers. This is the meaning that can be given today to Marx's statement that "Whenever we speak of production, then, what is meant is always production at a definite stage of social development – production by social individuals."[158]

There is a second, politically relevant aspect that my victimhood overshadowed. Those who first spoke of the "total mobilisation" of the social fabric dreamed of a totalitarian state,[159] in which every soldier would behave like a

---

[156] M. Ferraris, "From Capital to Documediality", in Romele, Terrone (ed. by), *Towards a Philosophy of Digital Media* cit., pp. 31–50.

[157] Arendt, *The Human Condition* cit.

[158] Marx, *Grundrisse, Foundations of the Critique of Political Economy*, cit, p. 18.

[159] E. Jünger, (1930) "Total Mobilization", *The Heidegger Controversy*, ed. by R. Wolin, MIT Press, Cambridge (MA) 1993, pp. 119–39.

worker, as already seen in the First World War, and every worker would act like a soldier, without giving in to weaknesses and civilised pauses, saturating every area of their lives with work, or more precisely with their willingness to be mobilised. This mobilisation did not take place at the time, so much so that, as we have seen, in February 1943 Goebbels was still complaining to the Germans that they were not sufficiently prepared for total war. One might conclude that if not even Nazis were able to achieve total mobilisation, and if this remained largely a dream or a nightmare, a kind of Turk's Head for subsequent analyses of the microphysics of power, then no one ever could. And yet this was not the case. Half a century after the end of the Second World War, and in liberal countries characterised by a strong emphasis on individual rights, the Web and the smartphone appeared, and total mobilisation began: the demand to respond at all times; the disappearance of the difference between work and life; the elimination of occupation, replaced by mobilisation. The most interesting thing about this mobilisation, promoted by the documedia revolution, is that it is only marginally coercive; it has some unique features that deserve to be understood, because they reveal our origins and explain the ongoing transformation.

Thus, mobilisation took place in a context of very strong demilitarisation of life – or, if you prefer, dissemination of war through terrorism –, without the central decisions of a totalitarian state and without coercive measures, but simply through access to services mediated by mobile devices. If the production of traditional commodities is turning into dead labour, humans are performing an enormous and irreplaceable living labour: the production of document commodities. This production merely records what humans already did before, which is to live with all this entails – curiosity, entertainment, boredom, despair, and so on. Except that now those great futilities have become useful because they are recorded, and they are recorded because they are useful, so they are quantified, unified, transformed – that is, capitalised.

The idea of coercion must therefore be overcome. Mobilisation is the fulfilment of the idea that the world, as we know it, is about to disappear, because the present is already projected into the future, with its obsession and urgency. So being modern – that is to say, on closer inspection, being very old, with so much history behind us – means being fast, and speed is the old age of the world.[160] Hence the realisation of the Faustian character of capitalism, that is, the fact that everything can become faster if – according to the reasoning of Mephistopheles, who embodies the spirit of technology – I can replace my legs with a six-horse carriage. But the story has an even more remote origin. In the beginning was mobilisation, i.e. action increased and motivated by need, not work. For thou-

---

[160] P. Virilio, *Vitesse et politique*, Galilée, Paris 1977.

sands of years, *sapiens* was a mobilised individual, not a worker. And people continue to be mobilised even after the end of work in its classical sense. This is why the definition of *homo faber* is inadequate and contingent: not only because it does not apply to our not so distant hunter-gatherer ancestors, but because it will not apply to our not so distant inventor and consumer descendants.

Mobilisation seems to be one of the few specific traits of what we are, because it leads us back to our organic component. On the other hand, if by sleeping, eating, or fornicating, I am working, then it is not clear what it means to work, if "work" supposedly comes with toil and alienation. This path leads nowhere, except to offending those (who still exist) who do work under conditions of toil and alienation. To avoid ridicule and mockery, a philosophical analysis is needed to distinguish between two concepts of labour. The first is work as toil and alienation, which is disappearing due to automation and which will eventually leave us all jobless (this is the bad news), but also all without toil and alienation (this is the good news).

Once this explicit labour has disappeared, and the things that we have rightly stigmatised as inhumane – mines, forced labour, the assembly line, ar of extermination as industrial warfare of materials, and so on down to delivery people and warehouse workers (bad jobs, but preferable to mines and factories) – have gone, what will be left? A few explicit jobs that will have survived the ones we know today and other jobs that will have arisen in the meantime, but it would be unrealistic to think of a break-even balance. This is where conceptual invention must come in. At a time when every activity, whether work-related or not (i.e. paid or unpaid), produces documents, which in turn generate value, it becomes necessary to recognise that *mobilisation is a job*, just as important if not more so than canonical jobs. Of course, if one wishes to be a stylite, well, nothing prevents one from doing so. The fact remains that well-being and greater material wealth favour the development of ingenuity and culture, so much so that as long as we had to run through the savannah chased by lions, the Muses stayed clear of us.

Another point, which is substantial and ties in with the master/slave dialectic set out in 1.1.4, is the following. The fact of being mobilised does not necessarily imply servitude: rather, if correctly understood, it can be a source of emancipation. In fact, mobilisation conceals the essence of human power and freedom, i.e. our infinite superiority over machines. Since humans can live and die, it is they who give meaning and value to machines which, without them, would have none. Let us never forget this, because doing so means giving up our freedom. Automaton cannot run itself. Behind it or in front of it, as its primary cause and ultimate aim, there is always a soul, which is not a vague cloud or an single point, but the entelechy of a body that has life in its potentiality: in other words, simply put, that which animates a body, and vivifies it precisely because, sooner or later, it will stop blowing, throbbing, and aching. In turn, of course, and by a process of capitalisation that I shall analyse in detail in 2, a sort of flywheel is produced between soul and automaton, whereby the social world sets the souls in motion just as they had set the social world in motion. But it must be clear who is in charge at every level, even the lowest ones, because the wheel will never turn without the hamster, and the hamster may sooner or later become aware of this.

1.4.4. What does it mean to "work"?

In this context, it is not a question of *re*defining work, but rather of defining it, clarifying that it is anything but a definite and obvious notion. Indeed, the question of the nature of labour is no less problematic, albeit of much more general interest, than the question "what is art?". In the latter case, the disappearance of beauty as an identifying element has suggested the answer that art is everything that a community of experts, an artworld, recognises as such. In the case of labour, the disappearance of the identifying features of industrial production means that labour is, in the end, nothing more than what a community of experts – let's say a "labourworld" – defines as such. However, while the interest of the artworld in multiplying the number of works of art is very strong, the labourworld is driven by the opposite interest: to reduce as much as possible the attribution of the character of labour to activities which, if recognised as work, should be rewarded in some way.

But on what grounds? By what arguments? There is consensus that labour should be defined as "an activity that produces goods or services, that responds to an obligation, implies an intelligent effort, that is time-consuming and obligatory, that aims at the satisfaction of the needs and the realisation of the abilities and projects of the worker or the manager, and contributes to the reproduction and the organisation of the institutions of a society".[161] Certainly, work as the relationship between humans and nature does not apply to the documedia society; the relationship with nature is managed by machines, with which humans relate as masters, the machines being the slaves. For similar reasons, it is difficult to identify the employers in platform-mediated activities. Who would that be: the platform owners? The companies or individuals who benefit from the services? Or the workers themselves, who in many legislations may be self-entrepreneurs? If we distinguish between labour and work, where the former is a subspecies of the latter in which exchange values are produced,[162] we realise that mobilisation

---

[161] A. Cukier, *Qu'est-ce que le travail*, Vrin, Paris 2018, p. 32.

[162] J. Dewey, "Interest and Effort in Education" (1913), in Id., *The Middle Works, 1899–1924*, vol. 7, 1912–1914, Southern Illinois University, Carbondale 1979, p. 189.

has all the characteristics of labour except one, that of being paid; which, as can be seen, is an injustice.

Let's start with an elementary question. Shaking hands is intellectual work, which for example marks the signing of a contract. However, no one could deny that it is also manual work, because it is done with one's hands, and this applies to most intellectual work, which involves the use of hands and fingers typing on keyboards or using pen and paper. And of course there are good reasons to argue that shaking hands is often not work if it is done in a convivial context, although the case is tricky, since maintaining relationships can be work: for example shaking hands during an election campaign, or at the end of a business lunch. If it is so difficult to determine the nature of a handshake, there are few solid grounds for answering questions such as: What is work? What distinguishes the time of work from the time of life? What, specifically, are the sociology, psychology and philosophy of work when the latter becomes mobilisation?

"Work" is a single word, but one that stands in contrast to many others (idleness, unemployment, play, life), which should suggest that what is called "work" is meant in many, and often very different, ways. Faced with this complexity, we generally cut to the chase, and roughly identify work with what our parents did when we were little, which we set as the norm. Since we have never experienced the hardness of that work, and the past colours our memories with rose-tinted nostalgia, the idea underlying discussions on labour is often that we should return to that lost work, which was supposedly "real" and "whole". This is somewhat paradoxical, since for a solid half century, and with good reason, people sought to overcome the old organisation of work and, more generally, work as a whole. One might say that the definition of work involves two dimensions: an empirical one, made up of historical oppositions, and a transcendental one, aimed at recognising the notion of "labour in itself". The latter, however, is an essence that emerges precisely through the comparison of these oppositional notions, or rather it is a notion of work that emerges differentially.

Consider the difference between play and work. Play may have no purpose, work always does. However, it is easy to see that a great many games do have a purpose, and can therefore become work – which is recognised in many cases, for example in sports defined as "professional", i.e. such that the activity performed is paid for. The rest/toil distinction is also problematic. Work requires physical effort, at least if we follow the biblical characterisation, rest does not, and its attributes are defined in Western culture, again, by ritual prescriptions originating in the Bible. Let us come to the pleasure/duty contraposition. There may be a dutiful character to work, for example in bearing arms or divine functions, yet paradoxically, these activities do not seem to fall canonically within the definition of work, unless they are paid for: soldiers work, and are indeed defined by the fact that they are paid, but what about partisans? Or terrorists? Is martyrdom a job? Finally, with regard to the *otium/negotium* distinction, it is worth noting that it embodies a reversal of values that is generalised in the documedia world. In fact, *otium* (reflection and consumption, whether cultural or not) has a higher value than *negotium*, which appears to be a mechanical activity of the kind that can now be overtaken by machines, leaving the human pole of the production of value to *otium*, which must therefore be recognised in its labour and wealth-producing dimension.

At a time when classic work in the form of repetition and toil is carried out by machines – whether humans like it or not – humans maintain their centrality in another sphere, which proves to be indispensable: that of consumption. The more individuals develop, the more they evolve spiritually, the more they consume. This intuition has been followed up in the wrong way, by claiming that humans will eventually develop a collective intelligence. They will not. Rather, they manifest their primary essence, which is dependence and need. The metaphysical transformation, on which we should reflect much more than we do, is the automatic capitalisation of acts. In this sense, in the past, movie-goers did not work, unless they were film critics. However, in the documedia age, movie audiences do objectively work, because they produce documents, and thus wealth. It is therefore necessary to recognise this production of value, above all because the more time is invested in it, the more documents are produced, and the fact that so much time can be invested is probably due to the fact that many of us are unemployed, i.e. do not work and are not paid in the traditional sense.

It could be argued that work involves a degree of coercion: if you are having fun, it is not work; or it is not work if you are obeying a natural dictate, such as caring for your children. These assumptions are nonetheless highly contestable, leading to the absurd situation whereby those who enjoy working should not be paid, because they do not really work; or to the paradox of two mothers exchanging children for a few hours each day in order to be paid as babysitters. Once these conceptual limitations have been overcome, the assumption I am making becomes clear, namely that labour is everything that produces value, and that mobilisation is the labour of the future.

This is where we need reflection and a bit of imagination. Economic and political problems often conceal metaphysical ones, and the mystery of labour is certainly an example of what I mean. The challenge is not so much to ascertain that automation will take away traditional jobs and create new ones, though not enough to replace the lost occupations. Of course this is the case and, for the sake of humankind, it is to be hoped that this is the case, and that indeed automation is the manifest destiny of labour. What remains to be done is the conceptual work, which in this case consists in a reformulation of the relationship between capital and labour, proceeding from a deeper understanding of both, in the awareness that the outcome of a conceptual change must be an institutional, political and economic transformation. We must realise that, in seeking an answer to the question "what does it mean to work?", we are led back to an even more challenging question: Who are we? Who are we who have long identified as workers, and who today are losing that traditional ground?

2. Revelation: Who Are We?

The documedia revolution confronts us with problems that cannot be solved by an appeal to alienation. The latter concept has one major flaw: it refers to an ideal and chimeric concept of "human nature" as eternal, immutable and, what is worse, originally perfect and degraded throughout history. It is obvious that if political action is based on such a Pentecostal anthropology, the results are bound to be disappointing. It is not a question of nailing the human race to a cross made up of imaginary virtues, but of recognising its actual history. Not in order to expel humans from some imaginary paradise, but to allow for their genuine emancipation, which is only possible by answering the question "Who are we?".

And this answer could hardly come from some heroic and often insincere effort of self-consciousness. Looking for it on the facade of the temple of Delphi is already more advisable, not so much because of the ultimately impracticable precept we find engraved on it, but rather because that building is the manifestation of a human life form. Since we only know ourselves in action, it is not surprising that traces of the action of others can guide us in our knowledge of past existences. If you want to know what humanity is, you have to leave aside its proclamations and look at the technologies it uses every day. This is so obvious and true that there are entire civilisations of which we have nothing left but tools and furnishings, yet this is enough for us to conjecture something about them. Full knowledge began when humans developed the arche-technology of writing; more sadly, what we produce through technology are the fossils of the future, destined to outlive the human species itself.

My argument proceeds from a critique of the creationism implicit in any definition, even a secular one, of human nature, and is based on emergentism and retrospection. Let us begin with the critique of creationism. Humans have a very long natural history, and this is generally acknowledged, but it seems that their cultural history can be explained in a few millennia. It is precisely in this assumption that we observe a well-hidden kind of creationism. Why should we believe that life can be explained in millions of years, while society can be explained in a few millennia, usually from the birth of our favourite empire? In fact, the Pentecostal explanation of the genesis of social reality exhibits a strange anachrony. Usually, when referring to nature, we embrace Darwinism and therefore imagine that evolution took place over very long periods of time, inaccessible to conscience and legislation. But when it comes to society, we presume that the world, as was still believed in the 17th century, is no older than six thousand years – and then we have to resort to constructions and conventions: language, political roles, the economy, family relationships, and obviously society as a whole.

If, however, we consider that the world – in which both the species and society originated – is four and a half billion years old, and that there has been life for more than three billion years, there will be nothing surprising about the idea of a long unconscious emergence, which, in addition to the species as we know them, produced writing, the arts, documents, money, the gods, and obviously thought, which is never independent of these forms of capitalisation. These institutions were already there when we began to reflect on them and recognise them as such, just as language was already there when consciousness, which is a consequence of it, was formed, and all the more so when we began to investigate the origin of language.

If, as I believe, this is the case, in order to be able to respond to human needs and to today's problems, it is necessary to shift from a generic and timeless "human nature" (as I said, there is no such thing, and this is precisely the profound nature of humanity), but rather traits that have gradually emerged in history and in correlation with technology. First of all, we have the organic nature of humankind, its foundation, which likens it to all other organisms, except that humans display a systematic connection with technology that is not seen in other beings. Then we have the characteristics brought by the technological supplements. The latter play such a decisive role that technology must be considered a part of anthropology, and, reciprocally, anthropology is the other face of technology. Only after we have recognised the characteristics of the technological animal (that interaction between organism and mechanism, between soul and automaton that I call "responsiveness") can we come to the characteristics of the social animal, endowed with language because it is predisposed to capitalisation, and finally to normativity, the capacity to act with reference to values (both venal and moral) that is characteristic of the human. In other words, virtue is there, but it is at the end of the path, always in front of us and only sporadically behind us.

2.1. Foundation: Responsiveness

"Man is born free; and everywhere he is in chains. One thinks himself the master of others, and still remains a greater slave than they. How did this change come about? I do not know."[1] The demons that corrupt the Good Man and the Good Savage originated here. Never mind that, after Rousseau asks how it happened that man is born free and yet is in chains, his answer is "I do not know", and these are things that are enough to disqualify a writer. What is more worrying than the incorrectness of the analysis is the absurdity of the premise. If we see that man is everywhere in chains, what gives us the right to claim that he is born free? Let us put it differently. Humans accept their chains, for example assembly lines. However, just as they have accepted them, so they can break them, because they have nothing to lose except those chains. Here's a hyperbolic and very concrete doubt: if chains are all we have, what is left once we break them? In fact, mature Marx proposed a different strategy: the man-machine relationship is not simply servile, because machines produce goods that would have no meaning without human consumption.

It is in this direction that we must seek the path to emancipation: man is born in chains, which are, properly speaking, the set of devices that allow him to survive, to mitigate his own constitutive inadequacy, starting with the "leading-strings of the power of judgment"[2] to which Kant alludes in the *Critique of of Pure Reason* but on which he expands in *Lectures on Pedagogy* (and it is touching to see a bachelor lose himself in details on how to swaddle or not swaddle children, how to arrange cots, prevent them from screaming, etc.). Of course, man can – and must, from a moral point of view – become free, as long as he becomes aware of his chains and knows that in some cases the only thing to lose is indeed one's own chains, even though they may be difficult to do without. The first step in reconstructing the foundation of hominisation is purely negative, and consists in freeing oneself from a deeply mystifying (and, what is worse, politically fatal) image of human nature.

2.1.1 The Rousseau syndrome

Everything we know about the human being points to a creature which, unlike any other soul, is born a slave because it is not very independent. The idea that there is some authentic, pre-technological human being who is gradually corrupted (usually in the century before we write, so in the 17th century if we write in the 18th, in the 18th century if we write in the 19th, etc.) is perhaps one of the most deeply rooted beliefs we have, and looks like an updated version of the golden age theory. I christened this historiographic hallucination the "Rousseau

---

[1] J.-J. Rousseau, *The Social Contract* (1762), J.M. Dent & Sons, Ltd, London; E.P. Dutton & Co., New York 1913, I, p. 1.

[2] I. Kant, *Critique of Pure Reason*, A134–5, B 174, p. 269.

syndrome": according to it, there is such a thing as human nature "in itself", which is a condensation of all virtues. Then technology and society corrupted us, bringing greed, lies, oppression, exploitation and many other misfortunes. This is the syndrome that underlies the vast majority of discussions about technology: look at us, always on the phone, always on social media, always arguing (and before that always staring at the television), always writing instead of talking to our friends and family ... It's as if, were it up to us, we'd be anything but, it's just that some external evil has transformed and alienated us. We are perfect and autonomous, bearers of a Kantian morality in which the subject is also the legislator, as well as an autonomous source of law. Obviously this is not the case. We are always on the phone, but is it not because Aristotle defined man as an animal endowed with language? We are always on social networks, but is it not because Aristotle defined man as a social animal?

The first effect of the Rousseau syndrome is resentment, born of disappointment with constraints that are not foreseen in the monument erected by the anthropological mirage. This gives rise to the system of guilt, bad conscience and the like, which – unlike in Nietzsche's time, still in the throes of liberating the human being from original sin – manifests itself today in the use of the term "alienation". As I explained in 1.4, this word goes in the antithetic direction, postulating a human being who was originally perfect and was then led astray. This generates a number of mistaken polemical targets, and precludes the path to true understanding. If a person who is sad or frustrated for some reason takes comfort in ice cream, is it fair to conclude that the need for ice cream is a technically and socially induced alienation? Of course not. Both have responded, for reasons that are anything but humanitarian, to a need that would have existed independently and would have been satisfied otherwise.

Another example. A child is playing with a smartphone. In their time, his father would have played the first video games and his grandfather with model trains, and so on, further and further back, until we find statuettes from the Palaeolithic era that could have been dolls. When did alienation set in? And wasn't it Schiller who maintained that man is only truly such when he plays? Another hypothetical case. Two professors are passing around files in a potentially endless conversation, in the belief that they will be published and contribute to the critical consciousness of humankind. A little further on, two non-professors exchange messages that are not necessarily thought-provoking. It really takes a lot of patience (on the part of the non-professors) and prosopopeia (on the part of the professors) to claim that these professors are inspired by Critical Reason, while all other social media users and bloggers are driven by narcissism. To impute things like exchanging messages, purchasing goods, watching TV series, playing with smartphones, and eating ice cream or taking drugs to some occult persuasion that perverts human nature means being completely blind to human nature, and one wonders what benefit can ever come from such an unrealistic view.

Immodesty aside, the Rousseau syndrome manifests a misleading essentialism: there is such a thing as Human Nature, perfect in the beginning, deformed by history, and to which we must return. I repeat that this is not the case. What supposedly alienates us is what reveals us for what we actually are, and are becoming. The declarations of human rights, which began to be drawn in the 18th century – excluding a significant antecedent in natural law – are based on a subtly misleading assumption, namely that of a stable essence, called "human", to be protected or restored. That this assumption is problematic can be deduced from the mere fact that declarations and laws have become necessary, a sign that – apart from what we are like as animals – there is nothing wrong with human nature: characteristically, Newton's laws are not intended to protect universal gravitation, whereas the declarations of human rights are intended precisely to protect and promote those rights. This is not in the least to criticise declarations of human rights – as long as the enunciation does not take the place of action, according to the mechanism illustrated by Talleyrand: "principles are always good, they bind no one". The problem is rather to understand the human nature which those principles so effortlessly claim to presuppose.

Leaving principles aside, we realise that there are very few characteristics that the human animal does not share with primates. The genome of pigs is very similar to that of humans (which, again, explains more than we might like), monkeys have opposable thumbs, dolphins' brains have two hemispheres, just like ours. But why are their forms of life so homogeneous, while ours is so varied, unpredictable and often absurd? The anthropological difference is first and foremost an inadequacy, a maladjustment to the environment, probably due to an original mobilisation that has taken humans out of the territory to which they were suited, namely southern Africa. This mobilisation has nothing Promethean about it, and should rather be seen as akin to the urbanisation of seagulls in search of food. The fact is that, unlike seagulls, for mobilised animals adaptation has required the adoption of technological supplements. The difference between the human animal and non-human animals lies outside and not inside: saying that the opposite is tantamount to postulating some intelligence gap. Indeed, it is outside of us, in the capitalisation I discuss in 2.3, that we find what makes us human.

The conviction we should abandon is that there is a pleroma, a "whole" of the human being and the world, present from the beginning and more or less fully manifested in history. This is not the case. That which is not manifested simply is not, or at least not yet; appearance is necessary for essence, so that what is given in the world is not the descent of the gods above, but the ascent of a multitude of organisms and underworld mechanisms. Humanity has been endowed with spirit through a path that starts from our past[3] and progressively arrives at the present, at what there is, what we are and what we want. Essentialism is therefore also exceptionalism, which is the secularised version of the view that humans are created by God in his likeness, through Pentecostal processes that are totally different from those used for other living beings. Those who might be tempted to claim that this view no longer has any followers, with the exception of a few fundamentalists, should consider that the prevailing political views of the last two centuries have largely endorsed the Rousseau syndrome. This is clearly a re-proposition of the very same exceptionalist perspective, with the aggravating circumstance that, while religious exceptionalism mostly acts within the confines of conscience, political exceptionalism legislates in parliaments.

Now, removing humankind from the pedestal on which it had settled without embarrassment or resistance, as well as grasping the specificity and novelty of the present, is an indispensable condition for effective cognitive and political action. In the same way, it is essential to understand that conjectures about the origin of the human race, and perhaps the dream of some original freedom, collide profoundly with the experience of each and every one of us, remote epigones of Adam or Prometheus. Unlike those imaginary progenitors, we are born full of needs in a highly structured world; authority figures, such as parents or teachers, loom over us (and if they do not, it is even worse); little by little, along with language and table manners, we also learn to respond to those who address us, with that fundamental structure of imputation which is our name. This is where our intentions, our ideas, our responsibility for our own actions come from, even without anyone forcing us to do so or judging us. This, of course, does not happen by default, but it *can* happen, just as it can happen that we are able to act freely. However, this is the rare and challenging point of arrival, not the point of departure.

2.1.2 The misfit animal

The second step in the reconstruction of our foundation consists in redefining what we are based on an assumption that is diametrically antithetical to the Rousseau syndrome: humanity is not corrupted by technology and society, indeed humanisation could not take place without them. The soul became a human soul not when it reached a certain cerebral mass, but when, through a series of fortuitous circumstances that we can retrospectively read as the manifest destiny of humanity, it was provided with technical supplements capable of overcoming its

---

[3] E. Avital, E. Jablonka, *Animal Traditions: Behavioural Inheritance in Evolution*, Cambridge University Press, Cambridge 2000; Epstein, *The Ant Trap* cit.

deficiencies. The essence of the human being as a non-stabilised animal[4] is therefore to be sought on the side of the technical supplement rather than on that of its
organic foundation. The latter makes us the same as any other animal, whereas
the supplement (what we are not and do not have) makes us what we are.

It is difficult for an animal to prove inadequate for the task assigned to it,
whereas for humans this is quite normal; it is unusual for animals to adapt to inhospitable climates, because they have normally chosen an adequate habitat as
their home. Likewise, it is difficult for an animal to cultivate colonialist or capitalist instincts, because it is fine as it is; human history, on the other hand, is a tale
of failures, migrations, colonisation, and capitalisation. This is not because the
non-human animal is good and the human animal bad, but simply because the
former is generally adapted, and the latter is generally maladapted. What in
non-human animals is solved through natural selection, in human animals is
solved through culture. Submission and empowerment are not counterbalanced
by anything natural, but by a cultural element intrinsic to the modern concept of
"society": the assumption that social reality can be transformed is a constitutive
part of social reality itself. We would not be willing to consider something immutable as "social". Transformation is co-extensive with social reality, just as
evolution is co-extensive with natural reality. The conferral or emergence of
meaning is a characteristic element of action in the social world, and there would
be no concept of "humanity" without the concept of "freedom" – which obviously does not mean that this ideal reference corresponds to a real state of affairs.

Thus, if humans had not been misfits right from the start, or if they had adapted to the short and meagre life, limited to a narrow environment, that nature had
assigned to them, I certainly would not be here writing this book (which would-
n't be so bad), nor, above all, would humanity have become what it is and what
it will be, in an indefinite but not infinite process. Without fire, the wheel, language and writing, we would not be better or more perfect humans. We simply
would be no different from non-human animals, and would soon become extinct,
as non-human animals have a far more adequate endowment to their lives than
we do: we are born naked, without nails or tusks, and above all take years to
become autonomous, if we ever do. Technological transformations, rather than
alienating us, reveal us, for better or for worse. Naturally defenceless, humans
develop later than other animals, and are therefore more in need of parental care

---

[4] Cf. A. Gehlen, *Urmensch und Spätkultur. Philosophische Ergebnisse und Aussagen*
(1956), Klostermann GmbH, Frankfurt a. M. 2004; Id., *Man in the Age of Technology* (1957),
Columbia University Press, New York 1980. Plessner, Darwin and Gehlen agree: we are more
skilled because we are weaker. An original take on this perspective can be found in T. Pievani,
Imperfezione. Una storia naturale, Raffaello Cortina, Milan 2019 and Id., *Finitudine*, Raffaello Cortina, Milan 2020.

and social protection than other animals: we continue to develop even after birth, and even when we have learnt to walk and talk, we keep on imitating others and depend on technological support. A cat is autonomous at four months, a human, if things go well, at forty years of age, but in reality never, because a human will always need society, language and culture. This explains why we are so highly dependent on so many things, even on the gaze of the other, including a cat's.[5]

Precisely because of this, far from being the sole owner of the world,[6] the human is the only animal that is dissatisfied with and therefore susceptible to any environment, except for those not-so-happy islands where humans, having no need to adapt, do not even develop as humans. Thus, as an unstable, unsuitable and unsatisfied animal, only the human is poor in world and in need of technology. One can certainly imagine a virtuous combination of circumstances underlying, for example, the survival of the Lapps, Samoyeds and Jakuts in their frozen, inhospitable moors: "But then we do not see why, generally, men must live there at all".[7] Atopic and out of place, humans have had to adopt sociality, writing, and symbolic thought, incurring the very real possibility, inconceivable for cats and white bears, of injuring themselves while opening a can, of not understanding a joke, or writing nonsense.

However, as we shall see at length later on, one of technology's greatest assets is its capitalisation, the possibility of transferring to the outside world something that, if kept inside, would clutter our minds and, what is worse, disappear with us. It is not without irony, therefore, that the most dependent, limited, displaced – and therefore least free – animal in creation conceives of freedom as one of its essential characteristics, which only increases its dependencies and frustrations. If humans are dependent for constitutive reasons, they suffer freedom as a phantom limb, to which they aspire but which at the same time pains them; that is, they do not see that freedom is a rare good, and is an ideal to be realised rather than a state in which we exist in a defective form, prey to pathologies of the will.

But every cloud has a silver lining: the mere fact of conceiving of a missing limb gives rise to an aspiration and stimulates an ideal, that of freedom, not as a starting point, but as a goal. The omnipresence of subjugation in the social world and in interpersonal relations was evident well before the documedia revolution, but was rationalised as the result of deeper and more sensible motivations. One submits to the gods for protection from an unknown and hostile world; one submits to power because it is stronger; one submits to the economy to obtain goods. But what exactly is the purpose that drives millions of human beings, myself

---

[5] J. Derrida, *The Animal That Therefore I Am*, Fordham University Press, New York 2008.

[6] As argued by Heidegger, among many others, see *Grundbegriffe der Metaphysik*, in Id., *Gesamtausgabe*, Klostermann, Frankfurt a. M. 1929–1930, pp. 29–30.

[7] I. Kant, *Critique of Judgment* (1790), Macmillan, London 1892, § 63, p. 271.

included, to mobilise themselves for free? Submission is not a historical contingency, but an anthropological constant. The essential way in which we relate to social reality is not that of a legislator, but that of the person subject to the norms, including the legislator as a concrete historical individual. This does not apply only to rules, but also to intentionality and conscience: if we admit (and this seems to be a common-sense consideration) that nurture, education, language and culture play a constitutive role in the formation of human conscience, then it will seem truly difficult to imagine a legislating and immune human being, a child of Pentecost and not of Emergence. More on this later; what we need to define now is something simpler and more essential, namely the difference not between human and animal, but between animate and inanimate.

2.1.3 Soul and automaton

The third step in the reconstruction of our foundation is therefore to define a basic polarity. The misfit animal has a priceless advantage over supplements: it can die, just like any other organism. And that changes everything. If there is something that the human as maladjusted animal is naturally inclined towards, it is technology (or, more precisely, technology is the only thing that makes the maladjusted animal human), and therefore also the *pharmakon*, the improvement of its state.[8] There are two poles in this relationship. On the one hand we have the soul, the organic process, which is irreversible (this appears as a huge limitation, but at the same time, as we saw in 1.4, it defines the mastery of souls). On the other hand, we have the automaton, the mechanism, which obviously has an end, like everything else, but is marked by a constitutive reversibility, that is, by the iterability typical of technical devices. The division of tasks between soul and automaton is clear. The soul, which will die and is therefore in a hurry, has history, needs, and ends to confer. The automaton, which has none of this, is built to respond to the soul's ends, and without those ends it is perfectly useless.

Let's think about this for a moment. A hairdryer is morphologically very similar to a duck. The difference between the two, however, is that the hairdryer can be turned on and off a large number of times (depending on its quality), whereas the duck, once turned off, cannot be resurrected. The same thing that happens to the duck happens to a professor, except that he can elaborate the doctrine of being towards death, and this is because he has equipped himself with mechanisms (from pens to books to language) capable of iteration. While a hairdryer and a traffic light turn off and on, a duck and a professor only turn off. However, one

---

[8] [7] J. Derrida, *La pharmacie de Platon* (1968), in *La dissémination*, Seuil, Paris 1972.

could argue, computers, cars and hair dryers also use energy and release heat: are they living beings? Of course not, for at least three reasons.

Firstly, organisms, unlike mechanisms, are able to use energy not only to exist, but to grow. Many radios have been known to switch off when the power goes out, but we have never seen a radio become a television set simply because it has been supplied with power. The machine only runs one programme and is already built; the power supply only serves to allow the programme to function. In the case of organisms, instead, energy serves for the passage from embryo to adult.

Secondly, organisms are in most cases capable of reproduction, i.e. of giving rise to other organisms similar to themselves. Again, it would be futile to put banknotes in the same environment hoping to obtain new ones, whereas it is entirely reasonable to expect something like this to happen in a trout farm. However, if we refer to the technological sense shared by both nature and artefacts, and if we think of bioengineering, we realise that even this difference is not absolute and insuperable. The point is another one: the fact that souls can reproduce gives rise to social needs that are inconceivable for an automaton, and which instead determine a basic continuity between animal and human sociality.

Finally, and above all, organisms, unlike inorganic matter, are characterised by the fact that they die. As in the case of the duck and the hairdryer, anyone who leaves banknotes in a drawer for a year will find them the same as before. On the other hand, if you left trout in a bucket, a year later you would find something considerably different. Of course, there are organisms capable of slowing down their metabolism to a state of apparent death, and some grains of wheat found in the pyramids have been able to germinate centuries later; but, in the end, every living thing dies.

A machine can switch on and off over and over again, that's what it is made for. We, like all other organisms, are only given two options: on, and then off, forever. The soul, whether of a plant, an animal or a human being, once off, is off (if the grain in the pyramid had been reduced to flour, it would not have sprouted). Therefore, even though there may be marginal problematic cases, the definition of the living appears clear-cut: alive is only that which can die, and only once. "Giving up the soul" and "drawing one's last breath" are equivalent. The breath or wind – *anemos*, the wind in Greek, and *animus* (whence the words *animate*, *animal*, etc.) have the same origin – which gave us life (for the ancients) is proof of the fact that we are alive (for the moderns). It is precisely the breath that is taken away by Covid-19, which apparently causes the sensation of death by drowning, as happened to Ophelia. This is the weakness of the soul: its dependence on a metabolism, because the soul is not the breath we give off, but the organism that gives it off. This is precisely where we tend to get confused, mistaking the last sign of life for something independent and immortal, and transforming the last breath into the birth of the spirit. Unfortunately, or fortunately, it is only the expiration of the body, so that only an organism, that is, a mortal – be it plant, non-human animal or human animal – possesses a soul, which upon dying returns to the earth and not to heaven.

The hypothesis of a vital impulse that would distinguish the organic from the mechanical is no less problematic or tautological than the idea that there is something special and unique about natural thought (or, more precisely, human thought) that would distinguish it from the artificial thought of a computer. This is very true, but it does not detract from the specificity of the soul compared to the automaton: it is not that the former expresses meanings and the latter merely manipulates signs, but that the former is an organism and the latter a mechanism. This is what life is: neither a mysterious fluid nor some occult power, but a certain end. There is no mystic force. It has been said that organisms are oases of order in an ocean of chaos. It would be better to say that they are engaged in a battle against chaos which when lost, unlike with mechanisms, is lost forever. This is how the flesh lives and dies: a cruel dependence on the intake of energy, a losing struggle against disorder, and at the same time a fundamental sense and direction that underlies all subsequent meaning. This means that a duck and a professor have much more evident and pressing urgencies than a hairdryer, and at the same time that their behaviour appears more motivated, more unpredictable, and all in all more interesting, if only because their purpose, unlike that of the hairdryer, is anything but evident. There is nothing in the professor that necessarily predisposes him to write *Being and Time*, nor in the duck that leads it to end up on the professor's plate. Instead, when it comes to the hairdryer, everything about it is already written, which is a good reason not to engage in activism for the rights of hairdryers.

If having an immortal soul would expose us to the disadvantage of eternal tedium, having a mortal one might seem even worse, but it is not so. It is the first step towards the human form of life, although for the moment it is no different from any other organic life form. The duck (like the philosopher) does not necessarily know this and may not contemplate a skull on the desk, but, usually two or more times a day, it is hungry and has to do something about it. It is under the urgency of these needs that the soul sets the automaton in motion to help it delay the irreversible end as long as possible. Only organisms have purposes, times, and ends. And a virus has none of these, any more than a clock has the purpose of telling us what time it is. Our foundation is thus the essential difference between the soul and the automaton: the first has no reason, it is born and dies because it is born and dies, there is nothing else to it. The latter is programmed to generate iterations, from the roasting jack to the smartphone via the clock. And if these things run out of battery, no problem: they recharge and start working again.

Consider two memorable scenes involving deadly automatons, Hal from *2001: A Space Odyssey* and the replicant from *Blade Runner*. The case of Hal is quite simple. Just as it is dead, or rather rendered oblivious, it can also be resurrected, provided its memory is restored. That of the replicant is more complicated. "I've seen things, you people wouldn't believe, attack ships on fire off the shoulder of Orion. I've watched C-beams glitter in the dark near the Tannhauser Gate. All those moments will be lost in time like tears in rain. Time to die." The replicant becomes human the moment he dies. Now, can a replicant really die? What if he woke up the next year? After all, it would be possible to repair him, and he would not decompose, as long as he was stored properly. And conversely, if the scene moves us and allows us to empathise with the replicant, it is precisely because we feel that he is like us, that he will not be revived, and that indeed the things he has seen will vanish with him.

And yet, even in the case of the replicant, as in that of any other automaton, I can prove that not only he cannot die, but that his relationship with energy sources can in no way be compared to the consumption characteristic of humans. First of all, let's clear up any ambiguity: I am fully aware, from my own experience, that sooner or later light bulbs burn out, so I have never contemplated attributing immortality to automatons. I am also aware that a car without fuel will stop. However, I am reluctant to consider the first event as proof of the bulb's mortality, and the second as an indication of the cars' propensity to consume. Quite simply, all automata need a source of energy, at least for their manufacture. As for the fact that they die, I propose a simple thought experiment. Before dying, except for tragic circumstances, one usually grows old. Which, mind, is not what happens to the replicant: his death, from this point of view, seems closer to the case of a broken electrical appliance, which suddenly stops working or starts giving problems. But these things have nothing to do with ageing and death. An appliance or a replicant could be repaired, except repairing it would be more costly than replacing it. After all, an old lion can well be cared for even after it can no longer work in the circus, whereas there is no point in looking after a washing machine that no longer works.[9]

---

[9] A very interesting question was put to me by Stefano Lanterna, whom I would like to thank: "What do you think of synthetic/biological machines? Are they souls or automatons? This is not science fiction, or rather not only. I am referring, for example, to the invention of Xenobots (https://en.m.wikipedia. org/wiki/Xenobot [02/02/2021]), small multicellular organisms built to move particles, in total autonomy and cooperating with each other if necessary. They are sensitive to damage, to which they react with integrated 'healing' mechanisms. Obviously, they do not have a consciousness, but they do have a body, a mortal body. Are they machines, then? I suppose so, since they have a specific purpose that the engineer who created them has given them. Do you agree? If I may, I would like to go a little further and think about when the biological systems we can create will be equipped not only with rudimentary locomo-

Now, human animals are not too well disposed towards non-human animals, and old foxes end up in the furrier's shop just as old horses are sent to the slaughterhouse. As for older humans, instead, we have invented technical-social supplements for when they can no longer play their instrumental role in society. I am talking about pensions, which today we consider fair to pay to the elderly – the Native Americans preferred to abandon them in the forests, and the Callatians, if we are to believe Herodotus, ate them.[10] Now, while a minimally developed sensibility suggests that caring for elderly humans is the right thing to do, and that caring for old animals is desirable, caring for old machinery seems like a potentially reprehensible oddity, because it would take time and resources away from caring for things that we consider more important (and rightly so). It follows, I believe, that an automaton neither consumes nor dies, since it can be repaired – provided it is not an automaton of great quality, like Woody Allen's Beetle in *Sleeper*, which reboots a century later.

Computers merely signal that their batteries are running low, while humans try to feed themselves, change their behaviour, reach for goals and turn to actions – from stuffing themselves with sandwiches to drawing up a charter of human rights – designed to satisfy metabolic needs. A bronze buckle or a glass vase have never struggled to retain their shape, whereas human bodies are perpetually engaged in this endeavour. Yet when archaeologists find a Mycenaean tomb, the buckle and the vase are intact, while only bones, if anything at all, remain of human bodies. This may seem like a frustrating circumstance, but it is also the reason why, unlike the buckle or the vase, the bodies were able to harbour hopes, beliefs and well-founded fears. Similarly, a Machiavellian intelligence,[11] which is found, in more or less sophisticated forms, in primates, children, gangsters and charismatic leaders, is completely unthinkable in an automaton. And this is neither because the automaton is more stupid, nor because all forms of intelligence that I have just mentioned lack the mechanical elements that are found in computers, but simply because the aims and needs of Machiavellianism are conceivable only for mortals. Now, mortality is precisely where the specificity of organisms should be sought, especially if we aim at determining the difference be-

---

tor and epithelial systems, but also with a nervous system and, later, a central nervous system. I am convinced that the creation of an artificial consciousness, through the adapted reproduction of an animal brain, will take place long before we know exactly what it comes from or how it works. Well, when machines are self-conscious, will they still be defined as 'machines'?". No, they will be organisms, at a very low level; if they have time to develop they will become something, as long as they have hands or some such thing.

[10] Herodotus, *Histories*, with an English translation by A.D. Godley, Harvard University Press, Cambridge (MA) 1920, Book III.

[11] R.W. Byrne, A. Whiten, *Machiavellian Intelligence*, Clarendon Press, Oxford 1988.

tween organisms and mechanisms rather than at identifying the specificity of the human form of life.

Conceptualising the alternative between ducks and hairdryers, what emerges is a contrast between the soul, an organ subject to metabolic pressures and death, and the automaton, independent of both, yet capable of indefinite forms of iteration. A simple automaton, like a tool, can break (a hammer breaks), but it can always be repaired or replaced. A complex automaton is programmed for as long a series of phases as possible (such as the traffic light, the internal combustion engine, the computer), and the more complex the process, the more fittig the automaton proves to be to its etymology (*automaton*, self-moving). Yet, this constitutes an ideal tendency destined never to be realised, because full movement requires internal purpose, whereas the automaton's purpose always comes from the outside, for example from the soul that regulates the thermostat in the house.

Just as automata do not have an internal purpose, but manifest that of their creator, they do not have a life, but promise a mechanical survival to living beings that cannot be resurrected, first by making up for their shortcomings and prolonging their life, and then, after the end has come, by preserving their traces (writings, photographs, mummification or other treatments for the most demanding). As we shall see at length in 3.4, a complex dialectic opens up between having an end (final point) and developing ends (purposes). Firstly, the finalism of corporeality gives life a direction, since it follows an irreversible process: one eventually dies. Secondly, direction is also a sense of urgency, a *dead*line (indeed), which imposes needs, goals, desires and, sometimes, responsibilities on organisms, both human and non-human. Thirdly, organisms, and especially human organisms, give explicit purposes to mechanisms aimed, as I have just mentioned, not only at prolonging their existence but, in some cases, at trying to project something of life beyond death.

This point deserves to be considered before moving on to the examination of responsiveness, because it illustrates a distinctive characteristic of mechanisms as opposed to organisms: their tendency towards perfect iteration. From a knife, one expects it to cut many things, until it becomes blunt; from *Antony and Cleopatra*, one expects Cleopatra to tell Domitius Enobarbus that all they have left to do is think and die until the end of time. And from mummification, one expects that – by a process of iteration, the same one we observe in traffic lights – it will ferry a body beyond death. Hegel wrote that meaning is twofold[12]: and indeed, even without Hegel, in Germany they say "*Einmal ist keinmal*", and the postman

---

[12] G.W.F. Hegel, *Encyclopedia of the Philosophical Sciences* (1817–1830), Clarendon Press, Oxford 1971, § 458. Cf. also J. Derrida, *The Pit and the Pyramid: Introduction to Hegel's Semiology* (1970), in Id., *Margins of Philosophy*, University of Chicago Press Chicago 1982, pp. 69–108.

always rings twice. He also wrote that "sense" is a wonderful word, because it has two opposite meanings: purely physical sensation and, at the same time, meaning, thought, as when we say "the meaning of life".[13] So how does spiritual-isation come about?

With regard to technology, Hegel observed that iteration is the way to ensure the transition from the material to the spiritual. The technology to which Hegel referred was precisely the Egyptian practice of embalming corpses, which he interpreted as an early intuition of the immortality of the soul, but which is in fact the transformation of the soul into automaton. He observed that death, as sug-gested by embalming, occurs twice: first as the death of the natural; then, with embalming, as the birth of the spirit. But the death of what is simply natural is true death, one and unrepeatable. The mummy, on the other hand, is a machine, an automaton allowing for the hysteresis of a life that is now mechanical and no longer organic. The tranquillity of mummies[14] is just a variation of survival through writing: it has nothing to do with an actual resurrection; it is a way of looking at life through the eyes of death, not of restoring life. The point is, how-ever, that death, real death, only happens once, and that is why we are souls and not automatons. In this sense, the conception of the human being as a "desiring machine"[15] is only acceptable if "desiring" is a characteristic that goes beyond that of "machine". Indeed, the notion of "desire" is inconceivable in a machine (as are "frustration", "will to power", "boredom", "ambition" and "hope").

2.1.4. The responsive animal

The fact remains, however, that only humans have shown such a technological propensity to devise ways of transforming the soul into an automaton, with the promise of resurrection or the practice of embalming. These hopes are as noble as they are futile, but they nonetheless point to the fundamental human inclina-tion towards technological iteration and the capitalisation that comes with it. The fourth and final step in the reconstruction of our foundation therefore consists in reacquainting ourselves with the specificity of the human soul, which, *like* every soul, is subject to irreversible processes, but which, *unlike* every non-human soul, has the essential characteristic of systematically relating to automata in or-der to make up for its shortcomings. However different they may be, the behav-iour of a prokaryote, a beaver, or a professor has a single origin: the ability to

---

[13] G.W.F. Hegel, *Aesthetics* (1836–1838), vol. 1, Clarendon Press, Oxford 1988, pp. 129–130.

[14] G. Leopardi, *Dialogo di Federico Ruysch e delle sue mummie*, in *Operette morali* (1827), Rizzoli, Milan 2008.

[15] G. Deleuze, F. Guattari, *Anti-Oedipus* (1972), Continuum, London 2004.

extract energy from the environment and use it to grow, develop, and self-repli-
cate, which confers irreversibility on events, and thereby determines the genesis
of sense, which is primarily *direction* and then specifies itself as *sensibility* and
reason. The passage to higher stages is possible because human organisms are
empowered by mechanical and repeatable performances, inside and outside of
them: *idealisation*, a possibility of iteration that is almost indefinite and cheap,
since pen and paper can suffice to obtain it; *hysteresis*, which is the condition of
possibility of idealisation; and the consequent formation of language and culture,
social institutions.

This is how the defective animal defines its place in the world, as a synthesis
of soul and automaton. If the human had not had chains, bands, rules (including
grammar rules), norms and social prohibitions, it would simply have been a beaver with particularly weak teeth. Responsiveness, that is, the characteristic quality of the human being understood as interaction with technology, and only then
as moral responsibility, originates in a constitutive link between the soul (the
organism that we are) and the automaton (the mechanisms we use to compensate
for the inadequacies of our organism). Unified by death, human and non-human
animals are differentiated by technology. A duck, unlike a human, makes do with
its natural equipment, while a human uses sticks, hats, benches, and wallets. Here
we are drawing a complicated picture. Let us try to work it out.

Let us consider a beaver, a chainsaw, and a woodcutter. Just like a human, the
beaver is a soul, endowed with physiological needs, metabolic urges and a finite
life course that make it capable of intentions – in this case, that of cutting down
a tree. And to those who object that we have no means of accessing the beaver's
mind, it will suffice to reply that if the beaver does not intend to cut down the
tree, one wonders what on earth it is doing. Reciprocally, if there is one thing I
would say without fear of contradiction about the chainsaw, it is that it has no
intention of cutting down the tree. In fact, it has no intention, no hurry, and no
desire, simply because it is an automaton, designed by someone for a specific
purpose (cutting down trees) but which itself has no purpose, having no need or
urgency. Finally, the woodcutter, just like the beaver, intends to cut down a tree,
but, unlike the beaver, he does not have good enough teeth, which is why he has
equipped himself with a chainsaw. In his action we see the cooperation between
soul and automaton typical of the human animal.

Imagining a duck using a hair dryer or a bull with a chainsaw is little more
difficult than imagining a hair dryer using a duck or a chainsaw *willingly* slicing
a beaver; whereas a human with a hair dryer or a chainsaw in hand is one of the
easiest things to picture, and the essential thing is that the human in question does
not confuse one with the other. Like any other animal, the human being is sub-
jected to the urgency of metabolic processes, and for this reason develops final-

istic behaviours, whose ultimate meaning is to be found in the satisfaction of physiological needs. However, as animals that are ill-equipped in relation to their living environment, humans have developed a series of supplements to fill the organic gaps. This encounter between the metabolic needs of organisms and the supplements inherent to mechanisms is what defines the specificity of human nature. This specificity is responsive insofar as it arises from the encounter between irritability as a macroscopic characteristic of living beings and the ability to iterate and capitalise on performance as a distinctive feature of technology.[16]

The coordinates that define the place of humans in the world therefore consist of two surpluses that are called upon to fill two shortcomings. The surplus that human souls possess with respect to automatons is *animality*, i.e. finitude and the finality that comes with it; the surplus that they manifest with respect to non-human souls is the *technology* that compensates for the human's scarce natural endowments – or, more precisely, technology as the product of a so-called "second-order" instrumentality. Indeed, while other animals occasionally make tools, humans make machines, complex objects whose function derives not from their shape but from the interaction between parts. The soul determines goals and urgencies, the automaton realises them, enhances them, and ultimately determines them. This is the unbroken thread that leads from the cell membrane to the human brain, which is its direct derivation. Born of the convergence of a weak soul, subject to an irreversible course, and a relatively more durable automaton capable of iteration, the human being originated in a techno-social system, the foundations of which are to be sought in our past as pure souls, but whose true beginning only took place when we encountered the first automata.

This is all too obvious. Just like beavers, humans are what they eat, which means that humans are what they consume. Therein lies their structure. Sophisticated superstructures may or may not come, but the starting point remains the same. As an organism to be nourished, the human being thus constitutes the entelechy of the whole social system, made to respond to, and if possible promote, consumption. This is a circle which, on the one hand, prescribes an end, and the urgency that comes with it, and, on the other hand, proposes ends, objectives which consist as much in postponing the end as in giving rise to aspirations, desires and frustrations inconceivable in a short, brutal and meagre life.

The fundamental characteristic of responsiveness is therefore this encounter between an organism, with its needs, its times and its deficiencies, and a system of mechanisms called upon to respond to these needs and to make up for these

---

[16] In this sense, ecologism overlooks a fundamental aspect, namely that the mind as a human fact is not only nature, but technology. In this sense, cf. the remarkable analyses of G. Bateson, *Steps to an Ecology of Mind* (1972), University of Chicago Press, Chicago 2000; Id., *Mind and Nature – a Necessary Unity*, E. P. Dutton, New York 1979.

deficiencies as far as possible. Responding to the former and making up for the latter, rather than the faculty of reason or love of knowledge, are what constitutes the essence of the human soul.[17] It is therefore necessary to conceive of technology not only as plastic and steel, but as a capitalisation of resources – the condition of emergence of social reality and of normativity itself, as well as of the concept of "rationality", which arises from the encounter between souls and au-

---

[17] Moreover, the soul shares some fundamental characteristics with the automaton, first of all that of keeping track of sensations and thoughts, just as a computer does. As Aristotle points out in Chapter XII in the second book of the *De anima*, the essence of this tracking is that it accepts the form of sensation without absorbing its material. Aristotle uses a famous image to illustrate this point: it is as if the seal of a ring were impressed on sealing wax, which retains the form of the seal, but not the gold, bronze or iron of the ring. Far more than a principle descended from some hyperuranium, as in Plato, Aristotle's soul is the outcome of an emergent process. In beginning his discussion of the connection between sensation, imagination, and thought (427a 17 ff.), Aristotle notes that thought differs from sensation and imagination, but at the same time he is very careful to show the debt that thought owes to these lower faculties, as well as to memory. The latter plays the role of a superfaculty in the whole of Aristotelian psychology, underlying the permanence of impressions, their elaboration in the imagination, and the articulation of images and concepts in thought. The intellect itself, from this point of view, is thought of in the form of memory. Indeed, Aristotle observes that the mind is nothing actual until it thinks, much like a blank tablet (429b 30–430a 3). Finally, in the Aristotelian soul, thought depends on the lower faculties also for another characteristic: the action of imaginative processes within reflection. It is precisely in this framework that Aristotle proposes the partition between two intellects: one, passive and referring to the senses, needs the imagination; the other, active, can do without it (430a 18–19). Obviously, the passive intellect is a close relative of the imagination, which in turn is derived from sensation. How, then, does thought differ from the lower functions with which it shares so many aspects? Exclusively in terms of a greater degree of abstraction. Sensation is of the particular, demonstration is of the universal, and the collection of particular facts differs essentially from intellectual knowledge (according to the argument that will be developed in the *Metaphysics*, 981a 7 ff.). The latter is as such independent of its basis, since the triangle that serves for a demonstration is not as such the object of the demonstration, not only in the obvious sense that one is not exhibiting a characteristic of that particular triangle, but of all triangles with the same properties, but also in the stronger sense that, for example, the circumstance that the sum of its angles is equal to the sum of two right angles does not depend on the isosceles as such (*Posterior Analytics*, 73b 25; 31 ff.). If, then, at the level of abstraction, the intellect is capable of moving beyond the here and now of sensibility, and beyond the otherwise determined character of the image, it remains that the contribution of these subordinate faculties is central to thought. With a very powerful image, Aristotle writes (*De memoria*, 450a 1–2) that the same phenomenon occurs in thinking as in drawing a figure, so that, although thought can in principle dispense with reference to a given size, in fact, due to an inescapable psychological circumstance, it does resort to it. Thus thought depends on the image for reasons that are not essential but accidental. It remains that this is a very important accidentality and, ultimately – for Aristotle this is not a contradiction – a necessary accident. In other words, the Aristotelian soul is both the paradigm of the living being as the source of needs and intentions, and of a thought that finds its condition of possibility in the retention of traces.

tomata, and not before. It is in this framework that concepts such as "trust", "responsibility" and "decision" acquire meaning. These involve the presence of technological components, of which computers are only the latest version, since everything began with the lever, the wheel and writing; and biological components, i.e. the existence of at least two organisms that interact socially. To claim that a computer is responsible, and therefore a guarantor, is a conceptual error regarding the notion of "responsibility", and is not different from thinking that if you take a 10 euro note out of your right pocket and put it in your left, you are lending 10 euros. The same, though for other reasons – including the lack of pockets – applies to non-human animals. More on this later. For the moment, let us take our eyes off the soul and turn them towards the automaton.

2.2. Supplement: Prosthesis

What could drive the Greeks to build a temple? Certainly not the desire to carve the maxim "know thyself" onto it. Excluding this hypothesis, conjectures are obviously controversial, but it is not difficult to find the single thread linking a cavern or a canopy to the shape of the Greek temple. In this need for shelter we find the characteristic element of humans, their inadequacy to the environment, their need to protect themselves, the scarce endowments they receive from nature which put them at a disadvantage compared to non-human animals. This generates a circle whose primary motive is need: in the human animal, less adapted and more ambitious than other animals, this need requires the intervention of a technical supplement that, by offering new possibilities, generates further needs and produces a very specific dependency structure. The pithecanthrope that picked up the first stick was identical to the one without the stick, only stronger. Nevertheless, that stick, like a magic wand, generated a completely different and transcendent being, although the natural endowments and emotional dispositions remained the same as those of the pithecanthrope. This continuity between the savannah and academia should not come as a surprise, since one of the salient characteristics of the supplement is the variety of forms it can take. Its metamorphic tendency is based on a common basis of capitalisation: the supplement is the club, but also the book, the coin, the dress, the institution. In all these cases, a technical accessory makes up for a lack in a non-accessory way.

Through the intercession of the supplement, the misfit animal becomes an accessorised animal, i.e. a capitalist; it exceeds the frugality imposed by the environment, it learns to appreciate order and beauty, luxury, calm, voluptuousness. If, however, there were not that initial lack, which lies at the origin of humanity, no civilization would be truly accessible and conceivable. The surplus of humans over animals does not lie so much in what they have inside them, but rather in what they externalise. Even their internal higher organ, the brain, derives its efficiency from continuous exposure to the outside world, in a process that begins with education and language and continues until retirement. After that – not by chance – a rapid psychic decay takes place, unless countermeasures are taken – measures that would appear pointless for an object. Here too, as in the case of understanding the Web, it is a matter of operating a Copernican revolution, which posits technology as the condition of possibility of the human being within the framework of a constitutive cooperation between organisms and mechanisms. In order to illustrate this role of the supplement, I will use an image that is a little more dignified than the contrast between a duck and a hairdryer: the dialogue between the Sphinx and Oedipus.

2.2.1. Oedipus's stick

"Which creature has one voice and yet becomes four-footed and two-footed and three-footed?" The first element that must be highlighted, in order to understand the profound nature of the supplement, is that it does not come *after* homogenisation, but precedes it, albeit by an instant. Before the proto-supplement, the stick, the human was still a non-human animal; immediately after picking it up from the ground, an evolutionary history began (which has undoubtedly been repeated who knows how many thousands of times in different places and with varying degrees of success) at the end of which we have humans as we know them today. The riddle posed by the Sphinx to Oedipus could not be more eloquent in this respect: the adoption of supplements, in this case a stick, is part of the essence of humankind. The supplement is not an extrinsic addition to an otherwise independent human nature, but a constitutive part of it. Being endowed with language and *hands* is an insufficient essence, therefore it is not an essence; it is not the fundamental *eidos*, but the simple predisposition to a destiny that has nothing manifest about it. The technological prosthesis is part of the definition of the human animal, and in the case of the stick it shows the human's propensity towards a life that is in many ways more unstable. Despite having four hands, chimpanzees do not systematically use sticks. Of course, they may use them occasionally, but we have never seen an old chimpanzee leaning on a stick, for the simple reason that the species has not made a definite choice in favour of the upright posture, and may, if necessary, go back to walking on all fours. What would chimpanzees do with an old age stick? And, *a fortiori*, could they ever appreciate the frivolity of a walking stick?

Let's think about the scene of Oedipus and the Sphinx through any of the many artistic reproductions available, from the Greek vases to symbolist paintings.

Usually, on the left there is the Sphinx, sitting on a pillar or sensually wrapped around Oedipus. It is an evolved and hybrid animal, because it has the body of a lion and a human face. If the Sphinx had kept its lion face, the scene would have been worthy of a comic-book, because lions cannot speak and have a snout to carry out functions which, in humans, are carried out by their hands and technical devices. Let's now look at Oedipus. He is not old (he has dark hair and a beard in Greek paintings; in Moreau he has blond, long and wavy hair, something between Christ and a gigolo), but he always has a stick, in all the representations we have of the scene. Over time, this stick seems to grow longer than the singular stick of Pinocchio's nose, to the point of becoming a long red pole in Moreau's painting. The reason for this is simple: only the human being, who has chosen an upright posture at its own risk, really needs a stick. And since the upright posture has aroused dreams and needs completely different from those it had before, the human being has ended up using not only the stick, but an infinite number of technical devices of which the stick is the fundamental emblem.

Thus, the human being is the only animal whose definition necessarily involves a technical device. Man is the *zoon bakterìa échon*, the animal endowed with a stick (unlike any other animal, despite having bacteria just like any other animal), who is able to use the stick and the carrot with other animals, including other humans. Therefore man is the animal endowed with glasses, shoes, backpacks, smartphones, books, and receipts. In this transformation, one can see a new value of the human, which has always been there but only now can be put to use. The observation that some non-human animals make use of apparatuses, such as prostheses or nests, is undoubtedly pertinent, but in the human animal the prosthesis has different functions in both quantitative and qualitative terms. A bird's nest, in fact, serves to protect the offspring in the earliest stages of life, which is completely different from the use that humans make of dwellings, nor does it produce the same social consequences as urban civilisation. Similarly, the fact that anthropomorphic apes can use bones or sticks as weapons because they have opposable thumbs does not change the fact that these same apes could very well live without these supplements, which is not the case for you and me. The addition of a technical supplement seems necessary for the attainment of a human form of life, playing a decisive role within an evolutionary circle that is much faster and more complex than that which, in the absence of technology, characterises non-human animals.

Not the creation of syllogisms or norms, but the construction of technical devices is the first act of the naked ape, whose first concern is to eat, to be armed, to warm up (and therefore also to dress up). Knowledge and duty will appear long afterwards, *as derived from technology* qua *capitalision*. I wish to stress this point, as it will play a decisive role from the next chapter onwards and it will form the backbone of the policy proposal I am drawing up in 4. It has rightly been pointed out that the gap between the ape and the human being lies in the latter's technological capacity, which does not take place on the level of *homo sapiens* as the rational animal, but on that of *homo habilis* as the technological animal. In order to justify an advantage that goes against the laws of evolution, because it decrees the prevalence of the *less* fit, one must not only look at the hand, but also at what it grasps.[18] The real difference lies in the stick: the stick of our old age, but also all the other sticks that fill the lives of each of us, and the history of humanity as a whole. So much so that for hundreds of thousands of years humans remained just naked apes. The change took place the moment when they grasped a stick: that's when the history of humanity and the history of technology simultaneously began.

The relationship with technology therefore constitutes the second distinguishing feature of the human animal compared to non-human animals. Imbecility is wrongly considered to be an insult like any other, or a painful acknowledgement. Instead, it is literally and etymologically the condition of the human being in the state of nature: *in-baculum*, stick-less, and therefore devoid of the system of supplements that determine the difference between the human animal and non-human souls. There is nothing in the human spirit that makes it different from non-human animals. There are hands and the ability, through those hands, to build devices, including language, writing, culture, and the capitalisation of resources, which come back to humans by prolonging their lives, enhancing their experience, and refining their sociality. It is precisely for this reason that only humans can be imbeciles,[19] failing in their distinctive characteristic, however

---

[18] In *Heidegger's Hand* (1985) in *Deconstruction and Philosophy: The Texts of Jacques Derrida*, University of Chicago Press, Chicago, 1987, pp. 161–196, Derrida commented on a famous photographic book on Heidegger that triggered a great invective by Bernhardt (T. Bernhard, *Alte Meister* [1985], Suhrkamp, Frankfurt a.M. 1993, p. 88) and focuses on Heidegger's reflection on the hand in an arc that goes from *Being and Time* to *What Is Called Thinking?*. On the basis of what I said about Oedipus and the necessary supplement, it should come as no surprise that these analyses, which obviously concentrated on the technical, philosophical and political role of the hand, require additional investigation. Specifically, one ought to place more emphasis on the supplement, the stick, which is all-pervasive and nearly more ubiquitous than the hand. In the pictures, Heidegger takes the stick, raises the stick, rests the stick on a bench, hands an additional stick to his wife, uses the stick's knob to look as if he was sticking out his tongue, etc. And when the stick is at rest (but Heidegger leans on the stick even while sitting in an armchair and conversing with the journalists of the Spiegel), then the spotlight should go on the pencils and pens arranged in battle order on the desk. Indeed, in Heidegger's study in Freiburg, just as in universal history, the most decisive revolutions arose from sticks, pens or barcode bars.

[19] Which explains many things. One always has the unfounded impression that the prevalence of cretins is a current phenomenon, whereas it is eternal, except today this prevalence is

accidental its genesis may have been; and precisely for this reason only an imbecile could call a non-human animal such.

The defining choice is a definitive choice. Orlando throwing away the arquebus to preserve chivalry is not giving up on technology: he is simply choosing an ancient one, the sword, over a modern one. On the other hand, the steel storms and material warfare of 20th-century conflicts have taught us how fables and myths can be recreated from devices more modern than the arquebus. Nature and spirit, myth and *logos*, are not two opposing dimensions: they are two faces of the same reality which communicate through technology. One of the oldest forms of technology is precisely the composition of narratives, of plots whose origin is no different from that of the textiles that have accompanied us since the Palaeolithic era.[20] When we protest against a technology, we never really do so in the name of nature and bare life, but only in the name of another technology that is more familiar to us. We are beginning to understand the apparent paradoxes raised by the documedia revolution: since the introduction of technology, i.e. from the very beginning, the difference between work and life, as well as between artifice and nature, has been threatened by essential indistinction.

The first and most obvious feature of the supplement is precisely its prosthetic nature, the fact that it makes up for a lack as befits a prosthesis.[21] We are usually

---

better documented. The fact that imbeciles were widespread even in the age of libraries and newspapers is eloquent proof of the pantheon of radical thinkers ridiculed by Marx and Engels in the *German Ideology*; likewise, some parts of the *Dialectics of Nature* are perplexing. The prevalence of cretins, however, is only the sad side of a more general process, which is the progress of humanity, guaranteed by technology, and in which humanity is called upon to make its contribution not by regretting the blissful years of punishment, production and alienation, but rather by rethinking itself as a bearer of needs, desires, consumption, and therefore as a generator of ends. In any case, the greater visibility of this old phenomenon depends on the greater technological power of documediality.

[20] M. Cometa, *Perché le storie ci aiutano a vivere: la letteratura necessaria*, Raffaello Cortina, Milan 2017.

[21] Labour does not begin with effort, but with the supplement. In this, Marx was not original. A. Ressi, *Dell'economia nella specie umana*, vol. III, in Stamperia e Libreria di Pietro Bizzoni, Pavia 1819, pp. 217–218 noted: "Instruments [...] have three objects: 1. to shorten the time of production; 2. to save a great amount of manpower in production; 3. to perfect production. The first consequence of these three things is *lower prices for better goods*, which is the most important aim of the human economy. Now considering these ways of exercising labour, it may be said that co-operative means are as many supplements or replacements for the actual labour of man, and as many abbreviations of time. It is dead *labour* that replaces living labour." The dead labour is the supplement, but as is always the case, it is transformed in its essence. "Coming, then, to the origin of man's labour, we shall find that his first living labour was that of performing dead labour (that is to say, some kind of instrument): this first dead labour helped the subsequent living labour, and so on, always progressing. Having said this, we have two kinds of labour: living labour and dead labour; and all the goods of the earth exist by living and

accustomed to thinking of prostheses in an anthropomorphic way, with an inclination that we also find in more complex prostheses like artificial intelligence or
society, and to which we therefore lend anthropomorphic and evil physiogno-
mies. A prosthesis, however, is not only the hook making up for a hand or a
wooden leg: it is everything that makes up for the inadequacies of our strength,
memory, intelligence and character. The benefit of having hands is, on the one
side, the possibility of having practical supplements, such as a stick; on the other
side, the possibility, in perspective, of having symbolic supplements, such as
language. The reference to the stick should therefore not be understood in a limiting sense. Technology is any externalisation of capacities that are enhanced and
shared, and which generate new forms of capacity through pre-adaptation mechanisms.[22]

The dead labour that every instrument incorporates, the possibility of enhancing the memory of writing, the ability to make up for individual shortcomings
through social organisation – these are all transfers that individuals make of
something that is not sufficiently represented within them. External hysteresis is
both a relief for internal memory and the condition of possibility for a complex
society and an economy that goes beyond mere subsistence, as well as for openness to knowledge, abstract speculation, artistic or moral disinterest. As Rousseau said, before the invention of writing, time passed and humankind remained
young, i.e. ignorant, because each generation had to start anew. Obviously, writing is not necessary for hysteresis to take place, especially if by this term we
mean alphabetic writing. If, however, we do not restrict ourselves to the alphabet,
it becomes very difficult to understand when writing began, and this explains
why it is not literally true that a society without writing always starts over. Similarly, any elementary technique is transmitted and perfected over very long pe-

dead labour together. Goods thus composed and manufactured are the representation of this
labour, and a good is nothing else but a consolidation of labour; and if the good disappears, so
does the idea of consolidation. But these are more or less *permanent*. Permanent consolidations
are what I call *capital*."

[22] Externalisation is well represented by affordance, in accordance with the perspective in-
augurated by the American perceptologist J.J. Gibson in *The Ecological Approach to Visual
Perception*, Houghton Mifflin, Boston 1979, and developed in the ontological domain by the
British philosopher Barry Smith (cf. *Objects and Their Environments: From Aristotle to Ecological Ontology*, in *The Life and Motion of Socio-Economic Units*, ed. by A. Frank, J. Raper
and J.P. Cheylan, Taylor and Francis, London 2001, pp. 79–97, and Id., *Toward a Realistic
Science of Environments*, in "Ecological Psychology", 21, 2, 2009, pp. 121–130). Cf. M. Ferraris, *Estetica razionale*, Raffaello Cortina, Milan 1997, pp. 545 ff.; D.C. Dennett, (*Kinds of
Minds. Towards an Understanding of Consciousness*, Basic Books, New York 1997, p. 138)
notes that for an older person, the affordances offered by a familiar environment are an excellent remedy for reduced brain performance. Cf. also F. Recanati, *Mental Files*, Oxford University Press, Oxford 2012.

riods of time through the cooperation of memory (the making) and matter (the artefact), which acts as a partial vector of memory.

The instrument frees the hand, while hypomnesis, hysteresis – writing, annotation, artificial memory – frees the mind. The bars of bar codes, as well as pens or the other kinds of stick we call bacilli, manage to store a quantity of data that would leave no trace on a stick used as a club, except some accidental notch left from a fight. Likewise, the sceptre (another type of stick) requires the memory of the subjects, unlike the sticks holding the rolls of the Torah, which are the actual bearers of a memory that has been able to defy the millennia. What we consider distinctive of our interiority, from consciousness upwards, in this sense, is nothing more than the internalisation, through education and social interaction, of something that has been formed outside. If a chimpanzee is probably not endowed with self-consciousness, and we are probably sometimes endowed with it, it is not so much because of what we have inside us, but because of what we receive from the outside. We can easily realise this by spending a few weeks living with other chimpanzees.

At this point, it is quite evident that those without a stick would end up very badly off in society. They would be helpless, with no authority, no knowledge, unable to make and receive promises. They would be little more than poorly trained pets. And if humans develop all the social attitudes we know, it is not because they possess magical, shared or infused properties. It simply depends on the fact that humans, thanks to the possibility of keeping track of knowledge and duty, are able to build complex societies, characterised by stratified inclinations and desires, by conflicts and envy, that is to say, by things that would not be explainable – and on closer inspection, not even imaginable – by collective intentionality or specific altruism.[23] But before we get to this, we need to put down the stick and look at the hand that grasps it.

2.2.2. Handy, all too handy

"Man is the most intelligent of the living because he has hands." Anaxagoras' saying is one of the most profound and true sentences that have been handed down to us. If we had a smaller brain, our history would probably not be very different. But if we had no hands, humanity would be, so to speak, an *ou*-manity, a strange handless species of non-human animal. The fundamental predisposition of the human towards the supplement does not consist in a peculiar intellectual endowment, but in the possibility of opposing the thumb and grasping the sup-

---

[23] M.E. Bratman, *Shared Cooperative Activity*, in "The Philosophical Review", 101, 1992, pp. 327–341; M.S. Gazzaniga, *Human. Quel che ci rende unici*, Raffaello Cortina, Milan 2009.

plement. The benefits drawn from the achievements and complexifications of the supplement are what transformed the human form of life, making it capable of language, society and symbolic forms. And here too, as we shall see, the hand plays a central role.

Let us never forget this, and let us think about it every time we make a gesture: the hand is the essential link between the needs of a soul and the resources of an automaton. This is why the human world is full of handles, handholds, handpieces and handbooks: because the hand is the tool of tools,[24] the first universal machine, the closest ancestor of the computer. We are human first and foremost because we have hands. The words *man*oeuvre, *man*ager, *man*ual (from the Latin *manus*), hand of cards, handshake, handle and handcuffs are all examples of how language is fully aware of the central role of the hand in our lives. So too, in addition to *man*ipulating for practical purposes, hands fulfil symbolic purposes: hands shake, join in greeting or prayer, reach out in blessing or shake in cursing, are kissed at court or in a mafia film. Everything we do, from typing on a keyboard to greeting someone, to opening a door, passes through the mediation of the hand, because our world is organised by organisms with hands. If we had beaks our world would be very different.

It is futile to imagine humanity as surrounded by a cloud of infosphere. Of course there is a cloud, but at the end of that cloud are hands tapping on a keyboard. The notion of "act" as a characteristic feature of the biosphere also blurs the distinction between intellectual and manual labour. The division between manual and intellectual work disappears when the intermediary of all activities is a keyboard, or rather a sensor recording our activities without differentiating between movement, leisure, productivity, readings and references. Only one form of mediation is imposed: the keyboard or the screen of the computer or smartphone. This is not without philosophical interest. Asking what it means to think, Heidegger compared thought to manual labour, and more specifically to building a cabinet.[25] This is less odd than it may sound at first. There is no doubt that a high degree of manual work is an indispensable component of intellectual work itself. But, then, how do we distinguish work from thought or anything else? The documedia revolution brings to the forefront precisely this question, which had remained in the background ever since the expulsion from Eden. Let us reflect on this for a moment. In writing this book, I am doing manual work. Obviously, one could object that in order to move my fingers on the keyboard, I must have some ideas in my mind, but – apart from the fact that this is not always the case – most of the time, in order to form them, I had to do the opposite. My

---

[24] Aristotle, *Parts of Animals*, Book IV, 687a–687b.

[25] M. Heidegger, *What is Called Thinking?* (1954), Harper & Row, London 1968, p. 14.

hands have opened books, pages, underlined lines, written comments on the side. In fact, it is difficult to find an activity of comprehension and expression that does not involve the use of the hands: even English gentlemen who considered it disreputable to talk while gesticulating had to make up for it by handling snuff boxes, sticks, fans, and handkerchiefs.

Why all this handling? Psychoanalysis has a favourite answer to this, and articulates it with a great number of examples and observations: neglecting the hands means repressing the body and its destructive and self-destructive potential, its longstanding uncontrolled desires. We clench our fists or suck our thumbs even before being born, and our knowledge of the world, as babies, passes through our hands and mouths, but it never leaves them for good, even when we reflect on the most abstract topics. We gesticulate when we speak, and not to make ourselves understood – we gesticulate even when we are on the phone and no one can see us – but because it is part of our expression. And when we listen, we take notes that we will probably never read, or we scribble, or we light a cigarette (at least, we used to). If we are psychoanalysts, we do even more. Freud used his hands a lot in his early days as a hypnotist, and later limited himself to caressing the idols he kept in his study. His daughter Anna had a whole weaving loom in her studio, and Lacan spent his time tying knots, in which he esoterically believed he had condensed his theory of the subject. So, the hand discharges and manifests the forces and desires of the body,[26] and at the same time allows them to be realised, by grasping things, writing, and technically transforming the world. The definition that someone gave of the last days of Proust, barricaded in his cork-lined and fumigated room, would fit perfectly into this framework: "a writing hand". However, as soon as we summon technology, we must admit that the hand is not only an outlet for our drives, but a tool. This is literally the case for potters, who use their hands as if they were spatulas, or boxers and martial artists who use them as weapons. The hand *par excellence*, however, is not the hand pressing the red button in an atomic war (perhaps after having fought in vain not to make a Nazi salute, as in *Dr. Strangelove*), but the writing hand.

The question remains: why would the hand, in addition to expressing our moods, make us intelligent? The answer comes from the treatise *On the Making of Man*,[27] which dates back to the end of the 4th century. In it, Gregory of Nyssa establishes an essential correlation between the acquisition of hands and the development of language, which Darwin would re-propose without variations fourteen centuries later. If humans did not have hands, then their faces, like the snouts

---

[26] D. Leader, *Hands: What We Do with Them – and Why*, Penguin, New York 2016.

[27] Gregory of Nyssa, *On the Making of Man*, Christian Literature Publishing, New York 1892. Cf. J. Laplace, *Introduction*, in Grégoire de Nysse, *La création de l'homme*, Éditions du Cerf, Paris 1944.

of quadrupeds, would have an elongated shape, lips suitable not for articulating words but for grazing, and a thick, calloused tongue good for kneading food. We may hold a pencil in our mouths when our hands are full, but we would never build a nest by carrying stuff in our mouths, as birds do. So, the hand frees up the mouth, the teeth and the tongue, and makes them available for speech: a transition that should not be understood simply as a physiological development, but also as a technological, economic and social event.[28] Indeed, the hand, unlike the mouth, can equip itself with a stick, and proceed to a series of capitalisations that would be impossible if the mouth had to perform the functions of the hand.

Gregory of Nyssa agrees with Anaxagoras but, above all, he explains *why* Anaxagoras is right, and this becomes all the more evident if we compare Gregory's and Anaxagoras' version with Aristotle's view that man was endowed with hands because he is more intelligent.[29] This is tantamount to saying that all animals are equal, but that there are some animals that are more equal than others because they have hands. Imagine the scene: the Prime Mover, reviewing all animals, subjects them to intellectual and aptitude tests, finally assigning the hand to the human. The animals are lined up, each with its own characteristics and abilities, and man stands proud with his sociability and his language, which came to him from the sky together with his monstrous intelligence. God examines and questions them, and, pleased with the intelligence of the human being, rewards him with a hand – or better, two. But human beings would never have won the contest had they not already been equipped with hands, i.e. the very prize at stake. Aristotle's view presupposes an exact moment when the human being would have received language, as a Pentecostal gift and a foreign thing; after that, he would have decided to form society in order to simplify his life (which is not necessarily the case) and put an end to the war of everyone against everyone (which is not necessarily the case either). Instead, and in agreement

---

[28] Cf. M.C. Corballis, *From Hand to Mouth: The Origins of Language*, Princeton University Press, Princeton 2002. Corballis provides a periodisation according to which the earliest graffiti date back to 30–40,000 years ago and logograms, phonetic writing, to 5,000 years ago.

[29] The most significant contributions on the philosophical role of the hand include: Derrida, *Heidegger's Hand* cit.; C. Bell, *The Hand: Its Mechanism and Vital Endowments as Evincing Design*, William Pickering, London 1833 (the hand must have a divine origin, as in Aristotle). Cf. also J. Napier, *Hands*, Princeton University Press, Princeton 1980; Id., *The Roots of Mankind*, Smithsonian Institution Press, New York 1970; D. Johanson, B. Edgar, *From Lucy to Language*, Simon & Schuster, New York 1996; K.R. Gibson, T. Ingold, *Tools, Language, and Cognition in Human Evolution*, Cambridge University Press, Cambridge-New York 1993; F. d'Errico, L. Backwell, *From Tools to Symbols: From Early Hominids to Modern Humans*, Wits University Press, Johannesburg 2005; R. Tallis, *The Hand: A Philosophical Inquiry into Human Being*, Edinburgh University Press, Edinburgh 2003; Z. Radman (ed. by), *The Hand, an Organ of the Mind. What the Manual Tells the Mental*, The MIT Press, Cambridge (MA) 2013.

with Anaxagoras, humans are the most intelligent of living beings *because* they have hands, which allowed them to carry out practical manipulations and hence, in a process that can take millennia, to develop their intelligence.[30]

This is precisely what is concealed by the Panglossian perspective, which sees technological development as oriented by some epistemological providence. To my knowledge, the most recent example of this is Colin McGinn,[31] who makes the astonishing assumption that thought is far ahead of language and is present in animal species that have no language whatsoever.[32] This is a factually unprovable and legally nonsensical assertion. Indeed, it is not clear what evidence, in the history of the species, demonstrates the precedence of thought over language; what we know derives from the fact that in our ordinary experience we happen to think before we speak, even if this is not always the case, so much so that we sometimes have to apologise for having spoken before thinking. It is even more

---

[30] In this sense, Anaxagoras' definition of the human would sound like this: the human is the animal endowed with hands. Humanism would thus be a *handism*, the ability to handle, and to have at one's disposal handbooks, grips, clubs, missiles, and of course also coins, manuals, notebooks and mobile phones, which – with a telling Anglism – the Germans call "Handy". Only hands guarantee the development of language, which in turn eventually determines the usefulness of smartphones.

[31] C. McGinn, *Prehension: The Hand and the Emergence of Humanity*, The MIT Press, Cambridge (MA) 2015. The book offers a good demonstration of the fact that a Panglossian paradigm, i.e. hyperfinalism, goes hand in hand with a and a logocentric one, i.e. a prejudicial and unmotivated bias favouring the role of the concepts in experience (*contra*: S. Oppenheimer, *Out of Eden: The Peopling of the World*, Constable & Robinson, London 2004; D. Sagan, J. Skoyles, *Up from Dragons: The Evolution of Human Intelligence*, McGraw-Hill, New York 2002).

[32] McGinn, *Prehension* cit., p. 40. The hand is said to be the bearer of an innate programme (ibid., p. 4), just as language would latently wait to be liberated by the hand (p. 49). In reality, no destiny awaited the hand or predisposed language. To assume such a hypothesis is to think that sticks are made to be grasped by monkeys and to give birth to consciousness and language. This is obviously not the case. The predominant feature of this productive history is not latency, but emergence: an enormous amount of time, material and acts, from whose entirely contingent form emerged the use of hands, its technical enhancement through sticks, and the subsequent liberation of the mouth for language. Panglossism is also logocentrism. In this perspective, in fact, we would have a series of latencies whereby language is latent in action just as thought is latent in language. And once the virtualities have been implemented, the normal hierarchy would be established whereby thought governs language and action. Assuming we can establish how things happened, everything we know about the world around us suggests that they happened differently. The hand was underused when our ancestors first came down from the trees. At some point, and for a contingency that hardly concealed the realisation of a latency, the hand grabbed a stick from the ground, continuing a habit acquired up in the trees. The hand thus freed the mouth, the tool enabled the manifestation of meaning (the purpose of the tool) through language and writing, and this led to understanding, mental meaning, and consciousness.

complicated to understand how an animal can be characterised as radically de-void of language, in a book that insists, like a long tradition that is carefully ig-nored, that gestures are language to all intents and purposes.

If one were to take seriously the thesis that thought is found even in animals devoid of any language whatever, one would have to assume the existence of thought in animals that are much less complex than bees or termites, since these social animals have communication systems and therefore, according to a Pente-costal assumption, are obviously endowed with thought. The result of this exer-cise would be that only worms are excluded from the sphere of thought, which is comforting, because since earthworms live on after they have been cut in two, the absence of thought frees us from the suspicion that a worm cut in two would have a split consciousness. So, once panglossism has been ruled out, there is one point left to consider. In establishing a symmetry between the soul and the hand, determined by the fact that both grasp their object without identifying with it, Aristotle clearly indicates the link between technology and knowledge. This con-nection transcends the generic relationship between prehending and compre-hending,[33] which is decidedly insufficient since one can very well prehend with-out comprehending and comprehend without prehending. The Aristotelian im-age, in fact, introduces a further and decisive element: hysteresis and the capitalisations that derive from it. For the soul understands not by grasping with its fingers, but by recording impressions and thoughts. However, the full symme-try is re-established by the fact that Aristotle, like the whole tradition that precedes and follows him,[34] represents the soul as a writing surface, as a tabula, which therefore presupposes the intervention of a hand holding the stylus and inscribing sensations and reflections. When we touch our chin in perplexity, we are linking back to practices already present in a foetus.

---

[33] Broad uses "prehend" to designate what Russell means by "acquaintance". The hands comprehend, grasp (Hegel knew this very well when he recognised in the German *Begriff*, "concept", the heir of the verb *greifen*, "to grasp"), indicate and, indicating without grasping, making gestures, initiate the production of symbols.

[34] Plato, *Philebus*, 39b; in Aristotle's *De memoria*, 450a 29–30, mnestic survival is akin to a *zographema*, i.e. a portrait (Aristotle, *De anima*, 429b 30–430a 3); J. Locke, *An Essay Con-cerning Human Understanding* (1689), Hackett Publishing Company, Indianapolis 1996, I, ii, 15 and II, i, 2; G.W. Leibniz, *New Essays on Human Understanding* (1705), II, Cambridge University Press, Cambridge 1996, I; S. Freud, "A Note upon the Mystic Writing-Pad" (1924) in *The Standard Edition of the Complete Psychological Works of Sigmund Freud*, Vol XIX, The Ego and the Id, The Hogarth Press, London 1961, pp. 227–232. I have described in detail the functions of the tabula in the history of philosophy in *Estetica razionale* cit., especially on pp. 234–240.

2.2.3. Epimetheus and Prometheus

I can guess a possible objection. Many anthropoid apes also have hands, along with the essential function of the opposable thumb. Sometimes referred to as "quadrumans", they have two hands more than bipeds. If Anaxagoras were right, the chimpanzee should be by far the most intelligent animal. On the other hand, it would be a mistake to think, with Aristotle, that it was the increased size of the brain that determined the evolutionary advantage for humans, since Cro-Magnon had a larger brain than Sapiens. The point is the cooperation between the hand and its supplement, in accordance with the fundamental trait of responsiveness. We must therefore look at hominisation from an emergent and not a Pentecostal perspective: something emerged to fill a gap without anyone being aware of it.

The decisive aspect of this evolutionary history consists precisely in the ability of a particularly maladjusted animal to find remedies that turned its limitation into an advantage. This, obviously, only applies if we take the human point of view: it is all but certain that if a lion could speak, and in the unlikely event that we understood it, it would proclaim its admiration for our forms of life. Our nature is the opposite of titanism – or rather, the titan from whom we descend is not Prometheus. Everything we have around us, and which makes our lives longer and more pleasant than those of our ancestors, does not originate in Prometheus, who stole fire from the gods. It comes from Epimetheus, who wanted to roast a sausage, and who therefore felt the need to master fire. Not only technology, but the factory defect that forces us to resort to it – that's the distinctive trait not only of our great-great-grandfathers, but of each of us at this moment. Quite trivially, I would not be writing these lines, and you would not be reading them, if what we so obscurely call "nature" had equipped us with tusks or fur to protect us from external aggressors and bad weather, and above all if it had provided us with claws, which are very useful for tearing and wounding but disastrous when tapping on a keyboard. Had they had adequate equipment for their environment, humans would not have wasted their time carving flints, developing symbolic systems, boosting technological developments that eventually resulted in me writing and you reading.

Our shortcomings, and the remedy that was devised for them – a mechanism with an explicit purpose – generated a complication that could be transformed into a resource, and which in any case defines us for who we are. One cannot imagine an animal in need of a suitcase (if only because animals don't usually have opposable thumbs or require clothes) or a holiday (if only because they do not work). Humanity is only conceivable on the basis of technical aids, and technology acquires meaning primarily in a social context. The fact that that same society or culture can create outcasts or serial killers is simply part of the game:

you will never find a lazy beaver, because such an animal would soon be dead. Likewise, there is no such thing as a killer tiger because this definition does not make sense – killing is simply its way of life.

Hence a circumstance that deserves further consideration. Whereas in the case of non-human animals needs are easily recognisable, in the case of human animals it is very difficult to distinguish primary from secondary needs, use value from exchange value, usefulness from ostentation and waste. This fact illuminates the enormous degree of chimerical abstraction that surrounds the notion of "human nature" when considered as an essence. Excess, the adoption of supplements to the natural endowment, must therefore be understood as a constant, the only thing that defines human nature, albeit only in a negative sense. However – note well, because it yields unforeseen effects – the *body* is the driving force of this need, not the machine or intelligence. Unlike non-human organisms, humans systematically resort to mechanisms, be they cloaks or hats, scrapers or mobile phones, sticks or drones. This phenomenon is fraught with consequences, precisely because of its circularity, which defines a process of capitalisation. On the one hand, the organism confers meaning, urgency and purpose on a mechanism that does not otherwise have any. On the other, the mechanism enhances the organism's resources, and creates the conditions for the emergence of knowledge, the world of culture and spirit, consciousness, social relations and the economy. In other words, it generates other interesting ends for humans, a new sphere of motivation, with what is to all intents and purposes the first process of capitalisation.

If this is the case, it is worth pointing out an aspect that will play a central role in 4.4, when it comes to justifying the reasons for webfare. The most sophisticated technology that the human animal has succeeded in developing is education, that is, not too paradoxically, the creation of contexts, such as those of learning, in which it is very easy to feel imbecile, i.e. inadequate, and there is an urgent need to remedy this lack with increasingly effective sticks: reading, writing and calculation. That is to say, there is a need to compensate for our natural endowments, to complete them with something better in order to bring us up to speed with the disproportionate nature of our ambitions that takes us out of our environment, making us even more inadequate than we already are. Kant wrote that nothing perfectly straight can ever be made out of the crooked timber of humankind[35]: only humankind is made of crooked timber, and only humankind demands to be straightened. Therefore the stick, the technical supplement, can also be the guardian, the straight wood that can help the crooked wood to straighten itself a little, through education and culture.

---

[35] I. Kant, *Idea for a Universal History with a Cosmopolitan Aim* (1784), Cambridge University Press, Cambridge 2009, p. 16.

Kant begins his *Lectures on Pedagogy* by arguing that the human being is the only animal that must be educated. Other animals, at most, can be trained to entertain an audience of other (human) animals, who at least have a supplement (money), to pay for their tickets. The human being was first educated by the environment, as there is nothing more formative (for those who survive) than being taught by a sabre-toothed tiger. Then it was educated by war and work, both of which, in their traditional forms, carry discipline, rules and values. After the documedia revolution, other rules and new values will have to be found, as well as a new work of the spirit,[36] knowing full well, and right from the start, that education never ends, or more precisely it ends only with the end of the student. Yet this structural incompleteness is what keeps our world going, namely the network of the social world of which the Web is the most recent manifestation, which otherwise would never have started.

Education is thus a defining term of human nature, just like any other kind of stick or supplement. If calling a beast a beast is bestiality, lamenting the brutality and wildness of humans is not only sensible, but also all too justified. This has always been true, and it is comforting to think how much more education is imparted to any citizen of an advanced society today compared to a century or two ago, not to mention the millennia since the process began. From table manners to virtuosity, heroism and perhaps even sainthood, it is a succession of drives that are all the more effective the more they are gentle, with a gradation well expressed in the title of a work by Roland Barthes: *Sade, Fourier, Loyola*. When, as sometimes happens, the push is not gentle, the succession is reversed, and this explains many things. For this encounter gives rise not only to good, but also to evil as a properly human attitude. Rationality is not motivated by logic (which is the effect and not the cause), but by the encounter between biology and technology, which is why it cannot be considered as right or good in itself. Foolishness, error, sin, avarice, the deadly and everyday vices that Baudelaire lists in *Les fleurs du mal*, everything that is blamed in humanity is the human essence and the origin of reason, teleology, merit, punishment and reward, and free will, which is the capacity to do evil as well as good – this is what is born of the marriage between the irreversibility of the organism and the repeatability of the mechanism.

2.2.4. *Faust and Mephistopheles: the origin of capital*

Let us consider one more point. Very simple and primitive humans have capitalised livestock (*pecunia*), but the reverse has never happened, capitalisation being

---

[36] M. Cacciari, *Il lavoro dello spirito*, Adelphi, Milan 2020.

the proper human trait, the only and indisputable distinguishing feature of our species. Talking about "human capital" is the starting point for any discussion of capital, since there is no capital that is not human and, above all, there is no humanity without capitalisation of material and spiritual resources. We are born naked, but fortunately in a world furnished by the generations that preceded us. Our first gesture is to grasp the finger of an adult. In the long run, and not without difficulty, precisely because we are not naturally inclined to knowledge, it may even happen that we take pleasure in reading a book, or perhaps even writing one. This capitalisation is not just about treasuring the past, making it fruitful: it operates in the present, in the life of each of us. It is highly unlikely that in the absence of rituals, documents or sociability, a human being could attach significance to elementary kinship relations or nurture ambitions, for the very good reason that in nature there are no cousins or brothers-in-law, no shamans or rear admirals, no Immortels of the Académie Française or US presidents. No one could seriously aspire to win the Nobel Prize or to cultivate an extra-marital relationship in the heart of the forest and without ever having had any contact with other humans. Such people would simply be pursuing their own internal goal, nourishing themselves to stave off the end that will catch up with them sooner or later. This capitalisation is what Oedipus was so good at responding to. He was no fool: he was armed with a stick, which enabled him to perform things completely beyond the Sphinx's reach, just as another stick, a burning ember, enabled Ulysses to blind Polyphemus.

That stick was the origin of capital. This is suggested by Marx when he states that capital precedes industry, and is linked to mercantile activity and the documents that made it possible.[37] Even where there seems to be no capital, one can still recognise this principle that allows accumulation, wealth, progress, and of course exploitation. Kings become businessmen, wars are fought for money, ideals are camouflages of this fundamental reality, and of course philosophy itself is nothing but an ideological superstructure that cannot aspire to independence from the material conditions that generated it. Here Marx goes beyond Hegel, that is, he unmasks him, precisely through the construction of a social object alternative to the system. Knowledge cannot be absolute, i.e. free from all determinations, because it is itself an effect and reflection of capital; or, put differently, capital is the real system. The unmasking is thus twofold: Marx unmasks capital through the system, and the system through capital. It is in fact the system that allows him to recognise the process of capitalisation, the primary accumulation and the genesis of value. But on the other hand, it is the reality of capital that allows him to unmask the illusions of the system; and of course, unmasking trig-

---

[37] Marx, *Capital* cit., Book III, section IV, chap. 20.

gers an interminable process. Today, the system is the Web. If this is the case, there is nothing more intimately contradictory than the image or dream of a society without capital, which would be a society without society. However, remember, there is nothing more intimately inconceivable than capital without humans, which reminds us that the road to revolution is always open.

Let us retrace the stages of capitalisation. The stick allows the capitalisation of force. This capitalisation can be immediate, as when using a stick as a direct extension of the arm, or mediated, as when using a stick as part of a technical device, be it a hoe or a plough. One can also use a stick to capitalise on the strength of another stick, like the javelin propellers of which we have so much archaeological and ethnographic evidence. In this case, the capitalisation is direct, does not involve a time deferral, and is based on the physical principles of leverage. However, we must not overlook the fact that it is precisely this capitalisation – at once concentration and channelling – that lies at the heart of the enormous capitalisation of force and memory that is technology in general. The first machines capitalised force, the latest capitalise memory. There is no hiatus between the beginning and the end, but rather a profound continuity. In war, as Victor Emmanuel III is reported to have said in declaring his opposition to entering the conflict, one goes with two bags: in one bag are the sticks to beat with, in the other the sticks to be beaten with. Also, as the proverb goes, live by the stick, die by the stick. Humans soon realised this banal truth. Instead of being a simple tool for multiplying strength and acquiring authority, the stick can become the *symbol* of authority. Be it the direct and still effective threat of the policeman's baton, the king's sceptre or the conductor's baton, the stick is very well suited to the capitalisation of authority, both as a deterrent and as a charismatic element. The *baculum* thus disappears in the human penis, unlike all other mammals, to reappear in the sceptre, the staff, the walking stick, the pen, the brush, the wand and the sword.

It may be interesting to note that different supplements all tell more or less the same story. The same process that starts from reinforcing the immediate and arrives at capitalising mediation takes place in devices whose evolution has taken place under the eyes of my generation. Such is the case with the mobile phone, which started out as a talking machine, i.e. as a prosthesis and megaphone of an immediate function, but very soon became equipped with time-deferral devices, becoming first a typewriter, then a recording machine. This evolution led to the current externalisation of hysteresis, whereby the mobile phone is only the intermediary between human users and the platforms in charge of recording their mobilisation. What would wrongly be interpreted as a process of dematerialisation or spiritualisation can more correctly be understood as a process of deferral, ultimately resulting in the enhancement of hysteresis, the full extent of which I shall examine in 3.

Capital is thus a panoply of supplements, i.e. the set of instruments and weapons with which the human animal has equipped itself – at different times, in different ways, and in different places[38] – to make up for its deficiencies. Here's probably the most interesting aspect of the whole affair: these shortcomings would not have arisen if the human animal had remained in its place and accepted its biological destiny. In other words, what non-human animals can solve involuntarily and through very long evolutionary processes, the human animal solves through technology and society. Instead of an internal evolution, marked by genetic selection, we are dealing with an external evolution supported by capitalisation: the supplement fills a gap but generates new needs, the satisfaction of which pushes resources and lacks even further apart. On the one side, the hand relieves the mouth of onerous functions, and moreover equips it with tools of increasing complexity, which make human life more suitable for the capitalisation of goods, knowledge and organisations: protection from predators and bad weather, increased social interaction, longer life span, development of symbolic capacities and hysteresis systems, which allow further capitalisation. On the other side, the mouth not only enhances humanity through language, but also confers an additional symbolic and social function on technological devices.[39]

This evolutionary circle, which is very easy to describe in hindsight, would have been impossible without the essential condition of capitalisation, hysteresis, and in this case, keeping track. This capacity, at first implicit and hidden in more conspicuous functions, such as the manual repetition of procedures, soon came to the fore and took over the scene: notches, wedges, hieroglyphics, numbers and culture took centre stage, and humanity realised that what was at stake was right there, in hysteresis.[40] The technical enhancement of hysteresis in the documedia

---

[38] J. Diamond, *The Tasmanians: The Longest Isolation, the Simplest Technology*, in "Nature", 273, 1978, pp. 185–186.

[39] This capitalisation also applies to thought, which cannot survive once the world has been removed (Frege, *Der Gedanke* cit. In Frege, this is an epoché that anticipates Husserl, and is driven by the same intentions: to avoid the reduction of philosophy to psychology). Thought requires support and capitalisation, i.e. it cannot proceed without hysteresis. "Keep recording, the meaning will follow", to misquote an Italian saying. But can the activity of thinking emerge from the passivity of recording? In Aristotle the problem sounds like this: the soul is a passive, i.e. receptive, function on which impressions are deposited; this is also true of the highest part of the soul, the intellect, which is like a wax tablet on which impressions and thoughts are inscribed. However, since it is not permissible for the intellect to be pure passivity, because it manifests requirements of freedom, spontaneity, and uniqueness, Aristotle artificially divides the intellect into two parts, a passive intellect, embodied and individual, and an active intellect, impassive, immortal, and universal (*De anima*, 408b 29). The only solution seems to be to embrace the hypothesis of a supernatural infusion, but Aristotle rejects this view, so the question is not answered and the difficulty remains.

[40] An endless amount of case studies on the connections between trace, number and alpha-

revolution is thus no accident, but rather the revelation of something that was already in place at the moment when the virtuous cooperation, and the division of labour, between hand and mouth began. Since the retention of a trace has been possible – that is when capital began. Capital is a technology, just like the lever or the wheel.

In the archaic past of flint chipping we find the present and the essence of humanity, namely supplementation and capitalisation as the fundamental resources of an animal in need of a second nature.[41] It can rightly be observed that chipping a flint with a simple blow is the lower threshold beyond which it is no longer possible to speak of "technology". It is nevertheless worth noting that everything we call by that name is essentially related to that first chipping: suffice it to say that the binary codes on which contemporary technological development is based are the evolution of that track. And just as authentic military power is manifested not in swords and banners but in the war room,[42] so the greatest strength capitalised by the stick does not lie in its direct action, but in the context it can produce.

As a weapon or as a cultivation tool, as a prosthesis for walking or as a sign, the stick exerts an indirect action of incalculable scope: reducing immediate need and pressure, it frees the powers of intelligence and reflection. In this sense, the stick is the – negative but indispensable – premise of those other capitalisations (of authority, knowledge and duty) which characterise human civilisation. A warrior displaying his strength in tournaments, a leader guiding a parade, a priest blessing the emblems: these figures do not simply form a social trinity deeply rooted in the Indo-European tradition. On the contrary, external deference to physical strength hides the awareness of the fact that brutal force is the foundation and the condition of possibility of intelligence.

I'll end on a final note, fundamental to the philosophy of history I will develop in 3.4.4, and which I propose here in the form of a conjecture: the adoption of

---

bet can be found in G. Ifrah, *From One to Zero. A Universal History of Numbers*, Penguin, New York 1985. Cf. also W. Warburton, *The Divine Legation of Moses Demonstrated* (1741), Garland, New York-London 1978; P. Tullio, *La forma naturale delle lettere dell'alfabeto*, in "Bollettino della Accademia Italiana di Stenografi" 1931; I.A. Gelb, *A Study of Writing: The Foundations of Grammatology* (1952), 2ª ed., The University of Chicago Press, Chicago 1963; A. Kallir, *Sign and Design: The Psychogenetic Source of the Alphabet*, Clarke, Cambridge 1961; E.A. Havelock, *Preface to Plato*, Harvard University Press, Cambridge (MA) 1963; Leroi-Gourhan, *Gesture and Speech* cit.; G.R. Cardona, *Storia universale della scrittura*, Mondadori, Milan 1986. For a detailed analysis, I would refer readers to my *Estetica razionale* cit., pp. 469–514.

[41] The Aristotelian theme of human nature as "second nature" was re-launched at the end of the 20th century by J. McDowell, *Mind and World*, Harvard University Press, Cambridge (MA) 1994. Only, the second nature does not consist in reason, but in technology.

[42] E. Jünger, *The Forest Passage* (1951), Telos Press Publishing, Candor (NY) 2013.

supplements is the origin of history and capital as exclusively human facts. History is progress, and capital is accumulation and enhancement, and neither of these is possible without recourse to hysteresis objectified in a supplement. Every new supplement that enters history fills a gap but generates new ones. Hence the beginning of a spectacular run-up: the supplement makes up for a lack (my glasses allow me to read after I have become farsighted), but generates new ones (is there anything interesting to read?), and so on, not infinitely but indefinitely, according to a deferral process in which the singularity of human "nature" is determined. This is why we will never see a Faustian cat longing for happiness and completeness that forever elude it. Capitalisation, with the circle that derives from it, is the essential characteristic of the human form of life, and this suggests to us – furthering the Copernican revolution – that one can understand humanity based on capital, but not capital based on a supposedly unscathed and immune humanity that would suddenly turn out to be venal.

2.3. Document: Capital

People often describe capitalism as the world of *1984*, forgetting that Orwell's book referred to communism and, what is worse, that this is neither 1948 nor 1984, and that capital has not been Marx's capital for a long time. In fact, in recent years, with the documedia revolution, we have found ourselves in the position of capitalising on every single act of our lives, recording it and thus transforming it into a possible source of value. A new capital is born, which requires us to repair the unperceived injustice of a labour that is not yet recognised as such, and to remedy the all-too-perceived discomfort of the disappearance of the society of coal, steel and toil, securing occupation that, if not guaranteed, is at least possible. But in order to repair the injustice and alleviate the discomfort, it is necessary to clarify one point: capital existed at the time of Anaximander and at the time of Thales, who foresaw a frost and speculated on the price of oil. It still existed at the time of Hegel, who in the idea of "system" gave a perfect definition of an ideal becoming of capital as an all-encompassing totality. But Hegel did not fall into the anthropomorphism of attributing intentions to the system. And we would fall into an even worse anthropomorphism if we were to attribute any intention to capital.

This in no way means denying the exploitation or evil involved in capitalism: quite the contrary. It means understanding that old explanations, forged on industrial capital and only slightly modernised with financial capital, have lost much of their usefulness when it comes to describing this evolution, which cannot even be explained as a simple mutation in the structure of capital. In order to understand what is happening, and above all to give a new foundation and a new perspective to the idea of emancipation and to the master-slave dialectic that is its vector, it is necessary to view the capitalisation process as a particularly significant variant of the relationship between soul and automaton analysed in 2.1. The soul, as an organism, dictates objectives, urgencies and temporality. The automaton, as a mechanism, enables the externalisation of memory that allows for technological and social development and, by that means, the genesis and enhancement of consciousness and intentionality. If this is the case, we can no more get out of capital than we can get out of metaphysics, and the question is rather to understand what capital really is. I will start by clarifying the central and omnipresent process we call capitalisation.

2.3.1. What does it mean to "capitalise"?

The answer to this question is very simple: it means activating a process of hysteresis. A singularity has taken place: a here and now leaves a trace, first in nature and much later in culture, settling in rituals, external media necessary for the fixation of memory in a society without writing. Once there is a trace, be it the decay of carbon-14 or the acquisition of an iterable skill, capitalisation begins. For example, there are technological acquisitions, which allow memory to be reified in artefacts; similarly, I think it is no coincidence that there wall paintings appeared very early on in our history. Again, this is a case of proto-documentality which, as palaeontologists suggest, has both a descriptive and a prescriptive value, just like our documents: the show where and how to hunt, for example, or what the sacred animals of the clan are. With the development of writing, sociality evolves much more rapidly, and the role of the document is enhanced. Documents establish and coordinate actions and take on a prescriptive role, generating value rather than merely recording it.

This is why the Web is the heir to a history that is older than it appears, a network that long predates today, and which, as we have seen, explains why the documedia revolution has imposed itself so easily. It was neither an invasion nor a colonisation, but the response to deep needs that have characterised humanity from the beginning, and which from the beginning, interacting with technology, has produced a network that goes by the names of "society", "culture", "economy", "world of the spirit". This network has been stretched out for a long time, waiting for a providential technical upgrade. But what exactly was the need that the network was responding to? To keep track, not to disperse, to conserve, to accumulate – this is all the more important for an organism that, as such, is destined to die, to lose everything, and that postpones death by acquiring instruments to capitalise on its strengths. In short, it is all a question of memory and the supplements that strengthen it by repeating it in matter.

Capital is thus the general system within which supplementation operates and finds meaning. As such, it is a general process that also exists in nature, precisely in the form of hysteresis: the straw that broke the camel's back capitalises on all those that have come before it. I realise that this bears little resemblance to what we think of when we speak of "capital" and "capitalisation", but I invite you to think that perhaps we are running up against prejudices. Probably because capital came to the forefront of philosophy with Marx, as industrial capital, financial capital was considered a perversion of the former, a sort of sick sexuality as opposed to a normal sex drive. However, industrial capital is not the norm, and financial capital is not a perversion. Indeed, industrial capital is by no means the prototypical form of capital: a factory is not the prototypical form of economic action, a proletarian is not the prototypical form of worker, and, above all, a worker is not the prototypical form of human being. More precisely, industrial capital is merely a relatively recent historical form of technological capital dating back to the dawn of hominisation, and financial capital is simply the current manifestation of document capital, i.e. the value that documents, as social objects, have held throughout human history. All these forms of capitalisation are now subsumed in documedia capital, as the union of humanity, technology and documentality.

There is no reason to pinpoint the genesis of capital to mediaeval merchants or Renaissance banks. One has to go much further back, as far as protohistory and prehistory, which we call so only because there are too few traces left to form a clear picture of it. A resource is accumulated, for example with the transition from living to dead labour via the construction of a tool or weapon or with the accumulation of goods, from salted meat to the preservation of grain and oil. This leads to an opening, if not to future generations, to future actions. Just as a banknote incorporates purchasing power, so a chipped stone is dead labour that allows me to reduce my future living labour. Similarly, the pot capitalises the effort made to construct it, objectifies it, and thereby reifies the method of manufacture. This accumulation means that a given event, which at T1 would have been catastrophic, for example a predator's attack or famine, does not have the same effects at T2, i.e. after capitalisation. The event does not change, but its outcomes do, and this is due to the action of hysteresis, which accumulated resources and temporally deferred them from T1 to T2. Obviously, as soon as the process is initiated, it generates further forms of capitalisation, according to the logic illustrated by compound interest. Unfortunately, the process can also operate with negative magnitudes, producing a growth in debt instead of resources, or, going beyond the financial dimension, generating demographic crises, reductions in the resilience of a human group, and loss of skills.

Reciprocally, the genesis of meaning only takes place within the framework of a code, thus requiring capitalisation. Every deferral, in fact, involves a possibility of differentiation. A" is different from A, just as it is different from B" (the trace of event B). Different in origin, B" and A" nevertheless share the common characteristic of being traces, and of being able to combine in ways that were not possible for the original events, giving rise to sequences like A'B", B'A", A'B'B'A", etc. The transitions from occurrence to meaning, from rhyme to reason, from chance to law, from singularity to system reflect this principle, the basis of which lies in hysteresis and the production of differences (both in the sense of "diversities" and in that of "deferrals") resulting from it. This principle is not only of semantic value, since it is at work in the passage from singularity to system, from sound to rhythm, which retrospectively gives sound a meaning it previously lacked. In this case, we are dealing with well-known phenomena in which simple enumeration determines meaning, as in the infinity of lists and the pleasure of collecting.[43]

That is why capitalisation underlies the genesis of meaning, which does not precede expression (as posited by the Pentecostal perspective), but rather – in accordance with the emergent perspective – emerges from the combination of elements that, as such, have no meaning. Meaning is not heaven-sent, but emerges from the interplay of differences and the hysteresis that makes it possible. The vowel A, alone, does not mean anything. It only means something together with B and C, but this process of differentiation between arbitrary signs takes time, and therefore hysteresis. One must imagine a three-way process in which hysteresis allows for diachrony, and the latter allows for emergence. The really crucial thing is that the sign (and meaning) come *after* the trace, not before as claimed in the Pentecostal vision. The sign is the necessary, though not sufficient, condition of meaning. In order to create even minimal meaning it is necessary that the sign is iterated until it generates a code. Fingerprints make sense only when they can be compared with police records; DNA makes sense precisely because it is repeated in the body; and a word makes sense only because it can be repeated (a language made up of words that cannot be repeated, or that constantly changed meaning, would not be a language).

---

[43] Think of the power of lists and "etcetera" (to which Husserl rightly drew attention); the power of catalogues, even if they are erroneous or nonsensical (but which retrospectively constitute meaning, a circumstance stressed by Borges); or the power of rhapsody (according to Kant, that was how Aristotle mistakenly constructed his categories, but instead it is an excellent example of machine learning).

This is the beginning of that great process of capitalisation of signs that we call "intelligence", whether natural or artificial, to which I will return at length in 3.2. My sight is not thinking when it identifies a random succession of letters, for example XCZW: it is rather handling signs. But, then again, it is not thinking either when it reads POT (you can get to the end of an article without remembering what you've just read). Thinking only intervenes when an obstacle comes up and, for example, we find the word PUT where we would expect POT. At this point, and only at this point, does perplexity lead us to wonder whether the author meant "put" or "pot". Is this the birth of the spirit? No, it is simply a more complex manipulation of signs.

Let me explain. When I say "I'm referring to number 5", I'm saying something like "I'm referring to what they taught me in school is number 5". And when I say "six times eight is forty-eight" I'm repeating a nursery rhyme learned by heart, which in no way prevents me from understanding "48". I could get the same result with an abacus or a calculator, and I would have no reason to postulate some intentional *homunculus* behind the scenes. In order for "C" to refer to "cat" in a dictionary, or for the dot with the word "London" on the map to refer to London, there is no need for a little man transcending the dictionary or the map. On closer inspection (education is just that), the spirit that goes from "C" to "cat" or that from the sign "London" refers to the city of London has been formed precisely through maps, dictionaries, calendars, poems learned by heart and cartoons.

There is no intrinsic value in any of the letters that make up the word "cat", so much so that by combining them differently we have "act": hence the contingency of the sign and the compositional nature of meaning discussed by Saussure – who, however, was convinced that the sign has a spiritual part, while in my view spirituality (if we want to call meaning that) is rather the Gestalt outcome of the combination of signs. The words "got", "gut" and "git" mean what they do not because of the intrinsic properties of the letters that make them up, but because of the system of differences of which they are part; in the same way meaning can only take place within a system and thanks to a network of differences, i.e. hysteresis. Also in this case, hysteresis precedes meaning and makes it possible,[44] thus allowing for the system itself (a private language, i.e. a language that does not rest on an iterable code, is not conceivable). Hence two consequences.

The first is that, since there is no intrinsic sense, everything, taken in isolation, can be said to make no sense (whereas two things put together can begin to make sense: means and end, for example, or identity and difference). From this point of view, meaning lies neither in the mind nor in the thing, but in the environ-

---

[44] J. Bybee, *From Usage to Grammar: the Mind's Response to Repetition*, in "Language", 82, 4, 2006, pp. 711–733.

ment,[45] that is, in this case, in the system. The second is that if someone by metaphysical chance wrote a truthful book on the meaning of existence, existence would lose all meaning, as meaning does not consist in a hypostasis, but in a system.[46] This circumstance suggests something of the utmost importance. The machines that we fear will do us all sorts of harm, and above all will replace us, are in fact already *in* us, since intelligence is a mechanical element of the human, and human learning works exactly like machine learning. In both cases, indeed, there is a shift from meaning to meaningless sign and, vice versa, from meaningless sign to meaning, as long as there is hysteresis.

2.3.2. Human capital: ichnology and technology

The ability to keep track, along with the resulting empowerment, evolved in humans because of the responsiveness and defectiveness that characterise them; thus, they acquired symbolic and expressive dimensions outlining the cultural traits we are used to calling "human". It follows that, just as there is no humanity before supplementation, there is also no humanity before capitalisation, as both determine what humanity is. The latter may later project squandering on grasshoppers and skimping on ants, but the fact is that both concepts can only be thought of in terms of humans and their horizon of capitalisation. No grasshopper has ever squandered, no ant has ever skimped, and obviously the queen bee is not a real queen.

Capital, as should be clear by now, did not begin when some proto-capitalist fenced off a field, but rather when a much earlier proto-capitalist generated an instrument that incorporated dead labour by reducing living labour, and, even more so, when a descendant of that proto-capitalist realised that other things could be made from the resulting chips, be they arrowheads, jewellery, or stones useful for carrying out elementary arithmetical operations. This evolution of the trace is like a carpenter who discovers that the really important thing produced by the planer is the shavings. The person we are used to calling, with undue confidence, "our ancestor", the hominid who first split a stone in two and made it sharp, was the first deliberate producer of tools, and unintentionally of traces.

When a rock is chipped, its fragments first appear as non-capitalisable remains. Much later, in a process of thousands of years that seems like the blink of an eye on this time scale, the rock is considered a reserve of material, whereas the shards are considered a product: they can be used as the tip of an arrow or, at a

---

[45] H. Putnam, *The Meaning of "Meaning"*, in Id., *Philosophical Papers*, vol. II, *Mind, Language and Reality*, Cambridge University Press, Cambridge 1975, pp. 215–271.

[46] Consider for example the story of "gobbledygook", a term born to manifest nonsense and that through use has acquired a univocal and shared meaning.

higher level of technological expertise, as a long, thin and sharp blade. In short: the first chip was voluntarily the first tool and involuntarily the first trace, and only later did we realise that the real capital is precisely the trace. It always happens that way, and as we have seen with the Web, one realises that the true capital lies in hysteresis only later. Everything that can be tracked can potentially be capitalised: today's platforms know it, and so did the Palaeolithic hunters who sought the marks of their prey. Rather significantly, here we have capital that comes before labour in the strict sense of the term, just as documedia capital comes after labour as we have known it for the last ten thousand years.

Once a trace is inscribed, it can be iterated. The living labour that produced it is transformed into dead labour, reducing the force needed for the production process. Quite simply, when I make a tool, I perform an act that is projected onto future actions, which will never be the same again, precisely because they now capitalise on the advantage of a pre-existing tool. Let's go back a little further than the manufacture of the tool: the capitalisation of traces. Traces, be they footprints, notches, or traumas, persist and modify the environment; for many souls, traces mean information: a very first process of capitalisation. This ichnological[47]

---

[47] In Greek, *ichnos* means "trace, track, footprint", both in a material sense ("to follow traces", for example in Plato's *Phaedrus* and *Theaetetus*) and in a spiritual sense ("track of one's story": Aeschylus, *Prometheus*, 845). The term resonates with the Greek name for Sardinia, *Hyknusa* or *Ichnussa*, i.e. "footprint", the shape of the island seen from above. *Ichneusis* is the search for traces, *ichneutés* is the seeker, *ichnéuo* is to seek (Parmenides, 128c), and *ichnographia* is map or sketch. "Icnologia" was the title of an essay I wrote in 1995 ("Icnologia", in *Interpretazione ed emancipazione. Studi in onore di Gianni Vattimo*, edited by G. Carchia and M. Ferraris, Raffaello Cortina, Milan 1995, pp. 103–135) and of the concluding chapter of *Estetica razionale* cit. By that term, which I thought was a neologism (and I had to be careful not to have corrected to "iconology"), I designated a general theory of traces, a sort of semiotics extended to include thought processes, and with a strong preference for recording over communication. Years later, I discovered that there is a scientific discipline called "ichnology", which researches traces, tracks, footprints, grooves, tunnels and in general any kind of imprint that documents the way of life of recent and ancient organisms. In this sense, it is wrong to consider it as a simple branch of palaeontology, since in addition to paleotechnology, which deals with ancient traces, there is neoichnology, which deals with recent traces, often capitalising on the experience accumulated by the former. The discipline dates back to the early 19th century, but has undergone a major acceleration in recent years. There are courses in ichnology (mostly related to earth sciences), there are ichno-libraries, and for some years now there has been a publication called "Ichnos. An International Journal for Plant and Animal Traces" (now at its 15th issue). Scientists working in the field of ichnology are predictably called ichnologists. But the idea, and even the name, of "ichnology" long predates the establishment of the discipline. Among the many inventions attributed to Leonardo da Vinci there is in fact also ichnology (cf. A. Baucon, *Leonardo da Vinci, the Founding Father of Ichnology*, in "Palaios", 25, 2010, pp. 361–367), and Leibniz expressly mentioned ichnology (thus becoming its onomaturge) as one of the components of the encyclopaedia of knowledge.

or stigmergic[48] capital is as much of a capital as the royal treasury, industrial production, the stock exchange and culture, because it involves a process of accumulation, deferment and transformation.

A trace – i.e., any sign on a surface or impressed on the mind – has the characteristic of spatialising time (something that took place leaves a spatial vestige) and of temporalising space (what is seen in space can generate a temporal succession). Quite simply, the lines I am writing at the moment translate into a portion of space a process that has taken place in time; reading these lines, in turn, consists in transforming signs that occupy a portion of space into a temporal process. Similarly, the mark I may make if I press the tip of a pencil into the palm of my hand, an impression fixed in the here and now, and dependent on the action of the pencil, is transformed by the simple force of hysteresis into something different and antithetical. It can remain even after I have removed the pencil, as a red mark that will be seen on the skin for some time; above all, it can become fixed in my memory and can be recalled indefinitely; finally, as I am doing at the moment, it can be written down in words, overcoming the spatio-temporal finiteness of my memory. Or I can multiply the traces, creating complex figures or potentially significant sequences, so that the more traces there are, the more arrows (real or metaphorical) we have in our arsenal.

Not to mention that, none of us being Adam, traces precede us. We are born into a society full of books. How could all this come about? Once again, the Pentecostal perspective proves inadequate to answer this question. The line between an animal marking its territory and a president building a defensive wall is thin. The paradoxical aspect whereby at the beginning (of what? Let's say: of everything) there would be a trace, i.e. the most abstract thing conceivable, is easily explained by the fact that a trace is also the most concrete of things. Surprisingly, but not too surprisingly, humans learned to write, i.e. to leave traces, before they learned to paint. When the first human trace was made is a question that is bound to remain unanswered. Too many empirical and conceptual difficulties stand in the way of this goal. The fact that extremely old traces can be found does not mean that they are the oldest, nor does it mean that these traces can be clearly recognised as already human or as marks that date back to our animal past.

---

[48] Stigmergy is communication that takes place by leaving information in the environment, such as when ants communicate by leaving pheromone traces. Cf. L. Marsh, C. Onof, *Stigmergic Epistemology, Stigmergic Cognition*, in "Cognitive Systems Research", 9, 1–2, 2008, pp. 136–149 and G. Theraulaz, E. Bonabeau, *A Brief History of Stigmergy*, in "Artificial Life", 5, 1999, pp. 97–116. It is worth noting that traffic signs, books, and culture in general also make up a great stigmergic system, which from this point of view is an extension of very ancient behaviour.

One thing, however, can be stated with reasonable certainty: since action always precedes contemplation, no one developed the concept of "trace" and then applied it. Before deliberately leaving traces, it was necessary to produce prostheses to compensate for human inadequacies, but at that point traces were already there, and they were acting. There is a vast literature on the role of culture in human evolution,[49] as well as common sense intuitions that are difficult to ignore; what I will add is that we are facing a process of capitalisation of traces, both in technology and in society. The prosthesis, through the way in which it is made, constitutes the model for other prostheses, which again can take advantage of the experience solidified in the first model: the process of making butter was passed down and the thread for cutting it was perfected. This capitalisation of know-how and tools is at the root of technological progress whereby, over hundreds of thousands of years, utensils are refined in a process in which the knowledge capitalised in the individual tool is as essential as it is undervalued: imagine a society in which each generation had to reinvent the hammer. As such, society is a general economy, since it is based on the same mechanism of promise that underlies the restrictive economy.

This is why, when we talk about society, sooner or later we end up talking about money. This does not in any way depend on the fact that human beings in general, or certain societies in particular, are more or less venal. It is simply that society is the technological device that humans, without any social contract but driven by a competence without understanding, have produced outside themselves by externalising their goals. Natural life, as depression teaches us, has no

---

[49] D. Bickerton, *More than Nature Needs: Language, Mind, and Evolution*, Harvard University Press, Cambridge (MA) 2014; R. Boyd, P.J. Richerson, *Culture and the Evolutionary Process*, University of Chicago Press, Chicago 1985; Idd., *The Origin and Evolution of Cultures*, Oxford University Press, Oxford 2005; R. Boyd, P.J. Richerson, J. Henrich, *The Cultural Niche: Why Social Learning Is Essential for Human Adaptation*, in "Proceedings of the National Academy of Sciences", 108 (suppl. 2), 2011, pp. 10918–10925; A. Acerbi, *Le trasmissioni della cultura*, in *Modelli della mente e processi di pensiero. Il dibattito antropologico contemporaneo*, a cura di A. Lutri, Ed.It, Catania 2008; A. Acerbi, D. Parisi, *Cultural Transmission Between and Within Generations*, in "Journal of Artificial Societies and Social Simulation", 9, 1, 2006; R.A. Aunger, *Darwinizing Culture*, Oxford University Press, Oxford 2000; R. Axelrod, *Evolution of Cooperation*, Basic Books, New York 1984; G. Baldassarre, *Cultural Evolution of "Guiding Criteria" and Behaviour in a Population of Neural-Network Agents*, in "Journal of Memetics", 4, 2001; J.H. Barkow, L. Cosmides, J. Tooby (ed. by), *The Adapted Mind: Evolutionary Psychology and the Generation of Culture*, Oxford University Press, New York 2002; S. Blackmore, *The Meme Machine*, Oxford University Press, Oxford 1999; R. Boyd, P.J. Richerson, *Not By Genes Alone: How Culture Transformed Human Evolution*, University of Chicago Press, Chicago 2006; J.M. Epstein, R. Axtell, *Growing Artificial Societies: Social Science from the Bottom Up*, The MIT Press, Cambridge (MA) 1996; D. Sperber, *Explaining Culture: A Naturalistic Approach*, Blackwell, Oxford 1996.

meaning outside itself, whereas social life has too much meaning, constantly imposing goals and deadlines on us. Hence the progressive resources of civilisation as acculturation. On the one hand, already in the framework of a restrictive economy, only concerned with give and take, it offers the possibility of giving and taking without dependence, making use of a universal intermediation. On the other hand, as we shall see in the next chapter, the restrictive economy must be read within a general economy, in which giving and having are part of a gift perspective: in the end we will all die, and this circumstance is what gives meaning and time to the economy in general, which serves the needs of an individual life that, as such, is and cannot but be purely loss-making.

Transgenerationality[50] is thus the essential vocation of capital, its origin and purpose, with respect to both nature[51] and culture. This is particularly evident in the genesis of the sciences, for example in the origin of geometry.[52] Thales had to record and deliver his discovery linguistically, so that oral language, which perfects the intellectualisation in place since perception, could free the object from the subjectivity of the inventor, fulfilling the criteria of objectivity and ideality as characteristics of science. However, oral language still limited the discovery to the synchrony of the original community. Only writing – that is, the most empirical and inanimate of elements – would be able to confer perfection on this ideality, rescuing it from the spatiotemporal finiteness of the proto-inventor and his contemporaries, achieving the independence of meaning from the current community in which the perfection of ideality consists.

On the other hand, it is difficult to think of myth or religion as individual and voluntary constructions, as knowledge that precedes being. In myth, being (the tale) precedes knowledge (meaning); and obviously, instead of myth, we can speak of "society", "history" and "technology", as well as of "writing", which

---

[50] T. Andina, *Ontologia sociale. Transgenerazionalità, potere, giustizia*, Carocci, Rome 2016; Id., *Transgenerazionalità: una filosofia per le generazioni future*, Carocci, Rome 2020.

[51] In nature we do not only have to deal with natural selection (which explains the disappearance of unfit species), but also with the transmission of advantageous traits, with a Baldwin effect. Cf. P. Godfrey-Smith, *Postscript on the Baldwin Effect and Niche Construction*, in *Evolution and Learning: The Baldwin Effect Reconsidered*, ed. by B.H. Weber and D.J. Depew, The MIT Press, Cambridge (MA) 2003, pp. 210–223; Id., *Conditions for Evolution by Natural Selection*, in "Journal of Philosophy", 104, 2007, pp. 489–516; Id., *Darwinian Populations and Natural Selection*, Oxford University Press, Oxford 2009.

[52] J. Derrida, *Edmund Husserl's Origin of Geometry: An Introduction* (1962), N. Hays, New York 1978. Heidegger is far from indifferent to the role of technology: he generally refers to tools that are not very sophisticated, like the hammer, but there is a passage in § 69 of *Being and Time* (p. 409) that perhaps anticipates Husserl's *Origin of Geometry*, and certainly contains *in nuce* all of Derrida's *Grammatology*: "even in the 'most abstract' way of working out problems and establishing what has been obtained, one manipulates equipment for writing, for example".

emerged separately in many different environments and at different times: in 3400–3200 B.C. in Hierakonpolis, Upper Egypt-Lower Nubia; in 2900 B.C. in the Indus Valley; in 1800–1500 B.C. in China; in 100 B.C. in Mexico; in 200–400 A.D. in Peru. As usual, just as these emergences took place, they might very well not have occurred: in all these cases we are dealing with emergent meanings and not with Pentecostal ones.

Something similar took place with the emergence of industrial capital, which once again should not be seen as the paradigm of capitalisation, but as a case that is historically close to us. At a certain point, steam, water and wind made automation cheaper than slave labour. By harnessing power other than human energy and with more complex machines, capital extended the domain of dead labour. And I am not just talking about breaking stones with a sledgehammer: anyone who has ever had to follow a protocol or fill out a pre-printed form will have noticed the benefits of capitalising on bureaucratic work that they would probably not have been able to do themselves. Also, the fact that documents have prosthetic value is proven by the fact that they are traditionally referred to as "legal instruments". The rest is modern and contemporary history: the worker, who was initially the main actor in the process, for whom the tool was only a prosthesis, becomes the prosthesis of the tool. This is where the future begins: to stop producing does not mean to be excluded from the cycle of capital – quite the opposite. The fact that the same principle governs the origin of geometry and that of industrial capital suggests a point that deserves the utmost attention. Just as there was capital before industrial production, there will be capital when all production is automated, which will make redistribution and socialisation much easier, not least because there is no analytical link between capital and private property or between capital and labour.

2.3.3. Document capital

As we have seen, traces are not only outside souls. They can also be deposited in memory, giving life to experience. And then, with a shift that constitutes the most important event in human history, internal traces can be transferred outside, overcoming the finiteness of memory and the mortality of its bearers. This is the birth of writing in the broadest sense of the term: let us not forget that the paintings of Lascaux, and in particular the fifteen black dots painted under the elk, are already a form of writing: we do not know what they numbered, but it is difficult not to associate them with enumeration and capitalisation. Numbering can have a purely individual meaning, as in noting the number of Siberian tigers killed just for the sake of it, but most of the time it has a social value. *The full deployment of technological and ichnological capital takes place in social, i.e. document, capital*, from caves to offices and Internet platforms.

This is where the stick turns into the stylus. In the current representation, a bureaucrat simply implements provisions decided elsewhere. However, on a less superficial examination, owning the means of recording is a power just like owning the means of production, and allows bureaucracy to override or anticipate the authorities – be they politics or law – to which it is formally subordinate. Documentation is first and foremost treasuring. An action without tools and witnesses can disappear forever, and never be capitalised on. An action with witnesses and documents can be treasured in the latter, and the action can be adopted by others, multiplying that first occurrence over the centuries.

In this capacity, documents are both the necessary and the sufficient condition of social reality.[53] On a first level, documentality, i.e. social capital, is a necessary condition of social facts, but it is not a sufficient condition, as individual intentions are also needed to give meaning to social objects and to create them with a minimum of understanding. In short, without documents – which can also be understood simply as shared memories in people's minds – there would be no promises, bets, marriages, money or wars, which all require some form of hysteresis to exist. This is also true in the case of mechanical agents: the algorithm that buys or sells has no awareness of its actions, yet its transactions are effective, whereas the operations of a hypothetical stock exchange of which human agents had awareness but no record would be null and void.

On a second level, however, documentality is a sufficient condition for individual intentionality, the emergence of representations, desires, intentions, and identity. Without documents, there would also be no intentions, social actions, aspirations, i.e. the sphere of motivations, collective intentions, and the forces that lead to the constitution of other documents. Indeed, in the social world, necessary conditions play the implicit role of sufficient conditions. Each generation enters the scene in a world that is already formed, and thus, in line with what I have just said, *receives a large part of the necessary conditions in the form of*

---

[53] Aristotle already emphasised the recording and inscribing value of money: "Now money serves us as a guarantee of exchange in the future: supposing we need nothing at the moment, it ensures that exchange shall be possible when a need arises [...] Money then serves as a measure which makes things commensurable and so reduces them to equality. If there were no exchange there would be no association, and there can be no exchange without equality, and no equality without commensurability. Though therefore it is impossible for things so different to become commensurable in the strict sense, our demand furnishes a sufficiently accurate common measure for practical purposes. There must therefore be some one standard, and this accepted by agreement (which is why it is called *nomisma*, customary currency; for such a standard makes all things commensurable, since all things can be measured by money" (*Nicomachean Ethics*, in *Aristotle in 23 Volumes*, Vol. 19, cit., 1934, 1133b 10–25).

sufficient conditions. Indeed, education, with a process that starts with bodily control, has always been considered a way of introducing normativity through outward and perhaps playful behaviour. Diet, clothing and posture are all external forms which, not without reason, exert a profound influence on people's inner selves. Our aspirations, our behaviours, our totems and taboos turn what is merely hysteresis into motivation. First we have competence without understanding, views that are not understood, rituals whose meaning is ignored, inherited social structures. Then, sometimes, and more rarely than we think, the spirit manifests itself, precisely as intentionality. Between capital and intentionality a circle is produced that echoes the circle of the supplement examined in the previous chapter. Capitalisation produces increasingly complex forms of sociality, which determine the conditions for the genesis of consciousness, responsiveness and intentionality. Let us examine this process in more detail.

First of all we have the documentation of social acts. In accordance with the principle that meaning comes from the sign and not the other way around, here intentionality derives from documentality. We do not need to wait for algorithms, the Web or artificial intelligence to ascertain this circumstance, which, moreover, reiterates a point that applies to the algorithm as well as to the bureaucratic machine,[54] namely that without souls there is simply no point. Thus, it is certainly true that documents do not speak on their own, and that humans are needed to make them speak. But it is equally true that humans do not speak on their own either: it takes other humans to activate the language module, and for this to happen, the mediation of a code is needed, that is, ultimately, a document. The language transmitted must be a public language, one that preserves its meaning and can be shared. In other words, it is not true that first there are intentions and then they can be laid down in documents. The opposite is true: we first receive document training (rituals, education, etc.), and only later can what we have received be transformed into intentions.

Secondly, we have valorisation. Documents arouse intentions and these generate documents with a mechanical iteration motivated by the urgency arising from organic interruption. As we have seen, the hallmark of humanity is not intelligence, but the externalisation of behaviour into documents that record it and produce new ones following a rhythm whose origin is metabolism. Here again I propose a reversal of perspective, abandoning the Pentecostal view that intentionality is cemented in documentality, and adopting an emergent view that conceives of a circle (the circle of capitalisation) between intentionality and documentality: documents generate intentions and values, which are cemented in

---

[54] Cf. C. Schmitt, *Das Problem der Legalität* (1950), in *Verfassungsrechtliche Aufsätze*, Duncker & Humblot, Berlin 1958, pp. 440–451.

other documents, and so on, in a circle that originates in our animal past and as such defines the ultimate teleology of the human being as a responsive animal.

Usually, what prevails is routine, the usual way, the road already travelled and much trodden – which is the etymology of "routine". This routine has the shape of a circle created between the technical tool and the users' intentionality. At the beginning of the circle we find neither value, nor norm, nor duty, nor utility, which are derived concepts, but only hysteresis. Archaic, instinctual behaviour is capitalised by the increased technological possibilities that are characteristic of the human being. This capitalisation gives new meanings to actions, for example to marking the territory, which acquires a legal value (definition of ownership), a religious value (separation of the sacred from the profane), an ethical value ("thou shalt not covet thy neighbour's house", "thou shalt not covet thy neighbour's wife"). And obviously, like a flywheel, these new meanings increase the sphere of intentionality, motivations and actions, which in turn increases the capital of recorded acts, and so on and so forth.

All this takes place in a self-referential system, a circle that gives meaning, through external goals, to a life that only has internal goals and therefore no meaning. Apart from the satisfaction of basic needs related to food, no human being is particularly motivated to do what they do. Motivation comes from imitation,[55] incentives, competition, boredom. Since it is easier to acquire habits than

---

[55] As noted by W.J. Verdenius, *Mimesis. Plato's Doctrine of Artistic Imitation and Its Meaning to Us*, Brill, Leiden 1949: visible forms are imitations of eternal ones (Plato, *Timaeus*, 50c), thought is an imitation of reality (Plato, *Critias*, 107; *Timaeus*, 47b–c), words are imitations of things (Plato, *Cratylus*, 423e–424b), sounds imitate divine harmony (*Timaeus*, 80b), time imitates eternity (ibid., 38a), laws imitate truth (*Statesman*, 300c), human governments imitate the true and divine one (ibid., 297e), pious men imitate their gods (*Phaedrus*, 252d; *Laws*, 713e). And indeed, it is not difficult to detect the omnipresence of mimesis (the derivation of intentionality from documentality) in a vast number of fields. In ethology, the centrality of mimesis in the animal world was already intuited by Darwin: cf. *The Expression of the Emotions in Man and Animals*, John Murray, London 1872. See A. Whiten, R. Ham, *On the Nature and Evolution of Imitation in the Animal Kingdom. Reappraisal of a Century of Research*, in "Advances in the Study of Behavior", 21, 1992, pp. 239–283. In psychology, see the classic J. Piaget, *La formation du symbole chez l'enfant. Imitation, jeu et rêve, image et représentation,* Delachaux et Niestlé, Neuchâtel 1945; and cf. G. Butterworth, *Neonatal Imitation: Existence, Mechanisms and Motives*, in *Imitation in Infancy*, ed. by J. Nadel and G. Butterworth, Cambridge University Press, Cambridge 1999, pp. 63–88. In sociology, reference must be made to G. Tarde, *Les lois de l'imitation* (1895), new ed. Les Empêcheurs de penser en rond, Paris 2001, p. 37. Theses and texts on imitation can be found in M. Lavelli, *Intersoggettività*, Raffaello Cortina, Milan 2007. The role of imitation and repetition was highlighted by G. Deleuze, *Difference and Repetition* (1968), Continuum, London 1997, to which I referred in *Proust, Deleuze et la répétition*, in "Littérature", 32, 1978, pp. 66–85. One should also not forget René Girard's thesis that society is driven by a mimetic desire (mutual envy) that cyclically vents on scapegoats (R. Girard, *Violence and the Sacred* [1972], Johns Hopkins University Press, Baltimore 1977). For a his-

to lose them, it is better to get used to praying than get used to drinking. Still, you have to get used to something, and most of the time it is not desire that generates habit, but, on the contrary, habit generates desire, just as documentality generates intentionality. Therefore, questions such as "what is money?" or "what is a driving licence?" have a trivial answer. However, if the question "what does this document depend on?" is answered by "other documents", we have a regress that, if not going to infinity, certainly gets lost in the origins of humankind and starts with scenes in which the distance between human and non-human animals fades, revealing the narrow gap between a Siberian wolf and the wolf of Wall Street.

I foresee a perfectly legitimate objection: isn't explaining documents through other documents like claiming that the world rests on an endless series of turtles? No. Because the pile of turtles is implausibly called upon to explain a physical phenomenon, whereas the dependence of documents on other documents is part of our most ordinary experience – the intervention of collective intentionality in the creation of documentality is not. The clerk and I may well agree on the absurdity of a bureaucratic procedure that wastes time for both of us: it remains that our shared intentionality is not enough to exempt us from the procedure, the inventor of which would be hard to find, as it results from a stratification whose original act cannot be found.[56]

2.3.4. Documedia capital

Hence a consequence that can shed light on the present. If talking about "surveillance capitalism" or "knowledge capitalism" leads to the problems I pointed out

---

torical presentation of the political role of mimesis, cf. B. Fuchs, *Mimesis and Empire: The New World, Islam, and European Identities*, Cambridge University Press, Cambridge 2003. Cf. also G. Gebauer, C. Wulnaf, *Mimesis. Culture, Art, Society*, University of California Press, Berkeley-Los Angeles 1995; Id., *Jeux, rituels, gestes. Les fondements mimétiques de l'action sociale*, Anthropos, Paris 2004; L. Rendell, R. Boyd, D. Cownden, M. Enquist, K. Eriksson, M.W. Feldman, L. Fogarty, S. Ghirlanda, T. Lillicrap, K.N. Laland, *Why Copy Others? Insights from the Social Learning Strategies Tournament*, in "Science", 328, 5975, 2010, pp. 208–213. Finally, as elaborated by Bourdieu within the framework of the notion of *habitus*, social hysteresis is a version of imitation as the foundation of society for Tarde. For imitation in aesthetics, with a contemporary outlook, cf. K. Walton, *Mimesis as Make-Believe: On the Foundations of the Representational Arts*, Harvard University Press, Cambridge (MA) 1990.

[56] S.P. Turner, in *Explaining the Normative*, Polity Press, Cambridge 2010, stresses the circularity of normativity and argues that it can only be overcome through an empathic transposition into the motives of others. I find it hard to imagine the empathic transposition of a Lager inmate towards his caretakers, and I doubt that a source of normativity can be derived from it. It seems more reasonable to think that the normative circle is not vicious, but virtuous, as I have tried to demonstrate in this section.

in 1.2.2, talking about "platform capitalism"[57] is entirely reasonable. The platform is the current form of capitalisation, i.e. the documedia capital that captures the acts of the biosphere and transfers them to the docusphere, generating new value and new intentionality and thus remobilising the biosphere. This capital is not yet evident. Just a few years ago,[58] the development of financial capital was recognised as the destiny of 21st-century capital, under the assumption that money would generate more profit than any other commodity and that finance is a space for the free play of the human imagination, a sphere of pure construction that enhances the supposedly hyperdynamic and fluid character of the social world. This idea, however, overlooks the fact that documents can now earn much more than money. If the general principle guiding my analysis holds, we are faced with the progressive revelation of an essence.

There is a solid thread from the earliest forms of humanity right up to us, and it consists of capitalisation. Capitalisation of resources through the accumulation of goods; capitalisation of energy through the development of working tools; capitalisation of social prestige, also through luxury and waste. In the mirror of documedia capital – which brings together all these aspects, constituting their *telos* – we can thus grasp the essence of human society in a nutshell. As ichnological capital, it is the result of human mobilisation, which in a highly automated society is predominantly oriented towards consumption rather than production. As technological capital, it essentially consists of the automation of production processes through the recording, archiving and profiling of human capital. As document capital, it is able to extend capitalisation beyond the narrow scope of production automation, and intervenes in every area of human activity, starting with knowledge. Thanks to documedia capital, it becomes clear that capital has value because it contains documents that are infinitely richer than money, as they keep track of every human act. Once again, it is necessary to think of value starting from capital and not the other way around, and this reflection takes us to the heart of normativity as a specific feature of the human form of life.

2.4. Monument: Value

While the scorpion in Aesop's fable can justify stinging the frog that is carrying it on its back by saying "it is my nature", no human will ever do anything of the sort. The scorpion's gesture is inadmissible both because it is morally reprehen-

---

[57] T. Scholz, N. Schneider, *Ours to Hack and to Own. The Rise of Platform Cooperativism, a New Vision for the Future of Work and a Fairer Internet*, OR Books, New York-London 2016; B. Vecchi, *Il capitalismo delle piattaforme*, Manifestolibri, Rome 2017.

[58] Piketty, *Capital in the Twenty-First Century* cit.

sible and economically foolish, as it is a choice whereby both will lose their lives. Among the meanings that capital gives rise to, a special place is given to value, both in a venal and in a moral sense. Understanding hominisation therefore requires that, as part of the Copernican revolution, we think of value starting from capital and not the other way around. Once again, it is a question of overcoming the prejudice of a naked and free man who would be burdened by the corruption of progress, which would be both a limitation of freedom and a vulgar barter whereby moral value is replaced by commercial value.

Now, it is an obvious but essential observation[59] that a single word, "value", indicates both economic status and moral virtue. This duplicity of meaning has obviously been interpreted in many ways, from simple homonymy to a genealogical derivation of economic value from moral value or vice versa. The hypothesis I would like to propose, arguing for documentality as a necessary as well as sufficient condition of intentionality, is that here too a single principle, the possibility of keeping track, constitutes the condition of possibility of both moral and economic value. However, this does not reduce one to the other, and above all does not preclude the possibility of categorical imperatives, independent of any empirical objective but not of the material possibility of keeping track of the commitment undertaken.

2.4.1. Gold and the categorical imperative

Without going too far, consider this: the lost war, which in 1946 at Nuremberg was the object of capital punishment, in 1919 at Versailles had been the object of economic sanctions. Of course, the faults were different, and in 1946 Germany was far more exhausted. However, this circumstance not only suggests the fundamental link between moral and economic value, but above all shows how erroneous it is to believe that the progress of capital consists in replacing moral values with economic ones (in fact, according to this philosophy of history, capital punishment should have prevailed in Versailles and economic sanctions in Nuremberg).

A strange moral romanticism leads even acute readers to postulate that the earliest writings, and the earliest traces in general, had an exclusively magical and mythical value. Only later on – it is said – were they bent to practical and especially economic purposes. The ancients were poets, heroes, and priests, not accountants or bookkeepers: they worshipped their gods and fought bravely. The same applies to peoples outside European history and geography. The earliest Chinese writings had some magical value, and the earliest Indian texts were the-

---

[59] A. Donise, *Valore*, Guida, Naples 2008.

ogonic and heroic epics. Native Americans did not care about money, so much so that bison were more important to them than gold. It is the white man who corrupted them, teaching them the use of dollars and the abuse of firewater. The Aztecs, it is true, had writing and artefacts made of gold, but their forms of capitalisation were aberrant, consisting for example in human sacrifices that by our standards would have the scale of genocide, with an accounting of death that recalls not the economy, but the madness of death camps.

Since this is a prejudice, empirical refutations do not invalidate the transcendental principle according to which the market comes after the temple, and the accountant comes after the bard. But since things (indeed) don't add up, and there is no shortage of examples of ancients or savages who were anything but economically disinterested, one is forced to hypothesise a hiccupping history of venality. Achilles was uninterested in money, while Ulysses was already more materialistic (not to mention the Proci), but this is explained by the fact that the *Odyssey* is about lower beliefs and customs. So be it. But then why were the Spartans so uninterested in money and rather seemed to cultivate the ideals of heroic communism? The Athenians already thought more about money, but never as much as the Phoenicians, which is often explained by their Semitic ancestry, triggering a process that in the long run would lead to pogroms. Also, it is not clear how, in the space of a single generation, moreover situated in times that are difficult to pinpoint and in any case remote, the Jews went from Abraham's sacred indifference to the trade between Jacob and Esau. It is also worth noting that *potlatch* and *kula* can only be defined as symbolic exchanges or gifts if we stand by the prejudice that only we modern Westerners are obsessed with economics.

An unbiased examination, instead, suggests that there is a profound continuity between those practices and phenomena such as the stock market, consumerism, the acquisition of status symbols, and so on. In this case, Abraham's sacrifice itself, his extreme and selfless gift, would not be foreign to economics. Perhaps his offering everything to God was a reckless financial manoeuvre: sacrificing everything to the Almighty in order to have even more. Indeed, it is far from certain that sacrifice is foreign to economics, as we are taught by the stories of tycoons who, by dint of sacrifice, have gone from pennies to millions. It is no more necessary to assume the magic of the spirit behind the emergence of normativity than it is to assume the magic of the spirit behind the emergence of intentionality. In both cases, before the spirit there is the letter,[60] and before value

---

[60] The idea of documentality as a promoter of intentionality is a general theorisation of what Foucault outlined in reference to the special case of norms: "The norm's function is not to exclude and reject. Rather, it is always linked to a positive technique of intervention and transformation, to a sort of normative project [...]. Finally, it seems to me that with the disciplines and normalization, the eighteenth century established a type of power that is not linked to ignorance

there is capital. We have forms – letters, pyramids, totems, bulletins as incomprehensible as totems, payment notices, notifications, decrees, unread emails, and then the crystal clear document we call a banknote. These forms share a single but decisive trait: they are hysteresis, they fixate an act, from the minimal one of a missed call to the more solemn one of a bank issuing money or a parliament legislating. This hysteresis, in turn, will conjure up something like the spirit, generating reactions to the email, the document, the law, and, in the best of cases, such as that of money, will create a collective intentionality that is far from mystical and very concrete.

One might add that social interaction is not reduced to account books and the records of assets, including the *Code Napoléon*, the Bible, and the Koran. Weber had posited religion as the origin of capitalism, but had failed to consider that the origin of that particular religion, Calvinism, was precisely the privatisation of consciousness generated by the printed dissemination of the Bible. In revelation as it is found in the Abrahamic religions, the letter precedes the spirit and constitutes it. It is something external, such as a burning bush, that manifests itself to consciousness, which, as a result of that revelation and the prescriptions it receives, becomes a believing consciousness. And this revelation, just like Thales' intuition, and following the same logic, is recorded in a book. Thus, it is no coincidence that religions of the spirit are also in the highest degree religions of the scriptures. In the religious sphere, writing contributes to the establishment and generalisation of norms, which are also conditions for the creation of universal religions: if, therefore, humanity can conceive of a cosmopolitan ideal of some kind, this does not depend on the spirit, but on the letter.

Please do not think that I am reducing everything to a question of money! I am simply trying to acquire a broader outlook, capable of understanding the meaning of value beyond the distinction between economy and morality. The confusion between economy and morality, politics and psychology, symbolic values and concrete interests, justice and revenge, selfless virtue and moral indifference, is not accidental, and therefore can be avoided in an ideal functioning of the economies of morality. However, it is still structural and constitutive. Just as a sign only acquires meaning within a system of signs, so value does not have meaning in itself, but only receives it within a system of value oppositions (pure/ impure, sacred/profane, etc., knowing full well that the two terms of the opposition can change places in different times and places). If there were only one action in the world, we would not be able to decide whether it was moral or economic, selfless or self-interested, and more importantly, free or necessary.

---

but a power that can only function thanks to the formation of a knowledge" (M. Foucault, *Abnormal* [1975], Verso, London 2003, pp. 50–52).

Hence my proposal: instead of thinking that capital stole our souls, let us consider that without capital, and without the network of concepts it is a part of, we would have nothing even vaguely resembling a soul – nothing with which to pay the psychoanalyst, but nothing to treat either. Far from leading to the destruction of morality, a correct understanding of capitalisation would provide a different foundation for the categorical imperative. Tables of values are not written by divine inspiration, nor are they the inner analogue of the starry sky above us. In a much more complex and, for that very reason, more interesting sense, they are the result of a capitalisation and technological evolution which, as I hope I have been able to demonstrate, rests on both dimensions of value. It is not surprising that Kant's morality ascribes dignity as an absolute value, a paradoxical value without value, that is, an absolute value that cannot be exchanged for anything, and therefore does not and cannot have any *Marktpreis*. Just as in the characterisation of the productive imagination, which Kant defines solely and exclusively in opposition to the reproductive imagination,[61] so the incommensurable value of dignity, its pricelessness, is defined solely and exclusively in opposition to market values. This does not fail to have significant consequences, because if this non-market or above-market value is only such in reference to the market, then there are good reasons to doubt the extraordinariness of moral value in relation to market value.

In this regard, there is a circumstance that deserves consideration. Speaking of the problematic character of the Kantian categorical imperative, Bergson wondered what would happen if an ant, for a brief moment, was endowed with consciousness and pondered about the work and toil it was undertaking, asking itself if it was worth it and why it should do it.[62] At the end of this brief flash of consciousness, Bergson posits, the ant returns to the darkness of instinct, and the last semi-conscious motivation that flashes to its mind is "I must because I must", i.e. the Kantian categorical imperative. We do it because we do it because we do it: the dogmatic description of duty is paradoxically identical to the critical hyperbole "you must because you must". When a child asks us "why" for the upteenth time, we answer "because it is so", inadvertently proposing a version of the categorical imperative, which describes morality as a fact as clear as the starry sky above us. Superorganisms,[63] of which termite colonies are an example, prove

---

[61] "Now insofar as the imagination is spontaneity, I also occasionally call it the productive imagination, and thereby distinguish it from the reproductive imagination" (Kant, *Critique of Pure Reason* cit., B 152, p. 257).

[62] H. Bergson, *The Two Sources of Morality and Religion* (1932), MacMillan, London 1935.

[63] B. Hölldobler, E.O. Wilson, *The Superorganism: The Beauty, Elegance, and Strangeness of Insect Societies*, W.W. Norton & Co., New York 2010; C. Detrain, J. Deneubourg, *Self-organized Structures in a Superorganism: Do Ants "Behave" like Molecules?*, in "Physics of Life Reviews", 3, 3, 2006, pp. 162–187; D.M. Gordon, *Ant Encounters: Interaction Networks and*

that a purely document-based coordination, the transmission of signals, is enough to confer a shared motivation. It is therefore not surprising that a higher social and mental complexity, such as we find in humans, can confer intentions, consciousness, responsibility, hypothetical imperatives and categorical imperatives.

What is certain is that these intentionalities and imperatives lose no value because of the emergent and non-Pentecostal nature of their genesis. Motivation does not descend from the heavens of the categorical imperative, it rises from the earth of hypothetical imperatives, originating in human practices and above all in that great vehicle for the transmission of practices and motivations which is imitation. In fact, those who argue that maybe the same barbarians who ate their own parents preferred beads to gold would hardly convince someone educated in a different way to exchange gold for beads. Why? Simply because we have been accustomed not only to conferring value on gold (a value that depends on the system, not on the subjects), but also to wanting it, or at least to considering the desire for gold a rational motivation. Jack London tells us an interesting anecdote in this respect[64]: he was drinking with a sailor, who after paying for the first beer kept ordering more drinks because he was sure that London would eventually pay for half of them, as was customary. London didn't and left the bar. Then, reflecting on the matter, London realised his mistake, went back inside, bought the sailor more drinks and got irreparably drunk. So, even a simple gesture like paying for a beer does not come by itself, but requires reflection. Let alone a categorical imperative that tells us that we must because we must, provided that the maxim of our will can be an object of universal legislation. Fortunately, in most cases, we are assisted by education, which makes us act in a certain way or another without requiring reflection and out of mere habit.

The birth of motivation is a technical process, related to the harrow which, in Kafka's *Penal Colony*, carves the words "Be fair" on the back of the convicted.[65] We are not born with innate moral principles – quite the contrary. We grow up in a system of education, prohibitions, and repression of instincts. And all these external pressures and repressions are reinforced by the fact that they do not come from a single person, but from the whole of society. It is precisely this

Colony Behavior, Princeton University Press, Princeton 2010; M. Resnick, *Turtles, Termites, and Traffic Jams: Explorations in Massively Parallel Microworlds*, The MIT Press, Cambridge (MA) 1997; W. Trible, D.J.C. Kronauer, *Caste Development and Evolution in Ants: It's All About Size*, in "Journal of Experimental Biology", 220, 2007, pp. 53–62; F. Tripet, P. Nonacs, *Foraging for Work and Age-Based Polyethism: The Roles of Age and Previous Experience on Task Choice in Ants*, in "Ethology", 110, 11, 2004, pp. 863–877; G. Flake, S. Lawrence, C. Giles, F. Coetzee, *Self-organization and Identification of Web Communities*, in "Computer", 35, 3, 2002, pp. 66–70.

[64] J. London, *John Barleycorn*, The Century Company, New York 1913.

[65] F. Kafka, *The Penal Colony* (1919), Transaction Publishers, Livingston (NJ) 2000.

sharing that makes the weight of society on morality so binding. We accept the laws of nature as if they were commandments, but this is only the other side of the process by which we tend to accept moral commandments as if they were laws of nature. By this I do not mean in the least that the commandments are actually laws of nature, or that they should be considered as such. Rather, I mean that moral duty, a very important element in individual and collective life, originates neither in the neurons of goodness, as the modern tend to say, nor in the infusion of a virtuous soul, as the ancients often said, but simply in the stratification of acts, behaviours, and regulations.

In Kipling's *In the Rukh*, the Englishman who, alone in his hut deep in the forest, wears a black suit for dinner, and does so out of self-respect, is poetic in its own way, but constitutes a stylised form of the otherwise elusive attribute we call "human dignity". An important consequence follows from this. We are used to regarding the moral world as a sphere of dilemmas and problems. This is nevertheless a kind of cognitive illusion not unlike the one that so often leads us to confuse experience with our awareness of it. In reality, although at every moment of our lives we perform actions that may be full of moral consequences, most of those actions are performed unreflectively and out of mere habit. The conception of the moral forum as an arena in which different principles clash at any given moment has the disadvantage of presenting us with an unlivable life, and is tainted by the unparalleled flaw of being false. We do not act in this way, and this courtroom-like image of the moral life captures true but rare moments – a form of ideal normativity that has nothing to do with the normality of our relationship with duty.

A tenacious thread runs between technology and spirit, between hysteresis and justice, the profane and the sacred, commercial capital and moral capital. This is what Hamlet's sentence "time is out of joint" suggests: injustice manifests itself as a technical defect, and justice, the remedy that Hamlet is called upon to provide, appears in turn as a technological operation, as a repair intended "to set it right" in which one finds both the material element of fixing something broken and the moral element of righting a wrong. And the economy? A radical injustice comes forward precisely through the condemnation of Shylock's greed and the misdeeds of capital, however it would be paradoxical if, after performing his reparation, Hamlet demanded to be paid. Though not because those who mend something broken or disjointed do not deserve to be paid, but simply because Hamlet was indebted to his father, he had made him a promise, and every promise is a debt. It is not surprising, then, that Hamlet wrote down his promise, so as not to forget it.[66]

---

[66] W. Shakespeare, *Hamlet*, I, v, vv. 93–113.

In this respect, money, banking and capital are technologies, i.e. devices that extend human resources. These devices, like documents, mobilise us first of all because they are hysteresis systems capable of generating individual responsibilities: we are called upon to *respond to*, under the action of an appeal that is addressed to us individually and that is recorded, i.e. is addressed specifically to us and cannot be ignored. However, this passive *answering to*[67] is at the origin of *answering for*, of responding in an active sense, as bearers of morality and freedom: it is precisely to the extent that humans are educated in the structure of *answering to* that they can formulate, later and in a derivative form, the structure of *answering for*, i.e. of being morally responsible. And technology, rather than requiring moralisation, is presented as a primary source of moral responsibility: we are called upon to respond, to give something back, and it is not surprising if we reflect on the logic of the supplement, which fills a lack and at the same time generates a need and a debt, triggering a system of exchanges. Hence my general proposal, which is to think of value, both venal and moral, starting from capital, and not the other way around. With this further Copernican revolution, we will be able to explain both why money has worth, and why there are values that transcend money.

2.4.2. Calculation

To this end, it is first of all worthwhile to take literally what Spengler said about "Faustian money", that is, the money of finance "as Function, the value of which lies in its effect and not its mere existence", first introduced by the Normans in England. This money is probably older than Spengler thinks. The first of the three functions of money, that of accounting, is obviously much older than money, and is the condition for its possibility. Just as we find it anachronistic to imagine Achilles smoking an electronic cigarette, we find it implausible to imagine Clytemnestra taking out a mortgage. However, it would be wrong to conclude that the first human concerns were heroic and sacred, while it was only at a later stage, which is always difficult to pin down in time, that economic interests came into play. There is no reason to view the ancients or primitives as simpletons or saints indifferent to capitalisation. The very fact that the first written texts are

---

[67] E. Lévinas, *Totality and Infinity: An Essay on Exteriority* (1961), Duquesne University Press, Pittsburgh 1969. The mobilisation I mentioned in 1 also relies on the power of documents and the fact that our actions are recorded, thus generating responsibility. On mobilisation, from different perspectives, see also F. Merlini, *Ubicumque*, Quodlibet, Macerata 2015; M. Wieviorka, *L'Impératif numérique*, CNRS Éditions, Paris 2013; E. Pasquinelli, *Irresistibili schermi*, Mondadori, Milan 2012; C. Formenti, *Felici e sfruttati. Capitalismo digitale ed eclissi del lavoro*, Egea, Milan 2011.

about accounting should prompt us to observe that, from the earliest records that have come down to us (but the fact that they have come down to us does not guarantee that they are the earliest!), interest in finance was as developed as it is today. Not to mention the fact that forgiveness of debts is a widespread religious precept as well as a Socratic philosophical principle (affirmed when Socrates remembers to return a rooster to Asclepius). This circumstance – which, dazzled by moral romanticism, one may be inclined to dismiss as ironic – suggests that many features that we are used to imputing to modernity are instead part of an infinitely more archaic human heritage.

If it is difficult to imagine non-human animals trading on the stock exchange, but not exchanging banknotes or shares, it is even more difficult to imagine that our forms of social organisation, such as structures of dominance, elementary kinship relations, and taboos, have no link of continuity with our pre-human origins. So, too, it is difficult to imagine a human social activity that is not decisively conditioned by its practical forms of implementation, as well as by the natural environment in which it takes place: for example, this is implicit in the notion of "geopolitics". The basic assumption of moral romanticism is easy to formulate: in times gone by, in the golden age – note the contradiction! – humans were more noble and disinterested than in the present, whenever that may be, with the paradoxical result that any time is both disinterested, when past, and interested, when present. This makes morality both very easy, since it is enough to go back to the good old days in order to be moral, and frustrating, because one can always find a more primitive and more moral age with a regress that ends with animality and that often does not perceive its inherent contradiction (what is the point of endowing a beaver or a tapir, or more often a dog, with a moral sense?), flatly concluding that animals are morally better than humans.

Like all prejudices, moral romanticism does not stand up to the test of facts. Sooner or later, tourists discover that the locals think about money at least as much as they do, or even a little more, since they are not on holiday. Anthropologists realise that the natives would like to be paid for the information they provide. And archaeologists have to admit that the similarity, in form and function, between the pyramids and banks is anything but external, with the aggravating circumstance that the Egyptians did not only undertake the journey to the afterlife with a good financial endowment, but even went so far as to take their debtors with them, so that the latter could not think of getting away with the death of their creditor. The fact that the classical pantheon contemplates not only mystic and heroic deities, but also a god of theft, commerce and writing, speaks volumes about the difficulty of finding an age of the world that is immune to calculation.

Moral value and economic value are but two sides of the same coin. Moral romanticism projects a huge theoretical prejudice into history, like an enormous purloin letter which conditions our examination of morality. This prejudice states that economic value is in no way comparable to moral value, that there is only an inexplicable homonymy between the two. Likewise, it is due to chance, if not to cynical fate, that the term "value" originated in economics, and that the first chair of moral philosophy in history was held by the author of *The Wealth of Nations*. Here too, in short, Pentecostal prejudice is prevalent. Just as (according to Aristotle) man has been endowed with hands because he is the most intelligent of animals, so (according to Kant) he is endowed with moral imperatives and a sense of duty that in principle assimilates him to any possible rational being, be it an angel or a Martian, but in fact differentiates him from any other being in nature, giving him duties that are moral only insofar as they are selfless.

In agreement with the Copernican revolution I was proposing, let us change perspective. I am holding a banknote, and I can pay the bill at the restaurant with it. I can do the same thing with a credit card, or with a smartphone, by photographing a barcode on the bill. What do these three operations have in common? The presence of hysteresis, i.e. of codes and – analogue or digital – memories in my pockets. So: money is a form of hysteresis, in fact all money can be traced back to this origin and function – and everything that can fulfil this function can take the place of money. In fact, it has been persuasively argued[68] that money is a form of bank debt and a substitute for memory,[69] and this hypothesis becomes more evident if we trace banks and money back to ancient times, where, as we shall see in a moment, money and banks coincided in the form of two sticks bearing notches. In order to understand the nature of money, some have proposed the image of a chessboard, on which the pieces record the positions of the two

---

[68] T. Lawson, *Debt as Money*, in "Cambridge Journal of Economics", 42, 4, 2018, pp. 1165–1181.

[69] Many economists have noted that money is a recording system (although they often speak of "information"; cf. J.M. Ostroy, *The Informational Efficiency of Monetary Exchange*, in "American Economical Review", 63, 4, 1973, pp. 597–610), a low-cost means of keeping track of prior resource allocations (R. Lucas, *Equilibrium in a Pure Currency Economy*, in *Models of Monetary Economies*, ed. by J.H. Kareken and N. Wallace, Federal Reserve Bank of Minneapolis, Minneapolis 1980), and that currency would be superfluous if agents had access to all their previous reciprocal interactions (S.R. Aiyagari, N. Wallace, *Existence of Steady States with Positive Consumption in the Kiyotaki-Wright Model*, in "Review of Economic Studies", 58, 1991, pp. 901–916). In the end, money is nothing but memory. This thesis has been developed, in particular, by the American economist Narayana Kocherlakota (N.R. Kocherlakota, *Money is Memory*, Federal Reserve Bank of Minneapolis, Research Department Staff Report 218, October 1996). Some have objected (cf. M. Bigoni, G. Camera, M. Casari, *Money Is More than Memory*, in "Journal of Monetary Economics", 110, 2020, pp. 99–115) that memory does not inhibit the rapacious behaviour of free riders, whereas money does. Which is true, but this only applies to individual memory, but not when the latter is recorded.

players in the course of the game[70]; the same can be done with an electronic transaction.[71]

In both cases, we are dealing with something concrete that represents something abstract, but the question is: what abstraction are we talking about? In my opinion, the most complete answer would come from representing social reality as a universal blackboard, on which all social acts are indelibly recorded and accessible to the whole of humanity. Obviously, there is no such blackboard, and even if there was, it would be of no use because it would be incomprehensible: the infosphere is an ideal, not a reality. What is certain, however, is that every human act will potentially be inscribed in the immense docusphere that makes up documedia capital.

So when did this calculation start? It is impossible to know, but let us consider an example. In 35,000 BC, in Africa, someone carved the Lebombo bone – and that someone was not called Lebombo, as that was the name of the location where the bone was found. For what? We do not know exactly, but one thing is certain: they did it to keep track of something. It is precisely at this time that we find the first traces of a human formative intent: small stone or wooden plates engraved with abstract motifs, with a rhythmic modulation. In the case of the Lebombo bone, there are exactly twenty-nine notches, suggesting that it was a lunar calendar. This is not an isolated piece of evidence. Another similar find, dating back to 20,000 BC, was also found in Africa, in Ishango. And recently, in Velletri, researchers retrieved a carved stone dating back to the Upper Palaeolithic, which, once again, appears to be a lunar calendar.

Now, it is undeniable that, for practical purposes, humans have used predictive systems whose importance they may have underestimated at first. This was the case with social networks when they took their first steps and we thought that their main source of income would come directly from advertisements and not from the collection of documents. This was the case with the inventors of double-entry bookkeeping,[72] which at first was simply a control instrument, but which nevertheless determined, with greater knowledge on the part of enterprises, the prevalence of the latter over the market, laying the foundations for the growth of financial capitalism that only in recent years (and thanks to platforms) has begun to decline, leading to a renewed prevalence of the market over enterprises. And this was the case with some of our ancestors tens of thousands of years ago.

---

[70] J.P. Smit, F. Buekens, S. Du Plessis, *Cigarettes, Dollars and Bitcoins. An Essay on the Ontology of Money*, in "Journal of Institutional Economics", 12, 2016, pp. 327–347.

[71] Without neglecting the circumstance, recalled by Leibniz, that Arabs could play chess by heart while riding a horse, without the need of a physical chessboard. (cf. Leibniz, *New Essays on Human Understanding* cit., II, XIV).

[72] Carpo, *The Second Digital Turn* cit.

Even if we leave aside the lunar calendar hypothesis, the basic idea is certain: something had to be kept track of, if only the passing of time, like Robinson on his island, because without this minimal retention, which takes place primarily in our memory and can be externalised in traces, everything would be an eternal beginning with no course. Even mathematics, which is seemingly the most abstract science we know, originates in these notches, because without an external medium operations would not last, and (even slightly) complex ones would only be performable by a mind endowed with the marvellous automatisms of an *idiot savant*. Even perceptual constraints count in determining mathematical ideals. We know about arithmetic systems that only had two numbers (which then came back in computer science): one and two. What we call "three" is a residue of the "trans", the "beyond" that goes over one and two – is it not still so in languages that include the singular, the dual and the plural?

Keeping track is therefore the basis of a series of functions that account for the origin of society: arithmetic, money, the calendar. Hysteresis, in the economic sphere, was born as an indication of ownership, and thus originated from seals or marks; marks were used since the Neolithic, imprinted on clay tokens called *calculi*, and acted as a unit of account for goods (oxen, sacks of grain), thus as a sort of currency; coins were then collected in cylinders, on which conventional signs designated quantity and content: and it is from this function that cuneiform writing arose, with tablets being the evolution of cylinders.

Let us take a closer look at the history of *calculi*, because it contains almost all that is necessary for the genesis and evolution of a society. Ten thousand years ago, in Mesopotamia, there were already symbols used for calculations: cones, cylinders, spheres, hemispheres, tetrahedrons, etc. Some represent the unit, others multiples of the arithmetic bases 10 and 60 (10, 60, 600). After almost five thousand years of undisputed use, a novelty occurred: as I have just said, around 3,000 BC, in the Elamite city of Susa, *calculi* were placed inside cylinders, indicating the corresponding quantity with signs that are the ancestors of cuneiform writing. A calculation is a simple manipulation, independent of what is represented by it, and herein lies its power. Originally, however, *calculi* were anything but abstract, indeed they were very concrete: they were the stones I mentioned – cones, spheres, marbles, discs, sticks, tetrahedrons, cylinders –, used to calculate. And the electronic calculator still has a relationship with those stones, which have evolved not only in an arithmetical sense, but also in terms of functions that have little to do with calculation: ornaments, religious symbols, amulets, tokens.

Take note of this circumstance: there were two parallel processes of symbolisation, not the passage from thing to sign, because the signs contained in the cylinder were in no way more concrete than those inscribed on top of it. Something similar happened in South America in 2,000 BC, concerning not arithmetics but chronology. The Olmecs used engravings or glyphs to indicate time units, such as days, years, and cycles. In 600 BC, the Zapotecs added signs for dates and for the characters involved in those dates. Here, the phenomenon of the construction of social reality by means of inscription is even more evident: if we did not keep track, time – on a psychological and social level, and even in terms of scientific theories – would disappear, and we would have no notion of the epochs we have left behind.

Technological compensation for our limitations does not only concern memory, but also perception. It is well known that we can easily discern up to three notches, but we already struggle with four and five (IIII), as we learn from experience whenever we have to fill in an IBAN by hand. This is why the Roman numeration indicates five with V, four with IV, and six with VI – and similar expedients can already be found in the notches of the ancient bones I mentioned. Carving from Cro-Magnon onwards, the bookkeeping of illiterates of all times, and using a cross as a signature – all this refers to a single sphere, that of traces. Just as fossils continue to live on in parts and organs of present-day animals, so the oldest traces are still found in the present, in wall markings, in dirty drawings, in scratches on school desks. And the 45,000-year-old painted boar in a cave on the Indonesian island of Sulawesi shows that even the most distant past is still with us, in our present.

2.4.3. Debt

The second function of money, acting as a means of exchange, also precedes currency and economics, and is located in a territory where moral value cannot be distinguished from commercial value. "Every promise is a debt", goes an Italian proverb that sums up the common source of economics and morality. Once the norm, and thus property, have been established, it seems natural to transfer resources, i.e. to grant credits. This mechanism initiates the first moral principle, which is not "do not kill", but "honour your debts". How does this principle arise? Human memory is short, and especially short when it comes to finances. This is why, since the Palaeolithic era, systems of external accounting, the basis of all other capital, have always been widespread. This accounting indicates a financial value, but also a moral value, because failing to meet a commitment is not only an economic offence, but also a moral wrong. This is how, from the very beginning, financial value has been indissolubly linked to moral value, even though many people persist in denying this with a tenacity dictated only by hypocrisy. In this context, claiming that some things are priceless is a euphemism for saying that some things have a very high price (which one is not necessarily willing or able to pay). But to conclude that venal value and moral value do not have the same origin is too bold and moralistic a step, and there is nothing to deny the evidence to the contrary.

Indeed, the idea that debt lies at the beginning of the social world has been criticised.[73] However, it is difficult to find good arguments for what is, at best, a complaint about the present, and not an understanding of the past and the basic structures of hominisation. Not to mention that private property, and hence the system that makes debt possible, is also attested in animal societies,[74] the theoretical argument amounts to saying that debt is a bad thing. Yet I suppose that a society in which it was not possible to incur debt would be even worse. Noting that violence is at the origin of debt, however, on the one hand displaces the problem, which is, if anything, that of abolishing violence, and on the other makes it generic and irrelevant: as we know, violence is also at the origin of the sacred, of philosophy, of court society and hence of dames and minstrels, as well as of the great international organisations that do their utmost to reduce violence through contracts.

It is therefore difficult to find a more characteristic element of the human animal than making promises,[75] i.e. contracting debts (since every promise is a debt): if a lion made us a promise, it would be the most senseless act in history, not so much because we would not understand it,[76] but because a lion is not a legal entity, nor can it become one – that is why we understand, even if we do not accept them unconditionally, the promises of children, which most often constitute the prelude to the excuses of adults. We do not hold other animals or automatons morally responsible; killing a lion that has eaten its tamer is not a punishment, but a precautionary measure. For humans, things are different: they have to account for their action as the result, not of necessity, but of an unpredictable contingency that takes the form of freedom. And even in this case, there are excellent reasons to think that this duty, this intention, this responsibility, is not

---

[73] D. Graeber, *Debt: The First 5000 Years*, Penguin, London 2011.

[74] R. Sacco, *Diritto muto*, in "Rivista di diritto civile", I, 1993, p. 689.

[75] "Have these genealogists of morality up to now ever remotely dreamt that, for example, the main moral concept '*Schuld*' ('guilt') descends from the very material concept of '*Schulden*' ('debts')?" (F. Nietzsche, *On the Genealogy of Morality* [1887], Cambridge University Press, Cambridge 2007, p. 39). "The feeling of guilt, of personal obligation [...] originated [...] in the oldest and most primitive personal relationship there is, in the relationship of buyer and seller, creditor and debtor: here person met person for the first time, and *measured himself* person against person. No form of civilization has been discovered which is so low that it did not display something of this relationship. Fixing prices, setting values, working out equivalents, exchanging – this preoccupied man's first thoughts to such a degree that in a certain sense it *constitutes* thought" (ibid., p. 45).

[76] As suggested by L. Wittgenstein, *Philosophical Investigations* (1953), Cambridge University Press, Cambridge 2010, II, 11, § 218.

something immediately human, any more than the mere fact of having hands can be a title of privilege for humans over other animals.

This spirit of promising, giving and taking, mind you, is not just a good-natured ghost or a benign memory. It can be – and in fact most of the time it is – an injunction, a monument: that is, a warning, requiring us to do something. Generally, it is an injunction to pay (bills, taxes, cheques, etc.), but let us not underestimate that paying does not only bear the principle of economy, but also that of justice. "You'll pay for this" rarely refers to a financial sanction. There is only one step between the paper on which Hamlet notes his oath and a savings book, and it is in this space that duty and responsibility arise: you answer an injunction, and then you answer for yourself, for your actions, for the justice or injustice that follow. This framework presupposes the existence of documents as a sphere for measuring and appreciating value, condensed in the notion of "monument",[77] which derives from the Latin *monumentum*, "memory, monument", which in turn comes from *monere*, "to remind". The system of models and principles underlying both economic value, which is based on the model of exchange, and moral value, which at least nominally rests on the model of gift, is generated here. Hence my conclusion: the notch comes before, if not debt, at least duty. If we are capable of making promises, it is not because of a natural instinct, but because of a cultural construct. We have simply grown up in a world where it is possible to write down promises, for example on a stick, and punish, with that stick or something better, those who do not keep them.

I realise that the idea of annotating promises on a stick seems far-fetched and ludicrous, yet that is exactly what has been happening for tens of thousands of years, until the time when barcodes became the new sticks and blockchains the new chains. In the beginning, in the Palaeolithic or even earlier, was debt: I owe something to someone. Nevertheless, in order for the debt to be such, it needs to be recorded, either internally or – better still, it is safer – externally.[78] But when did the circle begin? First of all, let's try to rewrite in technological terms the fundamental moral alternative: which came first, the debt or the technical tool to keep track of it? I imagine that, to many, this alternative seems a little less ridiculous than the comparison between the duck and the hairdryer, but the story I would like to tell you aims to show that it is equally well-founded. This fact,

---

[77] J. Le Goff, *Documento/Monumento*, in *Enciclopedia Einaudi*, vol. V, Einaudi, Turin 1978, pp. 38–48.

[78] P. Seabright, in *The Company of Strangers: A Natural History of Economic Life*, Princeton University Press, Princeton 2004, argues that it is our capacity for abstract reasoning that led us to develop economic instruments ten thousand years ago. The opposite seems more likely: it is economic technologies that have developed an abstract reasoning that today is beyond the reach of most economic actors.

which is well explained by our nature as unstabilised animals, and as such needy and dependent, is not surprising. Given that memory has a tendency to forget debts, or to get tangled up, and given that human relationships are manifold, debt comes with a technical problem: how is debt to be recorded in a reliable way that does not depend on the partial memories of the debtor and the creditor? This question is not extrinsic to the structure of debt. Just as science could not be formed and progress without a capitalisation of memory, so society cannot do without the hysteresis of social acts, and in particular of that paradigmatic social act that is the promise, of which debt, the "IOU", is the most obvious example. So how is debt recorded? (a question that turns out to be identical to "how is duty formed?", which is not surprising, since in both cases we are dealing with what is due).

A good system is to put notches on two sticks,[79] like carving (*tailler* in French, *intagliare* in Italian), which dates back to the Palaeolithic (or even earlier) but was still used a few years ago in the French, Italian, Romanian and Hungarian countrysides, as well as outside of Europe in agricultural and pastoral cultures and in the trades that have prolonged its legacy, such as bakers, milkmen, and shepherds. And these are not just marginal or residual uses. This method had been adopted in the English tax system up until the early 19th century (the *batons the taille* were called *tally sticks*), and, in the French Ancien Régime, it had given its name to the *taille*, the hated direct tax from which the nobles and the clergy were exempted. *Taglia* (from the same root) is also the Italian word for the reward that was given for bandits in the Far West – or at least in Western movies.

The way carving works (a notch for two, an instrument that is both primitive and sophisticated) is very simple. There are two sticks, which are both carved with a notch that stands for a credit; the creditor takes a stick, the debtor takes the other. The debtor cannot erase the notch, and the creditor cannot add one, because the tampering would be revealed by comparing the two sticks. From this point of view, I am not sure about the interpretation by which the three horizontal lines cut by a vertical one (the graphic element *jiè*) present in the Chinese ideogram *qì*, "contract", represent the stylisation of *one* carved piece of wood and a knife. Reflecting on the long-lasting use of carving, it seems more likely that it indicates *two* carved woods placed next to each other. You can do anything with a notch (barcodes are the best example of this), and it is no coincidence that letters and numbers continuously exchange roles, that both derive from traces and not from words, and that the same systems, for example South American nodes, have served as an alphabet, a calculation tool, or a calendar. The unfathomable antiquity of the notch, the cross in place of the signature, the knot in the handker-

---

[79] This was the case, for a long time, with the so-called *roboš* or *raboš* used by Slavs.

chief: all this refers to a single sphere, that of the trace, rather than that of the number as an ideal object. For its part, the blockchain is nothing more than a complex version of the notch, a notch that is brought to a world-wide dimension and multiplied countless times.

Now, let us look at the usual process through the lens of the Copernican revolution. The idea that a very remote ancestor of ours, embittered by the forgetfulness of their debtors, invented the tally seems no less far-fetched than the social contract, whereby that very remote ancestor of ours, fed up with giving and taking beatings, decided to enter into an agreement with his neighbours. Notch and debt were born together, and without the notch and its successors it would have been difficult to achieve a sophisticated economy, as well as a minimally demanding morality, unless we consider the latter the sublime science of simple souls. The tally is the origin of the *tâche*,[80] the objective, the end. In order to respond to a series of practical needs, notches were engraved on surfaces. This triggered a process that is part of the constituent endowments of the human being, and of which we have a growing historical manifestation today.

Nor is it surprising that the concept of "guilt" can be traced back to the concept of "debt", both expressed in German by the same word, *Schuld*. Following this very line of argument, Nietzsche contradicts everything he said before and everything he will say later about the absolute origin of the concept of value as force and virtue, and about the late imposition of resentment and herd instincts.[81]

---

[80] From Middle French *tache*, itself derived from High French *tache*, *taiche*, *taje* ("sign", "notch", "mark"), from Vulgar Latin **tacca*, **tecca*, and cf. Gothic ⰕⰀⰋⰍⰄⰔ (taikns, "mark, sign"), derived from Proto-Germanic **taiknaz*, **taiknǫ* ("sign, mark"), and Proto-Indo-European **deyḱ-* ("to show"). Influenced by forms of the Franconian **stakjan*, **stakkijan* ("to attack") and the Gothic ⵢⵜⰀⰍⰔ (staks, "mark"). See "*attacher*".

[81] In *The Genealogy of Morality*, after the premise – which sounds like an *excusatio non petita* in which Nietzsche hastens to say that the ideas set out in the book have nothing to do with those presented ten years earlier by Paul Rée, and that the latter is in part indebted to him for those very ideas – the dissertations follow an antiphrastic course, to say the least. In the first dissertation, as we know, Nietzsche illustrates the negative transvaluation whereby true values, the values of life and of the strong, have been slandered by the weak, who have ended up imposing their own values, those of resentment (here Nietzsche makes a long digression to explain that the resentment he is talking about has nothing to do with what is discussed in Rée's book, but there it is). In the second, which is by far the most interesting, he puts forward a hypothesis about the origin of morality that contradicts the hypothesis just presented. If in the first dissertation, the origin of morality is vital, aristocratic and heroic, in the second it is based in accounting from the very beginning. Moral guilt is nothing more than a financial debt, so (it is fair to assume, even if Nietzsche does not come to this conclusion) moral value and economic value are the same thing. In the third, finally, there is a further hypothesis about the origin of moral prejudices, which is nothing more than the analytical development of Rée's theory of resentment.

I renounce the futile attempt to get Nietzsche to agree with himself, but I would like to emphasise the acumen of his observations on the connection between moral value, economic value and cruelty, which places *The Genealogy of Morality* at the antipodes of *The Social Contract*: the contract precedes humanity and defines it for what it is: a race of sick, weak, indebted, problematic animals, interesting precisely because of this.

The blond beast is not yet human, it is indeed a beast, which has its advantages (a lion will never get a loan from Androcles, but will only ask him to take the thorn off his paw) but also its disadvantages (for example, with a lack of gratitude that depends precisely on indifference to the economy and morals, the lion might then eat Androcles). Nietzsche's thesis on the promise is not limited to the obvious observation that promising means taking on responsibility; it points to something more insightful and significant, namely that responsibility does not precede the promise, descending from some abstract moral heaven, but is derived from it. Education is training to promise, and therefore to give back. In order to do this, humankind first had to devise mechanisms for keeping promises. The fundamental social relationship is commercial exchange, selling and buying, being a debtor or a creditor, and thought is ultimately nothing more than a calculation arising from these exchange needs. All this is already evident in common sense and ordinary language, where economic terminology is frequently used to indicate moral value or disvalue: "Who finds a friend finds a treasure", "Is this how you pay me back?", etc. Not to mention mimetic desire, in which desire is increased precisely by the fact that others also want what we want, just as happens in the stock market. What we must do, therefore, is not deny the evidence, i.e. the link between economic value and moral value, but refine this correspondence, not seeing it in direct and simplistic terms (no one would ever seriously think that a rich man is more virtuous than a poor man), but by complexifying the relationship. So, what is it that makes us both partial and impartial, venal and moral? Capitalisation. This is what we must never forget.

First of all, we must note that morality also consists in a capitalisation: we are undoubtedly more morally sensitive than our ancestors, and this is because there is one great chain linking the first notches (with which, hundreds of thousands of years ago, hominids noted the passage of the moon through the constellations), the first credit devices, the names of the gods, and of course Mammon (money), the profoundly demanding deity that all religions have had to come to terms with it by establishing good neighbourly relations with it. Secondly, we must be clear that moral values are formed in a system, just like economic values, which again does not imply a direct correspondence between moral and economic values. No good deed is such in itself just as no metal is precious in itself, and the mechanisms that underlie both belong to the same sphere – that of valorisation. Thirdly, we must not fall into the fallacy whereby "If God is dead, everything is possible". This is not true, and we know it: the experience of obedience to divine laws is circumscribed in space and time, which has not prevented humanity from being moral or immoral, and above all from performing actions that it considered moral while to us they appear immoral, and vice versa. The idea of an ultimate guarantor of morality, in this context, is no less unfounded than the idea that the gold in Fort Knox is the ultimate guarantor of the value of the dollar.

Between the tablets of the law and the golden calf, the relationship is not one of opposition, but of mutual support, just as between the American monetary system and the deposits at Fort Knox. A precept such as "do unto others as you would have them do unto you" suggests a reciprocity that is fundamentally economic, as evidenced by the fact that criticism of the free rider, the one who does not do unto others as he would have them do unto him, optimising profit, often consists in the fact that this behaviour is inefficient in the long term – not from a moral but from an economic point of view. And conversely, when Plato and Aristotle observe that some form of fairness is necessary even among outlaws, they are referring precisely to the fact that one can be an outlaw while still obeying some law, and that this law can hardly escape a principle of give and take. Justice is an arithmetical relationship that regulates moral relations between humans, as well as between humans and gods, such as when Anubis weighs the hearts of the dead to assess their moral dignity. Not to mention the coin that the Greeks and Romans placed under the tongues of the dead so that they could be ferried by Charon.

To be human, and in this case Christian, is to ask God, for example, to "forgive us our debts as we forgive our debtors" (a prayer, the only one attributed to Christ, in which divine grace is called upon to take its model from the relationship between debtors and creditors, since it seems implausible that the Almighty needs our money); or to aim for a fiscal arrangement between what is owed to Caesar and what is owed to God. Moreover, we have irrefutable evidence that in the beginning was not the Word, but *logos* as relationship, as a relation of equivalence, and in particular as debt and credit. It is not surprising that Anaximander's saying is about giving and taking:

> Whence things have their origin,
> Thence also their destruction happens,
> As is the order of things;
> For they execute the sentence upon one another
> – The condemnation for the crime –
> In conformity with the ordinance of Time.[82]

---

[82] *Didonai gar auta diken kai tisin allelois tes adikias.* Translation by D. L. Couprie (*Heaven and Earth in Ancient Greek Cosmology*, Springer, New York 2011, p. 93). Nietzsche's 1873

In other words, he *also* describes an accounting process. Something has been given, generating a commitment, and that something will have to be returned, over time.

So not only can morality conceal economic interests, but above all – and this is no less important, though more often neglected – economic interests can conceal moral issues. It is not only for the sake of money, but also for the sake of revenge, i.e. to make amends for a real or alleged moral wrong, that money is so important in a divorce case; just as it is not only for the sake of money, but at least as much for the sake of excitement and risk that the wolf of Wall Street embarks on his financial adventures. When one rightly insists on the limits of economic rationality, one has to understand that what appear to be inconsistencies with an ideal rationality are precisely evidence of the fact that there is no such thing as a pure economic rationality unaffected by morality, politics and psychology, and vice versa.

2.4.4. Merit

The third function of money, the measure of value, continues and completes the perspective we have examined so far. Here is a classic fallacy that we often indulge in: gold is valuable. So, we start from gold coins, then we move on to silver

---

translation is a good compromise between possible ecological and moral values: "Whence things have their origin, there they must also pass away according to necessity; for they must pay penalty and be judged for their injustice, according to the ordinance of time" (*Philosophy in the Tragic Age of the Greeks*, Regnery Gateway, Chicago 1962, p. 45). In particular, the reference to time points to the close proximity between economic value, which only has meaning in time, and moral value: both an economic contract and a moral promise take place in time; an "IOU" without time specifications is just as nonsensical as saying "I shall redeem myself" without indicating when said redemption is supposed to take place. This perfectly honourable compromise between morality and economics is entirely absent from Heidegger's reading of the saying (*"The Anaximander Fragment"* (*Der Spruch des Anaximander*), 1946). I believe this is due to the combined effect of moral romanticism – which Heidegger, unlike Rousseau, does not attribute to the good savage but to the auroral Greek – and the stubborn opposition between poetic thought and calculating thought (almost as if triplets and hendecasyllables were not also calculus), which is also articulated as the opposition between morality and economics, between ancient and modern, between philosophy and science, and unfortunately also between Aryans and Jews. All of this takes place in roughly fifty pages in which the only thing that is clear is the intention to put every possible economic value out of play – it seems inconceivable to Heidegger, for example, that *tisis* should be translated as "penalty" and *didonai* as "to pay". This lengthy hermeneutical work produces a phrase that seems to be the product of an automatic translator: "… along the lines of usage; for they let order and thereby reck belong to one another (in the surmounting) of disorder" (M. Heidegger, *Early Greek Thinking: The Dawn of Western Philosophy*, Harper & Row Publishers, New York 1984, p. 57). Let's pretend we understand: justice is something no credit card can buy.

or bronze ones, up to paper notes, and eventually digital currency, but always with the promise that, at least in principle, one can go back to gold. But on 15 August 1971, American President Richard Nixon abolished the full convertibility of the dollar, without the latter becoming waste paper. On the other hand, and more importantly, there have been coins in history with no intrinsic value: shells, salt rendered useless for food, clay shapes. More radically, it seems quite obvious that, in order to claim that gold is valuable, one must already have coins – in this case as a means of calculation – recognising that gold is rare, which is precisely why it is valuable. Above all, there has to be an exchange system, as Winston Churchill pointed out ironically in a dialogue with a representative of Roosevelt:

A hint of Churchill's irritation with the United States might be noticeable from his only half-joking question to Hopkins at dinner the following night. He asked what the Americans would do when they had accumulated all the gold in the world and the other countries then decided that gold was of no value except for filling teeth. "Well," replied Hopkins, "we shall be able to make use of our unemployed in guarding it!"[83]

If we go back to what has been said above about computation preceding money and constituting its condition of possibility, we thus have: first *trace* as primary accumulation; then *computation* as capitalisation; and finally *money* as valorisation. It is precisely with this function that writing can contribute to the valorisation and mobilisation of capital, i.e. to the alienation of landed property in order to transform it into something else – in a sense, inscribed clay is the basis for the mobilisation of land. The emergence of the market economy, and *a fortiori* of shareholding, would not have been possible without writing – which, however, originates precisely from a sphere of document production, as a documentation of transfers, of property titles, etc. In other words, the meaning of money is distorted if we assume that its primary, if not exclusive, function is to replace barter.

What is more, barter as such has no need to be replaced by money. Money is not a refinement of barter, but the *creation of credit*. John received three sacks of grain from Dick and promised to give him a cow, which he did not have with him at the time; this formulation was fixed in writing, for instance by the tally system. This was a proto-cheque, which in turn was the origin of the banknote. Hence a consequence to which little attention is paid: the succession of metal coins → paper notes is no more fictitious than the succession of gold → money, or word → writing. Just as the genesis of writing is misunderstood by thinking that it is something imposed at a certain point for reasons of telecommunication and not of technical enhancement of hysteresis, so too the genesis of money is misunderstood by thinking that it is a *surrogate for value* rather than the *creation of value*.

---

[83] A. Roberts, *Churchill: Walking with Destiny*, Penguin, New York 2018, p. 397.

In fact, the succession that took place involved hysteresis from the outset, without it being guaranteed by any gold-value: act → document → debt.

To sum up, if you'll forgive my wordplay: counting counts, and must be taken into account. In Italian, paying in cash is called "pagare in contanti" (literally: paying in counting [items]). This expression renders very clearly the idea that the function of money is to number (think of the French *numéraire*) the value, thus determining it. It is naive to suppose that value precedes accounting, for example that gold is worth something in itself, and that it is then accounted for, and that finally, long afterwards, banknotes are invented to act as the paper counterpart of gold, which in principle can always be converted back to the latter. Obviously, this was not the case. First, there was accounting and the possibility of recording, and thus of establishing values primarily on the basis of comparisons and calculations of rarity. Then, in some civilisations and not in others, it was found that gold, due to its characteristics, i.e. malleability, resistance to chemicals and rarity, was an excellent candidate for condensing value. Value here is meant in every sense of the word, since, as we have seen, the golden calf, an indispensable element of any religion, is worth something because it is sacred and because it is made of gold, as demonstrated by the trade in gold cross necklaces, which does not seem to have suffered a setback in these alleged times of secularisation.

Thus, it would be absurd to think of settling Italy's national debt, which at the time of writing is 2300 billion euros, by selling the gold reserves held in the Bank of Italy – which are not small, constituting 10% of the world's gold reserves. This is because their worth is only 91 billion, a value that would obviously decrease considerably if one decided to put such a large amount of gold on the market.[84] This circumstance, as is well known, already occurred at the time of the great devaluation of gold following the conquest of Mexico and the consequent overabundance of gold. And it happened again in 1999, when the Bank of England had the bad idea of selling a large part of its gold, causing prices to plummet. After that, the central banks agreed not to sell their gold reserves in the future. And the value of gold, at this point, depends on one thing only, namely the document signed by the central banks.

Obviously, someone could decide to break that contract by putting the gold on the market, but even then the value of the gold would not depend on the gold, but on the market, i.e. on a system of hysteresis, capitalisation and valorisation. The metal is valuable not in itself, but because of a value that is attributed to it socially, based on a system of internal references. For the same principle, it would be bizarre to give someone the title of "archduke", "inspector", "professor" or "police officer" outside of a social and institutional network in which these positions

---

[84] S. Rossi, *Oro*, il Mulino, Bologna 2017.

have some value. To understand this, try to imagine what an archduke or a policeman might be in another solar system, or simply ask yourself whether an archduke or a policeman who were alone in the world, in the absence of the system of inscriptions and recognitions that constitutes the document capital, would really be an archduke or a policeman.

3. Speculation: Where Do We Come From?

In the first book I examined the ongoing revolution. In the second, I looked at the anthropology that made it possible, namely the complicated cooperation between souls and automata that we improperly call "human nature", and which systematically consists of a second nature. In both the first book and the second, I addressed a fundamental function, hysteresis, which is found, in the form of capitalisation, in both the technological revolution and the anthropological revelation. In order to dispel the impression that histeresis is some sort of *deus ex machina*, and to clarify what I am talking about, this third book will deal directly with hysteresis. It will do so in a way that is functional to the themes I dealt with in the first two books, and preparatory to the political development of my discourse in book four. For those who are interested in the purely metaphysical development of this theme, I refer you to my other works.[1]

Hysteresis thus constitutes the fundamental principle of what there is, i.e. *ontology*, as the permanence of substance in time through recording; of what we do, i.e. *technology*, as the iteration of the past in the present (for example through the dead labour which, in a flint, keeps track of the living labour that produced it, or the pedometer which keeps track of our physical activity); of *epistemology*, i.e. of what we know, as the alteration that transforms simple repetition, competence without understanding, into a reflexive concept; and finally of *teleology*, i.e. the interruption which, by putting an end to a process, simultaneously assigns it an end, a purpose, a meaning. The fact that every soul has an end (termination) – as we shall see in detail in 3.4.1 – is what gives souls an end (a purpose), a time and a rhythm, differentiating them from automata. This process is what draws up the sense, direction of flow, and meaning that account for the beginning, obviously not as a historical or physical necessity, but as a mere contingency – which is, however, necessary, i.e. it must take place, at least if we assume the hypothesis of meaning. Of course, we can certainly argue that life is meaningless, and indeed it is, considering that internal finality is nothing more than a lack of finality. If, however, as responsive souls, we articulate our biological destiny with its

---

[1] M. Ferraris, *Metafisica de la Web*, Dykinson, Madrid 2020; Id., *Hysteresis*, University of Edinburgh Press, Edinburgh (forthcoming).

technological supplements and the capitalisation that comes with it, then ends do arise, *en masse*: we must learn a language, we must integrate socially, we have ambitions to realise and frustrations to alleviate.

It is indisputable that all this metaphysics has an anthropic foundation, and that is why I deal with it after anthropology, and in the book on speculation, as a reflection on problems that are interesting only to us, if at all. But this does not mean that the answers, or at least the hypotheses, I will outline are arbitrary. A long time ago, Xenophanes warned us about the inappropriateness of considering the gods black, if we are Ethiopians, or blue-eyed, if we are Thracians. But, coming to us, this is also the story about every experience we have: even the smallest thing does not pass without leaving traces and, precisely for this reason, has consequences that can be referred back to the general principle of hysteresis. This anthropic trait should never be forgotten, but hysteresis is an omnipresent notion or (to put it better) function: it transcends all genera.[2] Indeed, in the case of the categories I use in this book, hysteresis crosses over into ontology (recording), technology (iteration), epistemology (alteration) and teleology (interruption). To put it in more traditional terms, we have being, nothingness, and becoming, to which I would add a fourth decisive element: ending as the destiny of all

---

[2] If you like, this is a transcendental notion in the classical sense, which relates to what Paul (2 Thess 2:6–7) defines as κατέχον, the obstacle, in the neutral, or κατέχων, the hinderer, in the masculine. It is the something or someone that holds back the Antichrist. When it or he is taken out of the way, then the Antichrist will unfold all his power, which for Paul is a good thing, because it will hasten the second coming of Christ. Closer to us, this is what Derrida expressed with the difficult concept of *différance*. Indeed, I have long wondered why it was necessary to join writing and difference as suggested by Derrida in a famous book; similarly, I have long found his reflections on *différance*, with an a, to be decidedly cryptic (J. Derrida, *Writing and Difference*, [1967] University of Chicago Press, Chicago 1978; Id., "Différance" [1968] in *Margins of Philosophy*, Chicago University Press, Chicago 1982). I believe that the very example of globalisation (and all that it contains) can illustrate very clearly, even banally, the reason for that somewhat disturbing pair, providing a concrete instance of the notion of *différance*, which I propose to call "differing", as indicating both diversity and (with a little effort) the act of deferring or postponing. To put it simply: what appears different ontologically, be it life or death, day or night, activity or passivity, is simply deferred chronologically: I am now alive, but my life is only a deferred death; it is now day, but day is only a deferred night; I am now proposing theses that have been actively worked out, but these, as the footnotes show, are the result of passively learned notions. Ontological difference is but a chronological deferment. This is a speculative thesis indebted to the Hegelian dialectic, and at the same time it is a common-sense consideration, which leads us to say that truth is the daughter of time. Let's give time to time (13.7 billion years), and space to space, let hysteresis run its course, and everything will eventually come out: this alone explains the omnipotence of imbecility (certainly not intelligent design!), the Black Death and the Lisbon earthquake, sodium chloride and Cromwell, the battle of Poltava and *The Thousand and One Nights*.

things. The three musketeers (recording, iteration, alteration) are thus four,[3] and the system, to be complete, needs a fourth element: interruption or teleology, which is the essence of speculation. Teleology, having a purpose, is the outcome of interruption: the fact that a process comes to an end. Interruption means crisis, the cessation of a system of recording, iteration and alteration. But it should not be conceived as mere negativity, because it is the condition of possibility of meaning and freedom.

This is an unusual addition in metaphysics, but one that I consider indispensable. While the merits of recording, iteration and alteration are indisputable, interruption is equally indispensable. And this not only in the mythical sense, where sooner or later Atropus will cut the thread of life that Clotho has spun and Lachesis has woven, and for Shiva to destroy what Brahma has created and Vishnu has preserved, but in the world. Let us try to imagine an endless (infinite) life. Its most interesting aspect would be that it would be a life without end (purpose). For one thing, the precept of not putting off until tomorrow what you could do today, which is based on the trivial fact that tomorrow you could be dead, would lose all meaning. The same fate would befall all calls for urgency, as well as the stimuli of metabolism and the signs of fatigue that remind us how a looming interruption gives meaning to life but, more profoundly, gives meaning to meaning. But let's not linger on the threshold and enter the exposition of the four pillars that make up the metaphysics of hysteresis.

3.1. Ontology: Recording

The first stage is ontology, or being. I am writing these words while sitting on a couch. I perceive its fabric, I see its colour, but beyond all these sensible phenomena, the couch holds me up, just as it would hold up my cat and it would fail to hold up an elephant. This support phenomenally presents itself to me with a kind of variation of touch, or rather with an explication of the two elements that are confused in touch, which are, on the one hand, the possibility of perceiving heat, cold, wetness, roughness, and, on the other, the fact of perceiving resistance. This resistance, for example the fact that I cannot cross the couch the same way I can cross the room, is evidence that the couch exists separately from my perceptual organs and conceptual schemes. The couch would continue to hold me if I fell asleep (in fact, that's one of the reasons I bought it) or if I were dead. What I think or know about the couch is completely irrelevant to the intrinsic

---

[3] R. Brandt, *D'Artagnan und die Urteilstafel. Über ein Ordnungsprinzip der europäischen Kulturgeschichte (1, 2, 3/4) (1991)*, Deutscher Taschenbuch Verlag, München 1998.

characteristics of the couch,[4] and claiming otherwise means confusing existence and knowledge in what I have called the "transcendental fallacy".[5] This sounds like mere common sense, but it is not so. When Kant describes the supreme principle of the intellect as the possibility that phenomena may fall within the synthetic unity of apperception, he is referring to the circumstance that knowledge presupposes hysteresis. The synthetic unity of apperception expresses an inter-

---

[4] Ontology is the doctrine of being, i.e. of what exists independently of what we know or believe. Its only condition of possibility, its principle of reason, is precisely hysteresis: permanence in time and space. Since permanence is the primary character of being, it is not surprising that its characteristic ontological function is recording, which brings into being what would otherwise be becoming. Now, what is recorded no longer depends on us, on our knowledge or ignorance, or on our will to power. It just is, and it manifests itself as resistance, that is, as unamendability (I refer you to my *Il mondo esterno*, Bompiani, Milan 2001, pp. 97 ff.). The fact that, for example, thought is unable to amend perceptual illusions does not mean that everything that appears is true, but rather that knowledge is unable to intervene on the plane of being, and that the latter is therefore independent of the former. Through our contact with the external world we do not know *what* the world is (*quid sit*), but rather *that* the world is (*quod sit*), according to Schelling's formulation: "in everything that is real there are two things to be known: it is two entirely different things to know what a being is, *quid sit*, and that it is, *quod sit*. The former – the answer to the question *what* it is – accords me insight into the *essence* of the thing, or it provides that I understand the thing [...]. The other insight however, *that* it is, does not accord me just the concept,but rather something that goes beyond just the concept, which is existence" (cf. F.W.J. Schelling, *Philosophy of Revelation*, 95, SW II/3, in *The Grounding of Positive Philosophy* ed. and trans. B. Matthews, SUNY Press, Albany 2007, p. 129). And, even without Schelling, the most obvious experience we have of this being is that it is more pigheaded than we are. A colour changes according to lighting conditions and proximity to another colour. On the one hand, this is something that happens at the mesoscopic level, which shows that quantum mechanics is not the only level at play. On the other hand, these colour mutations do not change the ontological consistency of the object. That I can experience an object, which undoubtedly happens, and that that object exists even without my experience of it, which is admitted by anyone who is not a correlationist or a speculative quantum relationist, are things that require permanence. In the case of my experience of the object, this permanence is the foundation of epistemology, while in the case of the existence of the object it is the foundation of ontology. Without epistemology there would still be lakes, mountains, epistemologists (in the sense of living beings) and odd numbers, whereas without being there would be no form of knowledge, which is always knowing something about something, *ti kata tinos*. This difference and precedence of being with respect to knowledge represents a material apriori stronger than any conceptual apriori, and is just as material, and no less apriori, than the axiom that there is no colour without extension. For if knowledge did not refer to something other than and prior to itself, then the words "subject", "object", "epistemology", "ontology", "knowledge" and "reflection" would have no meaning, or rather would be inexplicable synonyms. If this dependence of objects on subjects were true, there would be no difference between the dreams of metaphysical visionaries and physics, and then there would be no serious reason to put Newton before Manitou.

[5] M. Ferraris, *Goodbye, Kant!: What Still Stands of the Critique of Pure Reason*, SUNY Press, Albany 2014.

vention of the subject, not unlike the representations that – throughout the history of philosophy – depict the mind as a writing surface. The permanence and, at the same time, the becoming of the world are guaranteed by the subject (*cogito*, the I think), which becomes the invariant receptacle of all transformations, so that objects exist only in relation to subjects.

The solution is clear and linear, since we seem to have immediate access to our ego. Also, the very common experience of memory teaches us that past experiences can be recalled, and it seems intuitively obvious (let us leave aside whether this is true or not) that the sum of all those experiences constitutes the continuity of our ego. The price to pay, however, is high, since this perspective does not allow us to conceive of anything existing before the appearance of subjects, and specifically of historically situated subjects endowed with certain characteristics. Strictly speaking, its consistent outcome would be solipsism, a solution that is discarded for reasons of common sense. One of the very few things that we can assert with far greater certainty than *Ego cogito, ergo sum* is that Tyrannosaurus Rex never knew it was called that, which nevertheless did not prevent it from having all the properties it had. Likewise, to come to today's world, we have a good idea that the rabbit-duck contains both a rabbit and a duck, yet we can only see one at a time: the eye is not the docile servant of the brain, and this is the least of it. The most important thing is that this small resistance of perception heralds the enormous resistance of the world, which was there before us and will remain after us. This resistance, which is independent existence, manifests the first and most powerful ontological attribute of hysteresis, far more robust than our ego's weak syntheses on which the world is supposed to depend, according to modern philosophy.

3.1.1. Reflection

What applies to my couch – that's the speculative hypothesis – applies to everything else. The principle of sufficient reason for both the natural cosmos and the social world and individual psychology thus goes *nihil est sine hysteresis*: nothing exists without hysteresis. As black holes show, there can be recording without matter, but not matter without recording. Now, the question "why is it that something exists rather than nothing?" risks being futile, because it was formulated at a time when the search for a root cause of the universe appeared to be the essential task of science, philosophy and theology. But if we turn it into "why is it that something *persists* rather than nothing?" we do not even need the question mark. Everything that is exists because it is recorded, because it maintains a primordial level of hysteresis, whereby the universe derived from an original recording, from matter that was nothing but memory.

The reference to the couch is not dictated by laziness, but represents a precise methodological option. If the science of the 20th century took its cue from what by definition transcends the limits of experience and sought to describe the characteristics of experience through it, now we must follow a different, indeed inverse path: starting from something that is part of our everyday experience, enhanced by the technological revolution, we must go back from the given to its condition of possibility, a condition that transcends the possibilities of experience but does not contradict it. We must start from what is known to search for its conditions of possibility, with the awareness that what we are engaging in is reflection, or, if you prefer, abduction – even if, not being Sherlock Holmes, we have no guarantee of figuring out who did it. For philosophy, as for any other form of knowledge, the path of speculation must become once more practicable and necessary, as the only way not to hypostasise technology or absolutise anthropology. The revelations of technology follow an anthropic principle, but we can go further, which is why, starting from "*tout comme chez nous*",[6] we can speculatively reach something that is outside of us and before us – although, of course, it only makes sense to us.

In order to clarify this point, and since we are talking about ontology, I would like to exemplify the meaning of speculation precisely with reference to the question of being. The definition of being as permanence, illustrated by the couch as an entity, is undoubtedly dominated by an anthropic principle, because it is not clear what a couch could be for a proton. However, the fact that the couch makes sense to me and not to a proton in no way implies that the being of the couch depends on me. It is here that the path of speculation opens up as an alternative to the correlation of the last two centuries.[7] If correlationism asserts that being exists only in correlation with a knowing subject, speculation takes the diametrically opposite route. Being is certainly what I experience in every moment of my life, but it is, as such, what was before me and what will be after me, in a world of which I will never have direct experience for the very good reason that I was not yet there or will no longer be there.

I can talk about my couch to the point of maybe falling asleep, dreaming about the Big Bang or dinosaurs (depending on what I ate). These are all things I have no direct experience of – except, in the case of dinosaurs, for some bones I saw

---

[6] "C'est partout et toujours comme ici, c'est partout et toujours comme chez nous, aux degrés de grandeur et de perfection près" (Leibniz, *New Essays on Human Understanding* cit., I, I).

[7] Determining the so-called "correlationism", the idea that objects exist only in correlation with subjects. The term correlationism is found in Q. Meillassoux, *After Finitude: An Essay on the Necessity of Contingency*, Continuum, London 2008; but for a much earlier discussion cf. N. Hartmann, *Grundzüge einer Metaphysik der Erkenntnis*, Vereinigung wissenschaftlicher Verleger, Berlin 1921.

in a museum. I can also indulge in daydreaming and imagine the Earth four million years ago: all of this is fantastic archaeology, not very different from real archaeology, since what we know about those remote epochs is very little, and never derives from direct experience. Now, let us think about it: what is that archaeology oriented by? By teleology: it is I, as a mortal, and therefore as someone who is engaged in a short course of existence, who project myself forward, towards the moment when I will no longer be here, and backwards, to mysteries of when I was not yet here. This speculation is the reason why I do not assert the axiom "I think, therefore I am" (after all, an idle thought, because it has no practical consequence on my life or on the lives of others), but rather the principle that guides all my conduct towards the world: *being is*. And this is not because I perceive it at this moment (even if its resistance means something), but because I know, with a strange certainty that is not merely empirical, that being was before me and will be after me, and that for this very reason it is of interest to me and to others.

In the case of ontology, perhaps more than in any other, the archaeological question of "Where do we come from?" is linked to the anthropological question of "Who are we?" and the teleological question of "Where are we going ?". Only a human, as far as we know, and only one who has solved a large number of more pressing problems, can ask the question "Where do we come from?". They may even hope, over a very long time, to reach an objective and truthful answer, so much so that some answers, such as "we come from a cabbage patch" or "God created us three minutes ago with all our memories", seem less likely than "we come from a very long evolutionary history, which would not have taken place without hysteresis". But it remains that the question is one asked by humans wondering about their origin (or, more radically, it is a question that only a human could ask); and therefore the answer can only be speculative.

I wish to reiterate philosophy's right to speculate on this link between archaeology, teleology and anthropology. The science of the 20th century, as is natural and right, has advanced ambitious and controversial theories of everything, which contrast internally and with common sense (think of quantum mechanics) but which work, i.e., are technologically efficient. This technological efficacy is what resolves the conflict between the scientific image of the world and that of common experience. The table at which I am sitting is only slightly denser, atomically, than the surrounding air, yet for me there is an essential difference between the table and the air[8] and, above all, any scepticism about the existence of the table is overcome by the fact that the table is adequate for my use.[9] But hys-

---

[8] Cf. A.S. Eddington, *The Nature of the Physical World,* Macmillan, New York 1928.

[9] The same can be said of the controversy between phenomena and things in themselves,

teresis provides the principle that holds these two apparently antithetical experiences together and allows us to speak of a single reality, instead of various facets and levels, none of which can aspire to the claim of being ultimately real. For example, the words you are reading would not have been possible without a very long evolution that presupposes capitalisation and transformation, and yet the way you and I are led to interpret them is that these words are the product of my thought, that they merely express it, and that my computer is a simple instrument I use to record and transmit my ideas.

I believe that philosophy, precisely on the basis of equal opportunities, also has a right to speculate, though it should be handled differently. Science starts from the invisible in order to speculate on the visible, relying on the technological efficacy of its hypotheses. Philosophy, for its part, can start from the visible, from everyday experience, and in particular from a technology – the Web, which

---

which is dissolved at the moment of action, i.e. with technological performativity. When in high school they teach us that we only know phenomena and not things in themselves, after our initial perplexity this view seems obvious to us: the table I have under my eyes is white, but if I turned off the light it would be black; I like what I am eating, but if I were sick I would not like it. In short, we see things only as they appear to us, and we never have total or internal access to them. At this point, there are only phenomena and, as Schopenhauer put it, if a reasonable being reflects a little bit on it, they can only come to one conclusion: the world is my representation. But the question is: has this recourse to phenomena and representations really opened our eyes, or has it closed them instead? (cf. M. Ferraris, *Ding an sich*, in *Das neue Bedürfnis nach Metaphysik / The New Desire for Metaphysics*, ed. by M. Gabriel, W. Hogrebe and A. Speer, De Gruyter, Berlin-Boston 2015, pp. 119–132.) Let's consider a trivial but enlightening circumstance. As a paradigmatic representation, Schopenhauer proposes an eye looking at the sun, and seems to forget that a sun looked at is different from a merely represented sun: in particular, you can look at the sun only for an instant, while you can represent it as long as you want. Once it is clarified (which is not so simple) that the phenomenon of the sun is not the remembered or imagined representation of it, but the sun we are looking at right now (and only now!), many problems ensue. In fact, Schopenhauer speaks of an eye that looks at the sun and a hand that feels the earth. The use of the singular is not accidental, since if there were two eyes and two hands, as usually happens, the following doubt could arise: the fact that the sun and the earth give the *same* sensation to both eyes and both hands may depend on the fact that the relationship between the two poles of the representation, the subject and the object, is not perfectly balanced but leans towards the object. This doubt would become even more substantial if the one-eyed observer found themselves not only having two eyes, but in the company of other observers with whom to share their impressions and doubts. Further perplexities would come if the immobile observer switched to a dynamic situation, in which the objects (excluding the sun, of course) were manageable, and even better if these operations had no cognitive purposes but, as usually happens, were simple interactions with the environment. In fact, the appeal to phenomena and to the world as a representation, which works so well when we are the solitary and immobile observers of inanimate objects, loses much of its persuasive force when we are with others, in motion, and especially when objects are not static or we do not have a purely cognitive relationship with them.

has proved to be particularly efficient thanks to the power of hysteresis – to formulate a speculative hypothesis. Starting from what we know through the magnifying glass of the documentedia revolution, philosophy can postulate a unitary explanation for the common foundation of what exists (ontology), of what is iterated (technology), of what is altered (epistemology) and of what is interrupted (teleology).[10] Let me now continue my exploration, in accordance with the fate

---

[10] To those who would argue that magic can be performative, I would reply as follows: focusing on technology in ontology is not a renunciation of knowledge, but a way of taking it seriously. Simply put, the confidence – if not the credulity – with which I accept that the sheet of paper in front of me is a phenomenon changes completely if instead of the sheet there is a hungry tiger, but also, without going too far, if the sheet is a subpoena. Scepticism and contemplation give way to action, and the perspective changes. In fact, we discover that our relationship with the world focuses primarily on the interaction with artefacts, social objects, and people with respect to which the notion of "phenomenon" appears problematic. Is a corkscrew a phenomenon? You may believe it if you wish. Are our friends and relatives phenomena? It depends, but generally it is better if they are not. Is our troublesome colleague a phenomenon? We wish they were. If we like to think of a world of phenomena, we are welcome to do so, but only if we are able to explain what would change in our attitude and especially in our behaviour if instead of phenomena they were things in themselves. Everything is fine as long as it works, as long as it allows us to launch a rocket to the moon or cure a disease. Otherwise, nothing goes. But if this is the case, why don't we take this performativity as an example and build a metaphysics that ties in with physics, not on the controversial ground of foundations, but on the actual ground of operational possibilities? After all, from this point of view, the Web is a unique opportunity and a step towards a technology of everything, which is the general concept of hysteresis. Insofar as it can no longer claim to be the ultimate foundation of knowledge of reality, physics remains unchallenged in its technological presupposition, i.e. the mathematisation of the world: it is not the best conceptual scheme, but rather the most useful technical tool at our disposal. On the basis of what I have just said, I would like to make a point which, it seems to me, is not usually given the attention it deserves. Is epistemology primarily physics? Is the science of inorganic nature the yardstick of all cogent knowledge? Obviously not. If by "cogent" one means something obligatory and inescapable, it is totally false that "science", even in the vaguest sense, is the exclusive measure of cogent knowledge. The fact that Madame Bovary had a lover is cogent, much more so than any physical law (which can always be called into question by placing it in a more general framework). An infinitely more cogent fact than phlogiston is that Finno-Ugric languages are agglutinative, that in German the verb goes at the end of the sentence, that in French and English it is compulsory to make the subject explicit while in Italian and Latin it can be implied. Just as it is cogent that a sentence such as "Green is o" has no meaning, or that "I promise you that" is not a promise because it lacks an object. And all our scientific knowledge may change, but what can never change is a very simple law such as "there is no colour without extension". Conversely, is knowing that the Earth revolves around the Sun cogent? No, because humankind has been able to live very well and achieve high degrees of civilisation even in the age of geocentrism. Almost none of the laws of physics, except gravity, are cogent in the usual sense of the term, because we hardly ever find ourselves transgressing them, for the simple reason that they do not intervene in our lives. And even before we knew the law of gravity, people who threw a stone up in the sky would dodge it on its way down, perhaps explaining it with the law of natural places, or more likely not explaining it

of reason as a faculty burdened with questions that require going beyond what can be directly experienced.[11]

3.1.2. Independence

As the experience of my couch and, I imagine, your couch shows, the first phenomenological character of being is precisely resistance: the independence from the ego that I have called "unamendability".[12] This independence, denied by so many philosophers, is, once again, a mere matter of common sense. The only way in which the ego can really claim to give its own law to the universe is by positing *causal dependence*, whereby *the sufficient condition for the existence of X is its causation by a subject*.[13] Now, it is absurd to assume that only what is actually present to my thinking is real – which makes everything that is not real fall into unreality, starting with the theory of causal dependence and its proponents. If it were true that there is only what is actually present to the thought of a thinking subject, then there would be no difference between introspection and knowledge of the external world; all things past, from dinosaurs to Sumerians, would be present in exactly the same way as the thoughts that think them; all future things would be no less present than past things, and therefore the difference between possible and real would disappear; all things that the theorist of causal dependence ignores – and there is reason to believe that there are many – would be non-existent; on the other hand, everything that the theorist of causal dependence thinks of would exist, although only at the exact moment in which they think of it.[14]

---

at all and not even considering the problem. Obviously, the laws of physics are cogent when it comes to physics: I cannot claim with impunity that the Sun actually revolves around the Earth. But even on this point, a hermeneutic might skilfully point out that it is not a question of discussing the laws of physics, which no one has ever dreamed of questioning (let us rather say: no one has dreamed of questioning as long as they could – Galileo knew something about this), but simply of demonstrating that these laws have no bearing on our lives, that what really matters are death and taxes, two dimensions on which physics is ultimately not cogent.

[11] Kant, *Critique of Pure Reason* cit., A vii.

[12] Ferraris, *Il mondo esterno* cit.

[13] The only contemporary philosopher, to my knowledge, who subscribed to an uncompromising idealism, i.e. an effective and unambiguous dependence of ontology on epistemology, was John Foster (1941–2009). See his remarkable *A World for Us. The Case for Phenomenalistic Idealism*, Oxford University Press, Oxford 2008.

[14] Today quantum mechanics, or at least some of its philosophically motivated interpretations (C. Rovelli, *Helgoland*, Adelphi, Milan 2020), reproduces correlationism in the form of relationalism, which in philosophy had been proposed as a minor variant of correlationism by authors dealing with the philosophy of nature (cf. esp. A.N. Whitehead, *Process and Reality. An Essay in Cosmology. Gifford Lectures Delivered in the University of Edinburgh During the*

In order to avoid these contradictions, constructivists, who deny the independence of being from thought, turn to *conceptual dependence*, according to which *the necessary but not sufficient condition for the existence of X is its conceptualisation by a subject.* This thesis is one of the possible outcomes of Kant's saying:

Session 1927–1928, Macmillan, New York, Cambridge University Press, Cambridge 1929, for whom relations precede relata, which appears much more metaphysically sustainable than quantum relationalism, and is perfectly in line with the metaphysics of hysteresis that I am proposing). From the ontological point of view, unlike correlationism, quantum relationalism claims that objects exist only in relation to other objects; but from the epistemological point of view, as in correlationism, it claims that objects exist only in relation to the kinds of objects called subjects. Therefore for quantum relationalism, just as in correlationism, reality exists only in the relation between a subject and an object. So, just as in correlationism, when I do not see you and do not think of you, dear reader, you do not exist, and vice versa: if there is no existence without relation, there is obviously no relation without correlation. Now, there are five points to be made here. First of all, quantum relationalism, which conceives of existence as a reciprocal action (a category that was already envisioned by Newton without drawing the nihilistic conclusions of quantum relationalism) is from this point of view a simple reproposal of traditional correlationism. According to the latter, objects exist only in relation to subjects that observe them, whereas for quantum relationalism objects exist only in relation to other objects. If correlationism leads to obviously false statements, relationalism leads to the same difficulties, only less blatantly. Moreover, if what has been said so far is true, it is all but certain that these are novelties and difficulties brought about by quantum physics, whereby objects are nothing but relations. In this respect too, relationalism is but a version of correlationism, with which it shares all its weaknesses, and to which it adds a certain philosophical naivety. In particular, one does not need quantum mechanics to conclude that matter is a bundle of relations: Locke had figured it out in 1690, paving the way to what was soon to be Berkeley's radical idealism. Secondly, if it is argued that an object unrelated to another is the same as if it did not exist, then the existence of the object depends not only on correlation, but on knowledge of correlation, and discovering a new galaxy would be tantamount to creating it. Yet, after all, this is a lesser evil than the idea that America did not exist before it was discovered, from the Eurocentric point of view, while from the Native American point of view Europe, Africa, Asia, not to mention Australia, did not exist until the colonisers' arrival. Not to mention that, along with America, all the things we do not interact with, or *think we do not interact* with, would not exist. The obvious result of quantum relationalism then is that only what we know exists, and that what we do know does not actually exist, so we know nothing. If instead of "interact" you say "perceive" you'll have Berkeley's thought. If this is the case, then relationalism faces exactly the same problems as correlationism, only less obviously. "Relation", after all, is a less demanding word than "knowledge", but – if we leave the world of quanta and come to the world of tables and chairs – how do we know that something interacts with us? It may very well not interact with us at all, without us knowing it. This would change nothing in some cases, and everything in others (this is very clear in the human world: whether or not I know that my partner is cheating on me changes a great deal in the relationship). Thirdly, what is said about quantum relationalism is valid for the microscopic world, but not for the mesoscopic world in which our experience takes place, and in which, if quantum relationalism were to be asserted, we would run into manifest contradictions. The shift from the microscopic to the mesoscopic is a philosophical error that physicists sometimes make when speaking speculatively. To claim

"Intuitions without a concept are blind".[15] The latter, however, can be interpreted in two ways: in the weak form, whereby without the concept of "dog" we would not recognise a dog if we saw one, and in the strong form, whereby without the concept of "dog" we would not even see a dog if we were faced with one. When defending Kant, it is said that he meant the weak form: concepts reconstruct experience in general. But if he had really meant the weak form, his philosophy would only have been an epistemology, and not also an ontology, called upon to establish an anti-skeptical foundation not only for the possibility of knowledge, but also for the objective certainty of the known. If transcendental philosophy thought it had answered the sceptical objection convincingly, it was because it understood concepts to constitute experience in general. Defenders of conceptual dependence thus find themselves in an impasse. Strong conceptual dependence is traceable to causal dependence, and is subject to the same criticism. As for weak conceptual dependence, it is not a dependence at all. It is, in fact, trivially false to say that a dog depends on our conceptual schemata, just as it is trivially

---

that the world of quantum relationalism is our world – that is, the world we experience – is not very different from claiming that when we are thirsty we can first inhale hydrogen, then oxygen, and our thirst will be quenched. The fundamental elements do not equal the result, and our world is not at all the quantum world, so much so that the quantum world appears paradoxical and fantastic to us. Our world is nothing like that, and we do not experience quanta any more than we experience ghosts. As Schopenhauer said, life and dream are pages in the same book, only when awake we read it in a linear fashion, and when we dream we jump from one page to another. In other words: it is all too obvious that the subject experiences an object if and only if they experience an object; this is a solid tautology. But it is absurd to claim that the object exists only if the subject knows or experiences it. Fourth, quantum relationalism refers to the observation of objects in the presence and in the present, and does not ask any questions about the past and its existence and action in the present. This temporal limitation (which is initially an advantage, because it hides the difficulties, but in the end is a serious disadvantage, because it seriously limits the scope of the theory) has dramatic consequences for the possibility of making any sense of the world we live in and the way we live in it. Fifthly and lastly, since it is not a philosopher but a physicist who professes quantum relationalism, and since Bellarmine died long ago, his authority seems incontestable. Here too, however, a serious problem arises. The superiority of physics over metaphysics is said to lie in the fact that physics is based on sensory experience and metaphysics is not. If, however, objects only exist in relation to subjects, then there is no such thing as a meaningful experience, since there is no way of distinguishing dream from reality.

[15] In recent Kantian scholarship – and following the return of realism – there have been several non-conceptualist and realist readings of Kant (cf. for example R. Hanna *et al.*, *In Defense of Intuitions: A New Rationalist Manifesto*, Palgrave Macmillan, London 2013 and L. Allais, *Manifest Reality: Kant's Idealism and His Realism*, Oxford University Press, Oxford 2015). At last. However, it is worth noting the reason why the previous two and a half centuries worth of scholarship interpreted Kant in a conceptualist and idealist sense.

true to say that the *word* "dog" depends not on our conceptual schemata, but on our linguistic skills, which incidentally may be incorrect and are often sketchy.

So what is to be done? All that is left is to transform conceptual dependence into *representational dependence*, i.e. a weak conceptual dependence whereby *the necessary but not sufficient condition for the existence of X is its representation by a subject*. Compared to conceptual dependence, representational dependence has the characteristic of being programmatically vague. While conceptual dependence argues that intuitions depend on concepts, representational dependence suggests that our vocabularies exert some influence on the external world. In spite of this, the difficulties of conceptual dependence remain. For either "representational dependence" means that the name "dog" depends on us, and then it is not a dependence in any serious sense of the word; or it means that the being of dogs depends on us, and that is not philosophy, but magic.[16]

Indeed, here's a note on magic. I have no problem picturing a physicist reading a horoscope. But suppose this same physicist at some point has a severe pain in his left arm. There can be no doubt about the fact that he has a pain (if he really has it, of course), and even if, with some brain-imaging technique, it is shown that his pain centres are not being stimulated, well, we would be forced to admit that the superstitious physicist has a pain, for the good reason that he feels it. This and nothing else, I repeat, is the role I attribute to sensory experience: it is a measure of unamendability and cogency, not a measure of scientific truth. At this point, turning to epistemology, the superstitious physicist might ask himself whether his is just a pain in his left arm and not the symptom of a heart attack. If, as is to be hoped, he decides to have himself examined, there are two possible outcomes: it either was a heart attack or it was not. And unlike the brain-imaging example, here we can give an answer. If there are traces of a heart attack, then it was a heart attack. If there are no traces of a heart attack, then it was not a heart attack, but a pain in the left arm. Now, herein lies the cogency of medicine, which I would consider to be the paradigm of the natural sciences, constituting the example of an action that can, though not necessarily must, lead to knowledge – a knowledge that is extremely fallible and highly problematic (do we really know

---

[16] Rorty (*Nineteenth-Century Idealism and Twentieth-Century Textualism*, in "The Monist", 64, 2, 1981, pp. 155–174) pointed out the similarities between 19th-century idealism and 20th-century postmodernism: after all, both posit we have no direct access to reality, where the former thought it was mediated by spirit, and the latter by language. However, there is a fundamental difference between the two idealisms. 19th-century idealism played it straight: time, space and objects do not really exist, there is only the thought that thinks of them and makes them possible. In contrast, postmodernism follows a very different strategy, that of representational dependence (cf. R. Rorty, *Charles Taylor on Truth*, in Id., *Philosophical Papers*, vol. III, *Truth and Progress*, Cambridge University Press, Cambridge 1998, pp. 84–97), i.e. a pseudo-dependence.

why antidepressants work?), yet structurally linked to a connection between ontology and epistemology.

3.1.3. Permanence

As seen in 1.2, the social world shows that recording constitutes objects, and, as I was saying, the Web has amplified this law on a gigantic scale. And what about the natural world? Experience gives us evidence that does not contradict the role of recording in the constitution of objects in general: if there were no permanence in time, therefore a continuation of the effects even after their causes have disappeared (in this case, the carpenter's work), the table in front of me could fall apart. This is why the table is solid and has its specific properties. Is this enough to say that recording ontologically constitutes the table? Of course not. One still needs wood and a carpenter. Let's take a step back, and go back from the condition (the table) to the unconditional, or rather to its condition of possibility.

Another philosopher who enjoyed talking about tables, in his *Timaeus*, describes the action of a demiurge (a craftsman) who makes the world starting from ideas: ideas are the form, which is added to matter so as to give life to objects, and for that to happen there needs to be hysteresis. "Do you mean then, despite what you claimed at the beginning, that there is nothing outside the text? Is this what you mean by saying that there is nothing without hysteresis, and that nothingness itself, i.e. the end of all things, is an effect of hysteresis?" Obviously not. A text must have a meaning, express some kind of intelligence, have an author – otherwise the word "text" is just a metaphor. Instead, in hysteresis there is no author, no meaning, no intelligence, but only a great deal of time and space which, at certain moments and in certain areas, brought out meanings, intelligences, and even authors. Whether these authors have written sensible or intelligent things is up to the reader to decide, but that is another story.

In this sense, hysteresis is nothing more than a generalisation of the constitutive law of social objects O = RA (Object = Recorded Act), although, of course, the latter is to be regarded as a particularisation of hysteresis in the social world. It is necessary to keep track of events even after their causes have disappeared, and it is necessary to have a cycle whereby the sedimentation of the past produces new objects, which develop intentions, which in turn sediment into other objects, and so on, in an indefinite but not infinite process, since entropy will inevitably prevail in the end. In general, we are dealing with a large archive in which traces are preserved, and which allows for both genesis, whereby something originates and develops, and structure, whereby what has developed takes a form, relates to other developments, and can, in a long or short time, give rise to meaning.

3.1.4. Emergence

Moving from the couch or table to the Big Bang, let us now proceed to speculative generalisation, renouncing the comfort of common sense or at least of experience. There is nothing without recording, and even what appears to us as absolute nothingness, like black holes, are only a kind of recording so powerful that it does not let being manifest itself. When the recording weakens and is extroverted, we have a Big Bang, either once or many times, if we follow the Big Bounce hypothesis.[17]

First there was the Big Bag, at the end there will be thermal equilibrium, and in the middle there is time, which speaks of a growing tendency towards disorder and heat loss; but at each of these moments hysteresis is present as the possibility of keeping track of previous conditions. This, in the mesoscopic terms of our ordinary experience, defines the spatiotemporal characteristics of everyday objects: if the bookcase behind me suddenly disappeared, I would have good reason to believe that I was the victim of a hallucination, whether positive or negative. Therefore, the Big Bang hypothesis implies the retention of a trace, and a retention faculty that must have been present at the time of the explosion: this is the only way to explain why 10–6 seconds after the Big Bang the quarks joined three by three to form protons and neutrons. Hence the building blocks of the universe. The only precondition for this to happen is precisely the possibility of retention, and here the speculations of physics intersect those of metaphysics.

So, the question "Why is there something rather than nothing?" finds its answer precisely in hysteresis: there is something because the past is remembered by memory and repeated by matter.[18] A bent piece of iron recalls the action it has undergone in the best possible way[19]: memory manifests itself as time and matter as space, but the two poles are strictly interdependent, since we cannot conceive of time except as movement in space, and of space except as temporal presence.

---

[17] The big bounce is a re-proposition of the *tzim tzum* (צמצום), which in Jewish theology designates the self-limitation of God withdrawing after creating the world. This image is particularly significant because strictly speaking it is this retraction, the creation of an empty space (Khalal/Khalal Hapanoi, חללהפנוי) – that is, once again, of a *chora* (Ha-Makom, המקום, literally "the Place", is one of the names of God in rabbinic literature): in short, it is the work of hysteresis that gives life to the world. Creation, thus, is retraction and hysteresis. The Big Bang alone would have been a pure, timeless glow: it is only by retention that everything begins, which means that everything begins with a retention, and everything proceeds by successive retractions and retentions, in an image that coincides with the Neo-Platonic doctrine of hypostasis.

[18] H. Bergson, *Matière et mémoire. Essai sur la relation du corps à l'esprit* (1896), PUF, Paris 1965, chap. IV.

[19] E. Mach, *The Analysis of Sensations and the Relation of the Physical to the Mental* (1903), Open Court, Chicago and London 1914, p. 17.

It's a bit like Proust's image in the last pages of *La Recherche*: the guests in the Guermantes living room are magnified, as it were, because the small space they occupy in the room must be integrated by the enormous lifetime embedded in them, which turns them into dolls immersed in the colors of the time and giants walking on the stilts of the years past.[20]

If we start from the hypothesis that matter is made up of memory, both phenomena and the metaphysical foundation are preserved in a non-reductionist monism. That the past is repeated in and by matter means that there can be no inscription without a material medium, and that every material medium bears the traces of this inscription and is, in its essence, an inscription itself – the memory of its genesis. Conversely, and in the light of what has been said so far, the most important function of the universe and of life is emergence. Given a sufficient amount of time and space, complex structures can emerge from simpler ones; or, if you prefer, more superficial and recent structures can emerge from deeper or older ones. Thought emerged from action, the world from the Big Bang, society from animal life forms that predate humanisation. Is there still a need to postulate some intelligent design to account for a world that owes its very contingent emergence to an indefinite, yet not infinite, wealth of time, matter and energy? And conversely, does the fact that there is no meaning at the beginning imply that there can be no meaning in the instant of the end, of every individual end as well as in the end of all things?

The universe is therefore the reminder of the Big Bang, and every entity since the Big Bang bears in its structure the traces of its genesis. This space-time surplus is the best theodicy, and solves a problem that is as old as the *Ecclesiastes*: if the world is so perfectly ordered, there must be a Creator (God) or a Constructor (I). But if this order includes evil, we must conclude that either God is not the creator or he is evil; or else that the I is trivially masochistic. But things are otherwise. This perfection (provided it is such, given that we have no terms of comparison) is explained by the availability of time and material. This same availability, on the other hand, accounts for evil, since we have by no means reached the end of the process. Strictly speaking, it also accounts for God, but puts him at the end of evolution: God is not yet, and hopefully was not at the beginning, otherwise he would be responsible for far too much. Still, one cannot rule out that, just as amoebas and Tyrolean choirs, the pyramid of Cheops and VAT, Klee and garden gnomes, the day will come, maybe in a million years, when a very evolved

---

[20] M. Proust, *In Search of Lost Time*, vol. IV, Time Regained, New York, Modern Library 1993, p. 532.

colony of termites will come across a god. "As if You would truly come again / to judge the living and the dead. / And how many tears and semen shed in vain."[21]

3.2. Technology: Iteration

We now come to the second phase of hysteresis. At this point, we can ask ourselves where nothingness is, and we will find it where we least expect it: in iteration, in the repetition without difference that adds nothing, ontologically, to the being that is repeated. The Big Bang, the scratch on my mobile phone reminding me of when I dropped it, the compound interests of capital, learning a language, the genetic history that predisposes us to rheumatism and the neurosis that drives us to set up relationships with others by repeating patterns experienced when we were children are all manifestations of this. A medium preserves the trace of an action, i.e. it temporally postpones its presence, beyond the here and now of its production. This deferment is not an ontological doubling, but as it proceeds it can be associated with other traces and give rise to new objects that are more complex and qualitatively different. This process of capitalisation is the real secret of the whole technological movement, as we have seen in 2.3 and 2.4.

3.2.1. Capitalisation

Hysteresis certainly has the characteristics of recording: it is the permanence of a trace and of the surface that bears it. However, it is also the set of adventures that this trace can experience, constituting the metaphysical foundation of all that is and happens. From instrumentality to symbolism, in fact, there is only a short step: without great effort, a trauma becomes an externalised memory that records commodities, exchanges, debts and credits, as well as natural elements, by means of notches. And not only does the notch lend itself to keeping track, but shards become excellent reminders that work for numbering everything. A trace, once preserved, can be iterated; and even if the iteration is perfect, it takes place at a later time and in a different place, so it is potentially open to the creation of something new, to alteration, and finally to interruption, which gives meaning to the whole process.[22]

This iteration is characteristic of technology, and perhaps this is why, following an obscure and somewhat moralising intuition, technology has been associ-

---

[21] Vittorio Sereni. *Sopra un'immagine sepolcrale,* Italian original: "come se un Tu dovesse veramente | ritornare | a liberare i vivi e i morti. | E quante lagrime e seme vanamente sparso."

[22] For a more detailed exposition of this issue, I would refer you to my *Speculation* cit.

ated with nihilism, when in fact it is nothing new, indeed it reaffirms being. Indeed, it seems indispensible to find an intermediate term between being and nothingness for emergence to be possible. This appears to be a very abstract and artificial question, but it can easily be exemplified by aporias that are very much present in both the philosophical tradition and common sense: are time and space something or nothing? If time were something, it would lose its fundamental characteristic, that of flowing, and the past would be contemporary with the present. This is the aporia of time in Aristotle,[23] which can be further articulated in these terms: If the existence of the past depended on its observability, then there would be no past. However, if this is true, it is also true of the present: if the existence of the present depended on its observability, then there would be no present and no presence. The aporia of time, then, is potentially extensible to all ontology, and in particular to space: if space were something, there would be no place for objects that occupy it; but if it is nothing, would being float in nothingness? Now, this intermediate term between being and nothingness is precisely iteration: I am not multiplying entities, I am simply repeating them; I am not adding anything, and yet by doing so I am setting the condition for alteration, that is, for becoming.

What is given can not only be accumulated (preserve one, ontology) but also iterated (repeat one, technology) and, as we shall see, this iteration is the necessary condition for alteration (knowledge, epistemology). In order to pass from ontology to epistemology it is therefore necessary to pass through technology. What we call "technology" has many values, and there is undoubtedly a difference between a technical device such as a radio or a chainsaw, the ability to cook or to organise *coups d'état*, or the competence needed to compose a sonnet or draw up a law. What unites all these meanings, however, is the existence of a regular system of iterations: the radio is set up for certain functions, and so is the chainsaw. By following a preordained – or rather, prescribed – plan, one can pull off a *coup d'état*; by following another, one can make a pizza or write a law. When there is no order, then there is improvisation, a poor and clumsy relative of iteration. I do not think it excessive to maintain that iteration is the fundamental condition of technology, and that keeping track is, so to speak, the mother of all Muses. Rather, in view of what I have said so far, it turns out to be a necessary consequence of my discourse.

Technology does not seem to enjoy the same metaphysical nobility as ontology and epistemology. However, if we really try to understand what "technology" means, we realise that it is the quintessence of metaphysics, the common root not of sensibility and intellect, as was said of transcendental imagination in the days

---

[23] Aristotle, *Physics*, 217b 32–218a 8.

of idealism,[24] but of the intermediary step between ontology and epistemoogy, as should be recognised in the days of realism. Kant was so interested in a priori

---

[24] Kant, as we know, went in search of a "productive imagination", which appears as the analogue of Aristotle's active intellect, and is called to play a crucial role in Kant's theory of knowledge because it provides the schemes through which the understanding refers to intuitions. Kant insists that this imagination is different from reproductive imagination, that is, from a form of memory that recombines images of things, and the reason for this insistence is clear: if the faculty that produces the schemes was reproductive imagination, then the *Critique of Pure Reason* would become a critique of empirical reason – a theory of the association of ideas, an empirical doctrine in Locke's style that explains how general ideas are formed through the repetition of experience, and not, as is Kant's intention, a theory that explains how the pure intellect can determine experience through concepts and schemes that are foreign to it and precede it. However, all that Kant says to characterise this faculty which is so decisive for his system is that it is different from reproductive imagination, without providing further clarification. Reflections on capitalisation and hysteresis perhaps help to solve the enigma of transcendental imagination, a productive imagination that arises from reproductive imagination, that is, in economic terms, an activity that arises from a passivity. Just as passivity, once recorded, becomes the possibility of capitalisation, hence activity, so imagination from retention becomes production, of meaning, of ends and, strange as it may seem, of the origin itself, with a retrospective action. The imagination is not an absolutely creative faculty, a wonder and a mystery laid deep in the human soul, as Kant maintains when he speaks of the transcendental imagination. Indeed, Kant himself, who for structural reasons must hypothesise the independence of the productive or transcendental imagination from experience, is decidedly laconic about the nature of this surplus of creativity, and in fact characterises the productivity of the imagination as a different or deferred reproductivity. The good question now is why simple reproduction can present itself as production, to the point that, with complete naivety, Kant confesses that the only difference between reproductive and productive imagination is that the latter is different from the former. We can offer two interpretations of this statement. The first is ungenerous and suggests that Kant has no serious argument for drawing a difference between productive and reproductive imagination. The second, which is more generous and which I would be willing to embrace in this context, is that between reproductive imagination and productive imagination there is a recording-iteration-alteration system dominated by hysteresis, which reproduces the capitalisation processes described in 2. Schopenhauer stated that he could give up all of the Kantian categories except causality. But causality is a subtly epistemological principle (think of the notion of aetiology, where explanation is equated with the finding of causes: A caused B). But if we think about it, the only element with which we have ontological contact is the effect, that which remains even when the cause has disappeared. I therefore say that I can give up all the structures of Kantian transcendentalism, except hysteresis, which Kant himself refers to when he speaks of the syntheses necessary for the constitution of experience and when he describes the imagination as the common root of sensibility and understanding. This is particularly evident in the so-called "Deduction A", the one we find in the first edition of the *Critique of Pure Reason*. Here Kant argues that for an object to be known it must be perceived ("synthesis of apprehension"), contained in memory or, as Kant says, in the imagination ("synthesis of reproduction"), and finally conceptualised ("synthesis of recognition"). In his presentation, this is an epistemological argument, which isolates the minimal criteria of an experience. But Kant, as we know, also (and mainly) had an ontological objective, which manifests itself through the constructive action of schemes. These are forms of the imagination (i.e., of hysteresis) that

synthetic judgements because in his conception they derived from concepts that were independent of experience and therefore certain; indeed, rich and informative a priori synthetic judgements come to us from operations: we obtain the result of a calculation by applying the rule "add one"; and many other complex answers, for example what time it is or how much we weigh, come to us from machines that do not possess any concept, but only mechanisms that regulate their functioning. As we shall see more clearly in the next chapter, those performed by technology are synthetic acts of which synthetic judgements are a subspecies. In both cases, in fact, the result is not obtained from the analysis of the premises: 12 is not obtained from the analysis of 7 and 5, but derives from their sum, just as a statue is not obtained from the simple analysis of the marble of which it is composed. The same applies to the outcome of a battle, the taste of a cake, the signing of a contract, the making of a promise. The synthesis, the result, exists a posteriori, yet the conditions that make it possible exist a priori, in the sense that they cannot be anticipated or understood in the slightest by a simple analysis of form, acts or materials.

3.2.2. Competence

Let's examine the nature of these acts. "As rational metaphysics teaches that man becomes all things by understanding them (*homo intelligendo fit omnia*), this imaginative metaphysics shows that man becomes all things by not understanding them (*homo non intelligendo fit omnia*)".[25] That's what Vico wrote, and one could add: *machina non intelligendo fit omnia*. Artificial intelligence is possible because (just as in natural intelligence) one can make calculations, for example with an abacus, without knowing what mathematics is. The reverse is also true. One can be a great mathematician without having any aptitude for computation, and I'm not sure that Russell and Whitehead, after writing *Principia mathematica*, had become better and quicker at splitting the bill at a restaurant.[26] Obviously,

---

define substance as the permanence of something in time and cause as the succession of something in time.

[25] G.B. Vico, *La scienza nuova* (1744), in *The New Science of Giambattista Vico*, Cornell University Press, Ithaca 1948, p. 116.

[26] Consider the "Mary the Scientist" thought experiment proposed by Frank Jackson in the article *Epiphenomenal Qualia* (1982) and developed in *What Mary Didn't Know* (1986). The debate that followed is collected in the volume *There's Something About Mary* (The MIT Press, Cambridge (MA) 2004) featuring, among others, Daniel Dennett, David Lewis and Paul Churchland. Mary is a scientist specialising in the neurophysiology of vision, but she is confined to conducting her research in a black and white room with a black and white television. With her investigations she gets to know all about the physical and chemical processes that underlie the perception of colours. Then she is given a colour television: will she be able to

one can have both competence and understanding (or, at least, that's the hope). But such an understanding cannot do without competence, even the minimal competence of knowing how to use a technical instrument such as a pencil in order to make even simple calculations, which however exceed the possibilities of the human mind. The "setting itself to work of the truth", the fact that truth is the result of a technological process, as Heidegger claimed referring to art, religion, philosophy and politics, should also be extended to a sphere that he did not believe to be institutive, namely science and technology as a whole. Indeed, the condition of truth's setting itself to work is precisely technology.

This, given the current epistemological contingency, is a great resource. Physics and metaphysics got along well until Newton. Nobody was even aware that they were different professions, and we can be sure that if Descartes had not made a big mistake in his ideas about physics, he would have been remembered together with Newton and Galileo. Then Kant, who had elaborated a cosmological theory on the origin of the universe, had to set the pointers of metaphysics on those of physics. And this, mind you, was due to the successes and performances of physics, as demonstrated by the examples mentioned by Kant: Galilei, Torricelli, Stahl. Immediately after Kant, and especially with the theory of relativity

---

visually recognize the colours of which she knows everything in theory? The thought experiment has a certain degree of naivety – because it is not clear whether Mary is also in black and white (there are reasons to doubt it). The model of this experiment, which instead had no naivety, is the question that the Irish natural philosopher William Molyneux posed to Locke in 1693, quite some time before Mary and her author were born: "Suppose a Man born blind, and now adult, and taught by his touch to distinguish between a Cube, and a Sphere of the same metal, and nighly of the same bigness, so as to tell, when he felt one and t'other, which is the Cube, which the Sphere. Suppose then the Cube and Sphere placed on a Table, and the Blind Man to be made to see. Quaere, Whether by his sight, before he touch'd them, he could now distinguish, and tell, which is the Globe, which the Cube. Or Whether he Could know by his Sight, before he stretch'd out his Hand, whether he Could not Reach them, tho they were Removed 20 or 1000 feet from Him? If the Learned and Ingenious Author of the Forementiond Treatise think this Problem Worth his Consideration and Answer, He may at any time Direct it to One that Much Esteems him" (Locke, *Essay*, II.9.8). The experiment is difficult to perform because it is physiologically impossible to immediately recover vision, as this process requires long training (cf. A. Jacomuzzi *et al.*, *Molyneux's Question Redux*, in "Phenomenology and the Cognitive Sciences", 2, 2003, pp. 255–280; see also R.N. Shepard, J. Metzler, *Mental Rotation of Three-Dimensional Objects*, in "Science", 171, 3972, 1971, pp. 701–703). However, this supports the fact that there can be competence without understanding but no understanding without competence. In all likelihood, Mary would not be able to recognize colours on the basis of scientific notions (understanding) alone, though she would not actually need to because she would know that the sky is blue and tomatoes are red. Instead, Molyneux's blind person could perhaps trace the handled solids back to those he had seen, because he has had direct experience of them (for example, he will have experienced the edges of the cube, and this would be enough for him to distinguish it from the sphere, even if only visually).

and quantum physics, physics has increasingly departed from ordinary experience. And this meant the failure of Kant's project, which, by uniting physics and metaphysics, sought to provide a unified theory that would be valid for both science and experience.

Amid these shaken foundations, however, something has remained standing, namely the technological efficacy of physics. Science, and especially its traditionally fundamental core, physics, lost its traditional characteristics of certainty, replicability, and observability. This does not justify embracing the approach of "anything goes,"[27] especially because "anything goes" has a significant implication that is often neglected by its philosophical fans: "whatever works". Sure, as long as it *does* work, though – that's the point. You don't need to know what mathematics is to add things up, you don't need to know what politics is to draw up documents. A profound truth emerges from this. You don't need to know what you are doing to do it successfully; and, even more so, you don't need to know your physiology to feel the pangs of hunger. Understanding is a luxury, competence and action are a necessity, and in the transition from understanding to competence the documedia revolution implements a huge economy. Knowing the meaning of what one is doing, in terms of performance, is a great waste of time and worsens performance. As we have just seen, Vico, referring to the origin of the social world, said that humans act before understanding, generating gestures and processes.[28] Understanding, if and when it comes, is not the premise, as Cartesians think, but the result. It is a thought that is prior and external to thought,[29] that is, a *know how* and not a *know that*, a *können* and not a *kennen*.

We do most things without knowing, or knowing very little, and if it weren't so it would be a real problem, because we wouldn't get anything done. I don't need an theory of economics to get a coffee, and I can get one even if I have a completely wrong understanding of how the economy works. Yet to argue, again with Vico,[30] that we only know what we have created rests on a linguistic misun-

---

[27] P.K. Feyerabend, *Against Method* (1975), Verso, London 2002.

[28] For Gilbert Simondon, a technological operation is not a shaping according to a predetermined purpose but a movement or transition (cf. G. Simondon, *Naissance de la technologie* [1970], in *Sur la technique* cit., pp. 131–178). In an only apparently different field, Luigi Pareyson has defined artistic creation as a "doing" that, in the process of its own making, invents its own way of doing (cf. L. Pareyson, *Estetica. Teoria della formatività* [1954], Bompiani, Milan 1988).

[29] According to the definition of exchange in Sohn-Rethel; see his *Intellectual and Manual Labour* (1970), Brill, Boston 1978; *Warenform und Denkform*, Suhrkamp, Frankfurt a. M. 1971; *Das Geld, die bare Münze des Apriori*, Wagenbach, Berlin 1976.

[30] Phenomena such as recessions, economic cycles, and power systems show that social reality, albeit dependent on our beliefs and actions (i.e., "made by us"), is far from transparent to us. On this opacity see, for example: D.H. Ruben, *The Metaphysics of Social Reality*, The

derstanding, namely that a person who lives in society is also the person who created it. Now, if there is something that I do not feel I created, it is precisely the social world: I was simply born in an environment filled with rules, of which some I learned, some I did not, some I followed, some I did not. Instead, as for natural laws, I did just fine even without knowing them, for the same reason that a donkey goes straight to the hay without having read a line of Euclid. The best copyists in the Middle Ages were illiterate. Their proficiency in iterating signs was not hindered by their understanding of meaning, which could produce errors such as *lectio facilior* or *lapsus calami*. And it is rarely the case that the author, who hopefully understands the meaning of what they have written, is the best proofreader. Nonetheless, in most cases competence is a necessary step *towards* understanding. Technology, from this point of view, really acts as a bridge between ontology, a certain set of available entities, and epistemology, the knowledge that can be derived from it – not through the superimposition of conceptual schemes on perceptual elements (as in the binary system Epistemology → Ontology), but precisely through the ternary scheme Ontology → Technology → Epistemology.[31]

This is especially evident in the social world. We humans promise, gamble, and write wills, based on vague, and often mistaken, notions of law and society. If this is the case, there is nothing odd about the fact that we can be surprised by social reality, and that it can reveal hidden aspects to us. The form of society and its modes of production is no more transparent to its members than consciousness is transparent to its subjects. The fact that the two are the closest things we have is far from being an advantage: they are a little too obvious and ultimately escape our observation and consciousness. This is why answering the question of what gives value to a piece of paper does not mean showing any evidence, but

---

Free Press, New York 1985; A. Goldman, *A Theory of Human Action*, Princeton University Press, Princeton 1970; M. Gilbert, *Living Together. Rationality, Sociality, and Obligation*, Rowman & Littlefield, Lanham 1996.

[31] It will come as no surprise that the definition of a scheme, as offered by Kant, is that of a technical device. Kant rightly understood that to move from concepts (epistemology) to objects (ontology) it is necessary to use an intermediate term, which he called "scheme" and which he characterised with a strongly technological qualification – he spoke of schematism as a "hidden technique" (*verborgene Kunst*, which usually translates as "hidden art", but whose meaning is obviously the same as "hidden technology"). Cf. Kant, *Critique of Pure Reason* cit., A 141/B 180–1, p. 273: "This schematism of our understanding with regard to appearances and their mere form is a hidden art in the depths of the human soul, whose true operations we can divine from nature and lay unveiled before our eyes only with difficulty". However, conditioned by dualism and by his Pentecostal attitude (which in this case amounts to the idea of a priori as opposed to experience), Kant described schematism as a top-down process, although he did not exclude (and this is even clearer in the *Critique of Judgement*) a bottom-up process, which for Kant was the subsumption of the subcategories of the objects of experience.

rather solving a mystery, or at least trying to. What is the secular sacrament that transforms a human being into a doctor, into the holder of a driver's licence, into the recipient of a summons or a fine? After all, the track of the purloin letter is not a simple literary suggestion: all these cases traditionally involve a piece of paper, even if of course one can also use plastic, metal, tattoos, chevrons, distinctive signs of all sorts, or simply traces in a computer or smartphone.

The consolidation of competence does not occur through clarification, as in understanding, but through repetition, exercise, and capitalisation, which serve an essential function not only in physical activities, but also in spiritual ones like speaking a language or performing arithmetic operations. Capitalisation produces both reinforcement and facilitation. When I handle money, I do not apply any economic theory, or at least not necessarily a good one. I simply act, following the basic character of my relationship with reality in general. Little by little, through action, I may become aware of what I am doing, and in that case competence turns into understanding – that is, technology turns into epistemology.

In this framework, we have an immense sphere of competence underpinning life and surrounding a much more modest sphere: that of understanding. The latter is the sphere of consciousness and representation, which makes us say things like "here, I'll walk to the shelf, I'll take the *Critique of Pure Reason*, I'll open it and read it". This kind of proposition, or other more banal ones such as "I'm going to stop and fill up with petrol", very rarely accompanies our actions: from this we can infer how rare self-awareness is in our experience. And yet we represent it as perennially active, except when it comes to apologising for what we did or didn't do. Then it becomes very normal to lavish ourselves with apologies whose centrepiece is the idea that the "I think" did not accompany our representations at all ("I wasn't thinking", "I don't know what got into me", "I'm sorry, I didn't mean it"), or in which our representations have ambushed the "I think" ("It was a slip of the tongue").

Thus, first of all there is a technological development, which is an accumulation of competences without understanding: knowing how to chip a flint without necessarily having words to describe what one is doing, or, at the other end of that story, knowing how to make very complicated calculations by entrusting them to the knowledge placed in the paper, pen and inkwell, or in the computer, or in the particularly complicated technical instrument of the human mind. The latter, too, is made of competence without understanding: the real work of the mind is done by neurons that discharge without understanding anything[32]; understanding, the result of the sum of these discharges, comes later, if at all. Contrary

---

[32] D. Bray, *Wetware: A Computer in Every Living Cell*, Yale University Press, New Haven 2009.

to Kant's view, first there is the scheme, the method of construction; then the object and, if necessary, the concept, derive from it – though it is not strictly so: what is the concept of the *Mona Lisa*?

Art and war are canonical spheres of competence without understanding and, at the same time, they constitute the traditional domains of creativity, in that they lead to the emergence of something new, like a posteriori synthetic judgments. Von Moltke, the architect of the Prussian triumph of 1870, said of the art of war: "Only the profane see the course of a campaign as the coherent execution of an original idea, previously elaborated in full detail by the commander, who remained faithful to it until the end".[33] As for fine art, an artist does not know why they made the artwork exactly that way (Leibniz's doctrine of *nescio quid*); the work of art – since the invocation to the goddess in Homeric incipits – has always been conceived as inspired, as something that comes from the outside and is not fully mastered by the author. Most of the time, descriptions of the creative process appeal to unconscious elements or to some kind of automatism (the characters act as if they were alive, certain words – for example "Nevermore" in Poe's *Raven* – guide the whole composition). Even the most self-possessed of authors is at the mercy of their own work. Again, an obscure teleology guides the creative process, and only at the end, if anything, does the *telos* become clear, and retrospectively illuminates the archaeology: "This is what I wanted to do, this is what I've been searching for for years." The opaqueness of the process bridges the gap between artifice and nature.

Let us therefore leave the social world and come to the natural world. Following a powerful image proposed by Dennett,[34] compare a termite mound and the Sagrada Familia in Barcelona. They look very much the same – how can it be? Gaudí drew projects and had representations; termites have nothing of the sort. And yet, this is precisely the point: Gaudí's neurons had no representations either, they simply discharged electrochemical signals, just like termites. After all, the latter found a shortcut: instead of creating representations, like Gaudí's neurons, they went straight to creating the Sagrada Familia. It is true that termites are very different from Gaudí, but the difference is not found a priori, but a posteriori: Gaudí can have representations that termites cannot. However, compared to Gaudí's neurons, termites have the advantage of not representing what they produce and the purpose for which they do it. They just produce it, without needing a reason to.

---

[33] H. von Moltke, *Über Strategie*, in C. von Clausewitz, H. von Moltke, *Kriegstheorie und Kriegsgeschichte*, Deutscher Klassiker Verlag, Frankfurt a. M. 1993.

[34] D. C. Dennett, *From Bacteria to Bach and Back. The Evolution of Minds*, Bradford Books – The MIT Press, Cambridge (MA) 2017.

This applies to technology in general, which on the one hand appears to be the realm of iterativity, but on the other hand is presented as the sphere of an inventiveness that is interesting precisely because it is not animated by any preliminary intentionality. No one could have foreseen from the outset all the possible uses of the lever, the wheel or gunpowder. So, as we have seen, the functions that philosophers often attribute to a super-human faculty, imagination, should rather be attributed to the possibilities of hysteresis, exteriorisation and accumulation that are immanent to technology.

And let us not forget that social interaction is an eminent example of this competence without understanding. A human group performs a certain initiation rite that is linked to a pre-human past. The rite evolves into more complex and codified forms (scout promise ceremony, consecration, military service ...). Finally, it takes the legal form of adulthood, and creates a series of rules and laws, determining a sphere of truth. The image of human action as an unconscious *praxis* that becomes conscious only through historical becoming fully accounts for our intuitions. We do not know the reasons behind our actions, and only sometimes can we explain them. Reciprocally, knowing the principles behind our actions does not make us more efficient, otherwise, the professors of military academies would be the greatest strategists, which almost never happens. The answer, in short, is simpler than it seems.

To make a world, be it the cosmos or the Web, one does not need to know how to make it: absolute ignorance can take the place of absolute wisdom,[35] and chance can be stronger and more effective than intelligent design. This is why the notion of evolution – the biological counterpart of the metaphysical emergentism I am defending – could only come forward when we became aware of the temporal immensity involved in the history of the world. We are used to seeing understanding as a means of solving problems, yet we forget that most problems are solved without understanding (it is not through understanding that I learned how to ride a bicycle or tie my shoelaces). The salient feature of understanding is rather the ability to bring out problems, needs, stumbling blocks. Tradition taught Marcus Aurelius to use slaves, and it was understanding that suggested to him that slavery was unworthy of the human condition. The same is true for the environmental emergency, gender discrimination and social injustice. Here, understanding makes one sensitive to problems that were not previously perceived as such, and consists in identifying problems rather than problem-solving.

---

[35] The idea that wisdom could not arise from ignorance was the objection to Darwin found in R. M. Beverley, *The Darwinian Theory of the Transmutation of Species Examined* (published anonymously "by a graduate of the University of Cambridge"), Nisbet, London 1867, cit. in "Athenaeum", 8 February 2012, p. 217.

Turing's great discovery was that in order to calculate, it is not necessary to know what mathematics is, but it is necessary to have the technical skills that make the calculation possible. What I would add is that even the brain as a universal machine can do everything without understanding anything, since neurons do not think any more than computer chips do, and understanding – something like "ah, right, I get it" – is just a psychological effect that feeds the myth of understanding that I will discuss in a moment. Again, artificial intelligence sheds light on natural intelligence and helps eliminate unnecessary and misleading assumptions. An automatic translator functions when it satisfactorily carries out its task, and not when, in addition to this, it "understands" what it is translating. The same can be said of a human translator, who may well translate without "understanding" what they are translating and, perhaps even more interestingly, may believe they understand and end up translating incorrectly. A system of signs is made to interact for practical or ritual reasons. Over time, however, the signs become emancipated and sophisticated, and above all the operations that can be carried out with these signs become increasingly refined. In the end, the sphere of mathematics emerges as a collection of true theorems and operations. Thus, among the various products of technology, we must include *natural* intelligence.

But it would be a very serious mistake to conclude that intelligence comes before competence. The opposite is obviously true. It has been rightly observed[36] that gestures, understood as acts that have a beginning and an end and are endowed with meaning, have great philosophical relevance.[37] What is characteristic of humans as opposed to non-human animals is not that they have language, but – as we saw in 2.2 – the hand, and thus the ability to handle and develop technical devices. From this possibility, in a very long evolutionary time, we first developed the expressive use of the hand, through gestures; then, the possibility

---

[36] G. Maddalena, *The Philosophy of Gesture. Completing Pragmatists' Incomplete Revolution*, McGill-Queen's University Press, Montréal-New York 2015.

[37] A philosophy of gesture is the basis to understand technology as the production of thought and not as alienation, i.e. as a means of transition from ontology to epistemology. Referring to George Mead's pragmatist reflection, as well as to Derrida, in many important works (albeit with a different framework and with different aims from mine), Carlo Sini has proposed alternatives to this "Cartesian grammatology": cf. in particular *Filosofia e scrittura*, Laterza, Rome-Bari 1994 and *Etica della scrittura* (1992), new ed., Mimesis, Milan 2009. I discussed Sini's theses in *Estetica razionale* cit., pp. 495 ff. A more significant and original proposal is Maddalena, *The Philosophy of Gesture* cit. It is worth noting that even the so-called "Object-Oriented Ontology", which developed in the field of speculative realism, is in fact a reflection on technology that emancipates itself from the perspective that sees it as decadence, nihilism and alienation. Cf. G. Harman, *Tool-Being: Heidegger and the Metaphysics of Objects*, Open Court, Chicago 2002; Id., *Guerrilla Metaphysics. Phenomenology and the Carpentry of Things*, Open Court, Chicago 2005; Id., *The Quadruple Object*, Zero Books, Alresford 2010; Id., *Object-Oriented Ontology: A New Theory of Everything*, Pelican, London 2018.

of elaborating signs, both written and spoken; and finally, we reached the elaboration of complex linguistic codes, which manifest a *logos*. The latter is a real difference between human and non-human animals, but is a result rather than a starting point, and depends on events largely outside our organic endowment and on competence without understanding. How many people know why we shake hands in greeting or agreement? (the reason is well known: to show that one is unarmed, but most people do not know and do not wonder, as is right when it comes to social acts). And, a fortiori, how many children playing Ring a Ring o' Roses ("We all fall down") are aware that they are probably recreating the epic collapse at the end of the Titanomachy? The acts and technology that underlie our everyday expressions and actions incorporate, transmit and inform very ancient gestures. According to the famous anecdote, Sraffa embarrassed Wittgenstein by asking him what the logical form was for the Italian gesture of rubbing underneath one's chin with the index and middle finger to express indifference. Wittgenstein could have answered, following Bacon, that it was a transient hieroglyph, an expression whose origin had been forgotten and which, despite this, continued to exist in the social world. The same could be said about artificial intelligence and largely for natural intelligence as well.

Indeed, if the hypothesis of an originally gestural language holds true, *writing came before words*. In a chapter of *Tristes Tropiques*, Lévi-Strauss tells us this story.[38] He was studying an Amazonian population that was stuck in the Neolithic period, the Nambikwara. He recorded their habits and customs on notebooks, and this intrigued the natives, who eventually started writing too, except that – observes the ethnologist, who was also an ethnocentrist – being primitive, they did not understand that the essence of writing is to communicate. They simply drew lines (that was the name they gave to writing), and only the chief, who was a little more intelligent than the others, showed a glimmer of understanding about the communicative role of writing. Now, did the line-drawers really not understand what writing is? Perhaps they grasped its essence better than the chief and the ethnologist, because drawing lines is already writing, it is already recording and iterating a bar code: hysteresis precedes communication (this is the great revelation of the digital world, which manifests, so to speak, things that have remained hidden since the foundation of the world). The latter is by no means the sole and primal need of humans, as without hysteresis they would have nothing to communicate in the first place.

The definition of humans as animals with language is completely misleading, and not only if one understands "language" as "reason". Many other animals

---

[38] C. Lévi-Strauss, *A World on the Wane* (1955), Criterion Books, New York 1961, pp. 288–289.

have language,[39] before and better than humans. However much one may like to talk about "language instinct"[40] and "reading neurons",[41] it remains that neither one nor the other, provided they exist at all, would have been of any use whatsoever if there had not been a series of technological pre-adaptations and capitalisations that created the conditions for something like *logos* and culture to arise. Accordingly, histories of writing[42] insist that very similar symbols are found in different alphabets, though of course with different functions, and indeed the fact that the Semitic alphabets were taken over and transformed by the Greeks and Latins cannot be explained by a dismal lack of imagination. Is it because we have reading neurons? No, it is simply that there are not ten thousand different ways of leaving traces.

We have seen that gestures preceded language, and that writing preceded speech. Thus, the technological boom that, some fifty thousand years ago, liberated speech as a medium of communication triggered a first and decisive technological enhancement of hysteresis. I hope that, by now, the hypothesis that the system of thought → language → writing is the only plausible one has lost its lustre, or at least its unquestionable obviousness. All that remains is the third and final step, which must demonstrate the constitutive role of writing with respect to thought and language. I would be inclined to represent this passage as follows: hysteresis → thought → language → writing. Artificial intelligence makes good use of this circumstance. The fundamental social activity is not knowledge, but action, guided by competence without understanding, which can, but need not, become knowledge. The same applies in another great technological enterprise: Hegel's cunning of reason, by which humans believe they are achieving their own goals while working towards the more general goals of history. Islands of understanding and reflection are not the norm, but the exception in a sea of trou-

---

[39] S. Pinker, *How the Mind Works*, W W Norton & Co, New York 1997; Id., *Language as an Adaptation to the Cognitive Niche*, in *Language Evolution*, ed. by M.H. Christiansen, S. Kirby, Oxford University Press, New York 2003, pp. 16–37; Id., *Commentary on Daniel Dennett*, in "Mind, Brain, and Behavior", Lecture at Harvard University, 23 April 2009 (https://www.you tube.com/watch?v=3H8i5x-jcew [05/05/2022]); S. Pinker, R. Jackendoff, *The Faculty of Language: What's Special About It?*, in "Cognition", 95, 2, 2005, pp. 201–236.

[40] Pinker, *How the Mind Works* cit.

[41] S. Dehaene, *Reading in the Brain* (2007), Penguin, New York 2009, goes so far as to claim that if we have invented writing, it is because we have neurons that are designed for it. This is not very different from saying that if we have mathematics, it is because we have the "math bump".

[42] Tullio, *La forma naturale delle lettere dell'alfabeto* cit.; Gelb, *A Study of Writing* cit.; Kallir, *Sign and Design* cit.; Havelock, *Preface to Plato* cit.; Cardona, *Storia universale della scrittura* cit. For a detailed analysis, please refer to *Estetica razionale* cit., pp. 469–514.

ble, uncertainty and knee-jerk actions. In short, knowledge is not the premise, but if anything (if we are smart and lucky) the result.

3.2.3. Understanding

Hence a consequence that deserves some thought, also in view of the reasoning on epistemology that I will carry out in the next chapter. The point is to account for the passage from action to intelligence, and from competence to understanding. What is natural intelligence? On the one hand, it is in many ways identical to artificial intelligence (when I say that $2 + 2 = 4$ I am the same as an automaton, only slower and more fallible); on the other hand, it is completely different from AI, because it is located in a body that confers urgencies, motivations, deadlines and desires that artificial intelligence can never have. Let us start from Freud's old intuition[43]: consciousness arises in place of a mnestic trace, as if to say that the higher forms of psychic life are the reworking of phenomena linked to track-keeping and memory, exactly as happens in the comparison between the soul and a book. Thus, for Freud, consciousness is a superficial formation: located in the cerebral cortex, it would correspond to the outer membrane of single-celled organisms. So everything is already there: the automaton, the mnestic trace that is recorded on the tabula by iterating processes; and the soul, the organism subject to irreversible processes.

But what distinguishes the outer surface of the membrane from the wax tablet or perhaps the parchment (another outer membrane), the embodied automaton from the unembodied one? To answer this question, we must leave Vienna and go back to Athens.[44] On a hot summer's day, Socrates met Phaedrus outside the city, surrounded by the sound of cicadas. Phaedrus, who was fond of speeches, had just listened to Lysias and was eager to practise repeating the latter's speech by heart. Socrates, however, enjoyed frustrating people, and promptly pointed out that Phaedrus was hiding a written copy of the discourse under his cloak and suggested they read it instead. Over natural intelligence, Socrates thus seemed to prefer artificial intelligence – logography. Indeed, Socrates wanted to expose Phaedrus for using artificial intelligence, but then shamelessly chose to read the speech instead of hearing it from Phaedrus. But what's wrong with carrying a papyrus under your cloak? Why did Socrates care so much? For reasons that do not concern Socrates but his disciple Plato, the author of the dialogue. Indeed, Plato had gone into business and opened a school which, despite teaching how to be virtuous, still required a fee. In those years, however, writing was no longer a technology run by

---

[43] Freud, *Project for a Scientific Psychology* cit.
[44] Plato, *Phaedrus* cit.

a specialised caste of scribes: it was becoming popular and was being taught in the primary schools of Attica. So some people started saying that it was not worth going to Plato's school, because it was better to learn the alphabet and read books. What is the point of the Academy when we have Wikipedia?

Indeed, in the dialogue Socrates comes straight to the point and asks Phaedrus: would you ever agree to be treated by someone who studied medicine by correspondence? He could not have foreseen that people would waste hours on the Web trying to cure themselves, but there it is. This, in principle, is not an objection to artificial intelligence, but to natural intelligence. What Socrates is saying at this point is simply that the Web is not a universal panacea – can we blame him? The real blow comes shortly afterwards: not only does artificial intelligence not replace natural intelligence, it destroys it. Convinced that they can rely on machines, humans will stop exercising intelligence, which for Socrates, the theorist of anamnesis, is first and foremost memory. To make his theory more convincing, Socrates makes up a myth, the only other one created by Plato next to the Myth of Er. It's a fine creative effort set in Egypt: the land of writing and artificial intelligence: Theuth, the god who invented artificial intelligence, gets an earful from Pharaoh Thamus, who is very unhappy with the invention because it does increase the memory of the Egyptian people but will eventually wipe it out, making them inept. The myth is reminiscent of Asimov's story in which scientists of the future gather in admiration around a boy who is capable of elementary multiplication without using a computer simply because he memorised times tables.

De te fabula narratur. Don't trust artificial intelligence, not even the papyrus you've hidden under your cloak, because in the long run you'll become a slave to a master you don't even know. You will depend on a silent machine, and if you hope that it will speak, you are deceiving yourself no less than if you hoped that paintings could sing. Believe me, Phaedrus, come to my school and cultivate natural intelligence instead. And don't be like those fanatics who surf the Web and buy into all kinds of post-truths because the author is not there to correct them and explain what they really meant. As you can imagine, emojis are not enough to clarify things. And, assuming that the text really did reflect the author's intentions, and that you understood the words correctly, what if the author is a fanatic locked in an echo chamber, say, a flat-earther?

As you may have noticed, Phaedrus is easy to please, and he is happy with all of Socrates' arguments. And who can blame him? Naturally enough, after all this talk of artificial intelligence, Phaedrus is curious to know about natural intelligence, the good one, the one that you can improve by going to Plato's school and paying his tuition fee. The guru does not hesitate to do so and achieves a real *coup de théâtre*. After blaming artificial intelligence because it is artificial, because it is writing and technology, when it comes to defining natural intelligence,

Socrates says that it is "the word which is written with intelligence in the mind of the learner".[45] In short, he describes natural intelligence with the distinctive character of artificial intelligence. Should we conclude that Plato, or Socrates, is rambling? No. He is saying something very important, namely that we are no different from machines in terms of pure intelligence.

"I do solemnly swear that I will faithfully execute the Office of President of the United States, and will to the best of my ability, preserve, protect and defend the Constitution of the United States"; "Sh'ma Yisrael Adonai Eloheinu Adonai Ehad"; "Giuro di essere fedele alla Repubblica Italiana". Such statements are about as demanding as they can be for those who utter them. If they are so demanding, however, it is precisely because of the automatism, within the reach of a parrot or a recorder, that characterises them in a binding way. It is worth noting that the word *record* refers etymologically to the heart (*cor*), just as in the Italian "ricordare", the English "learn by heart", or the French "*apprendre par coeur*" – which demonstrates the proximity between external and internal recording. When I mentally perform an arithmetic calculation, I am handling signs just as I do when I solve it with pen and paper or with the calculator on my phone. Automation is within us and outside us. We find it in the smartphone just as we find it in the priest who consecrates the host, in the professor who explains Kant's ideas in class, or the child who learns times tables.

3.2.4. Embodiment

Plato is more modern than Searle. Contradicting Turing's thesis, according to which if a machine behaves in an intelligent way in its interaction with a human, then it is intelligent, Searle envisions a situation with a human in the place of the machine.[46] Imagine this: Searle, who does not understand Chinese, is in a room with a handbook. Someone slips him a piece of paper with some Chinese characters under the door, so Searle looks them up in the handbook, which – without translating them – says that the sequence of ideograms A must be answered with the sequence of ideograms B. And so on: to sequence C, Searle answers with sequence D, etc. Now, since the couplings have been made by a Chinese speaker, whoever receives the answer sequences will believe that Searle speaks Chinese, but obviously this is not the case. It follows that thinking does not mean handling signs but performing more complex activities, like interpreting and understanding. But is this really so? Let us suppose that in the Chinese room, instead of

---

[45] Ibid., 276a 5–7.

[46] J.R. Searle, "Minds, Brains, and Programs" (1981), *Behavioral and Brain Sciences 3*, no 3 (1980), 417–24.

Searle, there *had* been a Chinese speaker. They would have done exactly the same things as Searle, only they would not have needed the conversion manual on the table, because they would have had one in their mind. In both cases, thought consists in handling signs, and the difference lies, potentially, in the different emotional state – suppose one of the messages says that the subject will never leave the room again. Still, this second situation would not involve something metaphysically different from sign manipulation.

For Searle, thinking means interpreting: on the one hand there is the soul, which interprets; on the other hand there is the automaton, which manipulates. And only the soul thinks. This attitude recalls philosophers who are apparently at the opposite end of Searle's cultural horizon. For example, at the beginning of the 19th century, Schleiermacher argued that there can be no act of understanding without interpretation. Now, it is hard to believe that Searle was influenced by Schleiermacher. Rather, in both cases, and albeit with antithetical rhetoric, the same operation takes place: one assumes that thought is a given thing, which appears to be only possible for humans, and one concludes that only humans think. Yet this means not taking into account that the performance identified with thought – in this case, interpretation – may not be precluded to automata and animals, and is actually sporadic in humans who are honest with themselves. What follows is that automata and animals sometimes think, and humans spend most of their time not thinking anyway. Taking it for granted that "thinking" means "interpreting", and that "manipulating symbols" is not thinking, is problematic (or, if you prefer, dogmatic) to say the least. And Searle's experiment, even in the light of the four decades that have passed since it was proposed, appears to rest on two conflicting exaggerations.

The first is the strong AI hypothesis, which is criticised by Searle. For it, the computer is not a *tool* that can be used to study the brain (or hurricanes, or Roman history), but rather a *representation* of the brain, perhaps a miniature brain. The second is Searle's image of human thought, which underlies what I propose we call the "myth of understanding". As many critics of Searle's argument have observed, in these terms one can only conclude that computers do not think, and that only a fool would think otherwise. As usual, if you claim that Hal (which is more or less the computer Searle had in mind) can think, then you have to resign yourself to having your arguments proved only in science fiction. But this does not take into account that even expecting a philosopher, a preacher, a spouse, a president, or a taxi driver to think at every moment of their lives, instead of behaving like automatons, is a symmetrical exaggeration.

Driven by an unconscious mythology of understanding, Searle ascribes to natural intelligence properties that are intrinsically different from those of artificial intelligence, in that they cannot be referred to extrinsic aspects, such as the phenomenological fact of first-hand experience or the physiological fact that natural intelligence is embedded in an organism. This myth is much more widespread than is generally believed, and it was revived at the very moment when the first computers raised the problem of the difference between artificial intelligence and natural intelligence. Most of the time, this difference has been sought through the presupposition of understanding: an *ad hoc* faculty that would make natural intelligence metaphysically different from artificial intelligence, just as the *virtus dormitiva*, in *The Imaginary Invalid*, is summoned to explain why we sleep.

The "parity principle",[47] the functional equivalence between natural intelligence and artificial intelligence, fully applies to these considerations. There is no reason to place what we call "intelligence" in the head. If a certain cognitive process takes place in the head, and if the same process takes place outside the head, then both processes are equivalent. But there is one fundamental difference, which is not taken into account by the parity principle: the process that takes place outside the head only has a meaning for a human, but not the other way around. In other words, the extended mind is a phenomenon that is much older than artificial intelligence, since it coincides with the relationship between soul and automaton, which, as we have seen, is the constitutive characteristic of hominisation. However, just as the soul cannot take the place of the automaton, so the automaton cannot take the place of the soul, and this is precisely because the two poles are complementary in the phenomenon of responsiveness as a characteristic of human intelligence.[48] But in what sense and why would the soul be

---

[47] A. Clark, D.J. Chalmers, *The Extended Mind*, in "Analysis", 58, 1998, pp. 7–19. This led to the development of a controversial theory of the extended mind: cf. A. Clark, *Supersizing the Mind: Embodiment, Action, and Cognitive Extension*, Oxford University Press, Oxford-New York 2008; Id., *Natural-Born Cyborgs*, Oxford University Press, Oxford 2003; Id., *Whatever Next? Predictive Brains, Situated Agents, and the Future of Cognitive Science*, in "Behavioral and Brain Sciences", 36, 3, 2013, pp. 181–204; Id., *Surfing Uncertainty: Prediction, Action, and the Embodied Mind*, Oxford University Press, New York 2015. *Contra*: D. Marconi, *Contro la mente estesa*, in "Sistemi intelligenti", 17, 3, 2005, pp. 389–398; J. Preston, *Is Your Mobile Part of Your Mind?*, in *Seeing, Understanding, Learning in the Mobile Age*, ed. by K. Nyíri, Passagen Verlag, Wien 2003, pp. 313–324. Unlike the extended mind hypothesis, however, I do not claim that we are dealing with a real extension of the mind, but only with an externalisation of its functions. Cf., in this sense, J. Domenicucci, *Trust, Extended Memories and Social Media*, in *Towards a Philosophy of Digital Media* cit., pp. 119–142.

[48] Paradoxically, the parity principle works very well in an idealistic perspective like Plato's, whereby the body is an inert container for the soul, which is to the latter what a pilot is to a plane. For this very reason it seems paradoxical that, in this perspective, the functioning of the soul is described through automation. If, however, we do not embrace Plato's perspective, the parity principle does not apply, not because an automaton cannot perform the same cognitive operations as the soul – indeed, the latter always behaves like an automaton when performing them – but because the purpose and meaning of those operations require having a body. The

different from the automaton? Where does this difference lie? In the fact that the automaton is *res extensa* and the soul *res cogitans*? Nothing could be less true, and nothing could be more blatantly Pentecostal. Let's start from a point that is easy to agree on. It is hard to deny, or to reduce to mere appearance, the psychic state that characterises our saying to ourselves, "OK, I get it". It is difficult to overlook a vivid and common experience, which differentiates the soul from the automaton. The point is, however, that this phenomenological difference does not reveal a *qualitative* difference between natural intelligence and artificial intelligence, but only that intelligence takes place in a body, forming what I have called the "soul", the compound between a mechanism capable of iterations (for example, following the "add one" rule when counting), but subject to the urgencies and irreversibilities of an organism.

Let me explain what I mean with an example. Say we are in a supermarket, we get to the checkout, the optical reader cannot read the barcode of a product and the cashier manually taps the number written under the code. Has the cashier "understood" the price better than the scanner does when things go smoothly?

---

purpose of artificial intelligence is thus the inverse of that of natural intelligence, i. e., reason. While the latter seeks to allow the emergence of meaning through automation, artificial intelligence uses automation to produce correlations useful for profiling, regardless of their meaning. While human intelligence is thus a movement that, from technological competence, aims (at least ideally) at epistemological understanding, artificial intelligence follows an opposite movement, starting from understanding. In artificial intelligence, in fact, understanding is presupposed, because the programme it runs on embodies goals formulated by natural intelligence, and aims at technological competence and effectiveness. One possible objection, here, is that I am talking about things as they stand today, but may change in the future: maybe, sooner or later, an automaton will become so sophisticated as to generate a finalistic super-intelligence. Well, I don't think so, and I also think that the idea of an automaton perfectly identical to a responsive soul is untenable for five reasons.

The first is the *imbecility argument*. Human intelligence is not a superintelligence, and needs automata, whereas artificial intelligence is independent and super-intelligent. To conceive of superintelligence as an enhancement of human intelligence means overlooking the fact that human intelligence, in addition to systematically requiring the input of automata – which is obviously not the case for an automaton – is much less self-interested than superintelligence, which is explicitly intended for cognitive purposes. When Woody Allen calls his brain his second favourite organ, he captures a typical fact of human intelligence that no automaton will ever really have, except in a simulated form introduced from the outside. As for the finitude immanent to corporeality, even a technical device can definitively interrupt its own functioning, and this indeed happens most of the time. But while we complain about a hair dryer that lasts too little, no one could ever reproach a human being for dying prematurely, and if suicides are viewed negatively, it is precisely because they are definitive, that is, they have nothing in common with the stand-by mode of a computer. The second is the *finality argument*. Human intelligence is found in a body that provides internal purposes, whereas superintelligence, being an automaton, would only have external purposes. In this sense, superintelligence will always encompass the realisation of a human purpose, much like a hammer or a bottle of Bourgogne.

And, conversely, has the reader that translates a sequence of bars into a numerical series not "interpreted" information? In this sense, can it be said that the optical reader in the supermarket, when it works, reads the price per kilo of carrots with

---

Believing that automata will one day replace us is a very ancient fear, and one that has obviously increased with the progress of technology. However, the fact remais that finality is not a detail: machines can do everything, but they do not run on their own, they require the temporality and purpose that come to them from human needs. Precisely for this reason, and by definition, machines do not design themselves, because production entails a purpose that can only come from the system of needs and irreversible processes characterising humans. What is at stake here is not intelligence, but reason as the faculty of purposes. One may object that very many living beings have no awareness of death, and that most humans believe themselves, deep down, to be immortal. But it is not necessary to have read a treatise on metaphysics, nor to have practised some form of authentic existence, to feel the pangs of hunger and thirst, which remind us, with a clarity that would be uselessly sought in some manual of devout life, that we are dust. For this very reason we are interested in meaning, truth, and what is so obscurely called the "world spirit". All these things are beyond the reach of superintelligence, but are perfectly clear to Forrest Gump or a beaver. The third is the *disobedience argument*. Values are "loaded" into superintelligences much as they are with souls, as we have seen: from this point of view, the latter turn out to be identical to automata as regards the genesis of intentionality. But these values would be null and void if they did not foresee the possibility of transgression, which is true for responsive souls but not automata, however superintelligent they may be. When Leibniz describes the human soul as a spiritual but free automaton he is referring precisely to this circumstance. Indeed, we have no evidence that our actions are at all free, but we have endless evidence of how often we deviate from our plans, through boredom, weakness, or a whim. And it is indeed difficult to conceive of a superintelligence subject to akrasia – and, frankly, why should one conceive of such an intelligence at all? We humans are more than enough to apply the principle *video meliora proboque, deteriora sequor*. The fourth is the *simulation argument*. It could be objected that the brain in a vat – the one which, in Hilary Putnam's famous thought experiment, is placed in an organic tank by a mad scientist and stimulated with electrodes – still has experiences. That brain thinks and feels that it has a body, so, from the perspective of superintelligence, the creation of automata simulating the fact of having a body should not in principle be ruled out. And, in fact, Tamagotchis already simulated organic needs and requests for care, much like dolls that pretend to pee and require a change of nappies. But this is simulation, not reality. A computer that believes it has a body and organic needs dictated by metabolism and the finitude of life could certainly *delude* itself into wanting power, but its delusion would be even less grounded than that of a madman who thinks he is Napoleon – at least the latter would really *have* a body. Finally, there is the *discontinuity argument*. Of course, one can posit that the development of superintelligences may not be linear (and therefore reduced to a simulation of human intelligence) but may proceed with unpredictable discontinuity. But in this case we would have a change of state, and we would find ourselves speaking of "intelligence" in a radically different sense from the one we use for human intelligence. In fact, there already are examples of superintelligence – think of bats endowed with sonar, or dogs and cats with a superhuman sense of smell – yet no one would seriously think of speaking of "superintelligence" in these cases, but simply of intelligence different from our own. Otherwise we come dangerously close to statements like the Italian saying "If my grandmother had wheels, she'd be a bicycle".

more "spirit" (because it engages in interpretation) than the clerk when they have to take over from the reader because the label is illegible? Of course not. It is clear that the bar code reader does not have the same experience as the cashier, but this does not derive from the fact that the latter has a particular "understanding" that their automatic colleague does not have, but from the circumstance that the cashier has a body, senses, experiences, memories, and, presumably, emotional ties that the bar code reader lacks. The difference between natural intelligence and artificial intelligence does not consist in the quality of the operations performed, which are in any case a manipulation of signs, but in the fact that in the case of natural intelligence these operations take place *in an organism*, which is destined to interruption, while in the case of artificial intelligence they take place in a mechanism, which is destined to iteration. The differences we find in human intelligence – where it would be better to speak of "reason", as will become clear in a moment – such as the fact of having goals, of being conditioned by emotions, of showing social openness, depends not so much on the fact that humans have a brain, but on the fact that this brain is a living organ that is in turn part of an organism, which fits into a technical and social context. This leads to a crucial consequence, which follows from the difference between soul and automaton discussed in 2.1, and whose implications I'll describe in 3.4: the difference between natural intelligence and artificial intelligence is not so much about understanding but rather about purpose.

In production, artificial intelligence, designed to execute orders, works better than any human agent, just as a mechanical arm can score much more at basketball than any champion. But no human would be willing to watch a basketball game between mechanical arms. And building spectator robots makes no sense, and above all has no purpose. One can rightly imagine that if we watched a football match without knowing that the aim of the players (apart from the two goalkeepers) was to score goals, we would not understand the reason for all the excitement on the pitch. There is another and equally obvious aspect that deserves to be considered in times of increasing automation: a football match must be played by human players, a car race must involve human drivers, a parliament and a hospital must have human parliamentarians and human doctors. Sports, healthcare, and political representation require, in their most complete form, human intervention, that is, a responsiveness made up of a bodily component embedded in a technological context (providing a supplement to the bodily component).

Artificial intelligence knows better than anyone else that *logos* has become flesh, and it cares little for what we value most about natural intelligence, but appreciates very much the forms of human life – how they move their hands, how they speak, what they desire and what they fear – which depend essentially on the fact that this *logos* is embodied, inserted into a body that dictates times and needs, imposes ends and above all has a certain end. The fact that God became incarnate means that by this act he became an authentic player in human affairs, just as the Greek gods were, and as a pure spirit or soul would not be. Reciprocally, resurrection must be of the flesh to be meaningful. The mere resurrection of the spirit, besides making it impossible to identify the resurrected ("do you see that little cloud at the bottom? That's your grandfather"), would be to actual resurrection what a video game is to a car race or a football match. Which is to say that even in the afterlife it would be difficult to interact without bodies – let alone in life. This is a decisive point that is often underestimated. Personally, I doubt that we will be resurrected, but I am certain that here, in this world, the interaction between our finitude and the boom of hysteresis in the documedia revolution will be able to provide a new meaning for a new world, provided we have the patience to seek it out rather than retreating into nostalgia for times past – the only form of hysteresis about which I feel neither tenderness nor sympathy, and least of all, of course, nostalgia.

With respect to the orthodox version, I would like to add an important point, which is that *logos*, at least in our case, did not exist before the flesh[49] and technology. Technology makes up for the deficiencies of our bodies, building a system of tools and a society, which would not have come into being without a fire around which to gather, a language with which to speak, and a scripture or rites and myths to transmit to future generations the legacy of the previous ones. Incarnation changes a lot for God, but it would change even more for machines. Mechanisms receive their purposes from outside, so they have explicit purposes that, however, are not self-given. Organisms, on the other hand, only have internal purposes, which manifest themselves as the set of needs, volitions, movements, sensations and feelings that characterise the functioning of an organism as opposed to the functioning of a mechanism. And it is precisely this organic rootedness that underlies reason as the faculty of purposes. The brain is no more made to think than the eye is made to see and the nose is made to have glasses rest on it. Instead, the idea of a useless machine, of an automaton with no purpose, is nothing more than an intellectual provocation. It is perhaps also for this reason that very few contemporaries have taken seriously the vague hope of res-

---

[49] A. Noë, *Out of Our Heads: Why You Are Not Your Brain, and Other Lessons from the Biology of Consciousness*, Farrar, Straus and Giroux, New York 2009; Id., *Action in Perception*, The MIT Press, Cambridge (MA) 2004; A. Clark, *Being There: Putting Brain, Body, and World Together Again*, The MIT Press, Cambridge (MA) 1997; S. Hurley, *Consciousness in Action*, Harvard University Press, Cambridge (MA) 1998; H. Putnam, *The Threefold Cord: Mind, Body and World*, Columbia University Press, New York 1999.; E. Thompson, *Mind in Life. Biology, Phenomenology and the Sciences of Mind*, Harvard University Press, Cambridge (MA) 2008.

urrection,[50] which, on the other hand, if it were possible, would turn the soul into an automaton, an iteration device.

This circumstance deserves some thought. Because if it were possible to rise again, a mechanism would be activated that would destroy any possible meaning – this was noted by Augustine, who was indeed committed to finding a meaning and an end to history. Christ dying and rising cyclically is no different from an actor playing the same part every night. In order to have meaning, it is necessary to maintain the cooperation between an ending soul and an iterating automaton: that is why we find nothing strange or abnormal about achieving immortality through letters.[51] Of course, traditional immortality, with the body and all, appears more seductive, but only apparently so. Having a body does not only mean being situated in an environment, as is often claimed, but also entails *forward projection*, planning, physiological needs. Herein lies the origin not only of consciousness, but of intentionality. On closer inspection, those who militate for the hypothesis of intentionality do not do so for the need to explain the specifics of the psychical realm or of society and their respective genesis, but rather to account for the fact that in society and in life there are things that we call "intentions", "aims", "needs" – i.e., things that we do not find in inanimate matter. We understand this very well, except that to explain it there is no need for an *ad hoc* faculty: it is enough to recognise a series of functions belonging to inorganic, organic and mechanical elements, that is, what I have called "responsiveness": life as we know it is the result of the interaction between a body full of needs and technologies, including the symbolic structures which, in a responsive animal such as we are, are translated into consciousness, intentionality, will.

---

[50] An influential American analytical philosopher, Peter van Inwagen, in 1978, seriously argued for resurrection (you can find the essay in his book *The Possibility of Resurrection and Other Essays in Christian Apologetics*, Westview Press, Boulder 1998): at the moment of death, God – cremation permitting – would take our body, or a relevant part of it, for example the brain and the peripheral nervous system, set it aside to preserve it for resurrection, and put a simulacrum in its place, destined to decompose. Quite a convoluted theory, I think. More recently, another analytical philosopher, this time a French one, has tackled this subject (R. Pouivet, *De van Inwagen à saint Athanase, une ontologie personnelle de la résurrection des corps*, in "Klesis, Revue philosophique", 17, 2010, pp. 98–123) by rehabilitating the traditional Thomist interpretation: the soul does not exist before the body, it is generated together with it. When the body is recreated, through a miraculous intervention, on the day of resurrection, the soul rises along with the body. Pouivet told me about the scepticism and difficulties he encountered in dealing with these issues during a seminar for doctoral students. He said it was as if he were talking about zombies – a topic that, incidentally, has risen to philosophical glory over the last few years thanks to the Australian philosopher David Chalmers. And indeed zombies (i.e. the undead) seem more acceptable to common sense than resurrection.

[51] Cf. U. Eco, *Five Moral Pieces*, Penguin Random House, New York 2003.

It is not surprising that an automaton inserted in a body, a soul, is capable not only of setting goals ("I would like to reunify Germany under the Hohenzollern dynasty", if the soul is Bismarck), but also of illuminating an archaeology in the light of this teleology, that is to say, to achieve a retrospective re-reading of meanings. Freud considered this the prerogative of psychoanalysis (the *Nachträglichkeit*) but is actually found in the life of every person (there must have been a moment when Napoleon III realised that the insolent reworking of the Ems telegram was intended to provoke him). Indeed, this archaeology is the professional activity of historians, the masters of hindsight, just as good strategists are masters of teleology. It is this interweaving of archaeology and teleology that provokes the first-person effect differentiating the soul from the automaton; if one wants to call it "understanding", one can, provided that one does not see it as a difference between artificial and natural intelligence, but rather as the mortal illness of souls, which has nothing to do with intelligence.

What artificial intelligence teaches us about natural intelligence is the perfect identity, *from the point of view of intellectual performance*, between soul and automaton. In other words, it confirms that positing something qualitatively different between natural intelligence and artificial intelligence means postulating a spectre in the machine,[52] a *homunculus* animating mechanical performances. The performances are mechanical, yes, but there is no *homunculus* behind them. Certainly there is no *homunculus* understood as an incorporeal entity, and if we understand it as an organism, it is outside the machine and not inside it. We need to overturn the Pentecostal approach: the spirit, thought, is a mechanism. It is the most mechanical and document-based thing there is. Feelings and reason are about the soul, and the soul is not about intelligence, but about corporeality and the purposes that derive from it.[53]

This also applies to personal identity, which is not simply memory, but memory located in a body in accordance with the principle that there is no memory without matter. The body, before the part of it we call "mind" identifies us and tells us who we are, is a character we share with other organisms, but is inconceivable in reference to a mechanism. When I transfer my archive from one computer to another, the nature of the archive does not change along with the physical medium that hosts it. If, however, I transplanted my brain into the body of another person, it would no longer express the same intelligence, and we can be sure that my responsiveness (perceptions, tastes, preferences, and of course purpose) would soon be radically transformed. Above all, the organism is an impe-

---

[52] G. Ryle, *The Concept of Mind*, Hutchinson & Co, London 1949.

[53] Cf. Stefan George on the soul as opposed to the spirit. L. Klages, S. George, *L'anima e la forma* (1902), Fazi, Rome 1995.

rious master, it does not wait, it cannot postpone its hunger, thirst, or fatigue until tomorrow. And this is why humankind has resorted to hysteresis, inventing devices that do not simply make up for our shortcomings, but allow for accumulations of resources to cope with our urgencies. Human needs would make no sense for a mechanism, and indeed determine an ontological leap with respect to both non-human souls and automata.

This changes everything. Can you imagine an automaton itching to go on holiday, to change political or religious opinion, to get a new job, favourite restaurant, home, gender, or partner? No. Is it because it is more rational than we are? No, the reason is the opposite: it lacks reason completely, even though it is full of intelligence. Rationality is rightly considered to be the specific character of the human being as a responsive animal. The same cannot be said of intelligence, which is present not only in many animal species, but also and increasingly in a large number of automata. Instead of defending the indefensible, attributing some mysterious property to human intelligence, we must recognise that natural intelligence, unlike artificial intelligence, is related to reason as the faculty of purposes, because natural intelligence is the expression of a body, its needs, temporality and intentionality. As a result, while there are good arguments for attributing rationality to some non-human animals, there is no argument for the rationality of artificial intelligence. As far as the capacity for complex operations is concerned, natural intelligence is equal or inferior to artificial intelligence, which works in the same way: neither I nor the computer need to know the fundamentals of arithmetics to perform division or multiplication. The only difference is that, if the numbers are complex, I must rely on external devices such as pen and paper, and if the numbers are very complex, I will have no choice but to resort to a calculator. The soul therefore contains an automaton, a technical element. But in all three cases under consideration (simple, complicated, very complicated calculations), the meaning and purpose of the operation depend on the soul, on its urgency and necessity. From this, in conclusion, we can draw four arguments.

The first argument is that of organicity. If having intelligence is more or less equivalent to having an automaton inside and outside of us, being a body entails very different, and much broader, consequences. What is unquestionably different about the human compared to the automaton and to some extent to a non-human animal (e.g., having goals, being conditioned by emotions, showing social openness) depends not so much on the fact that humans have a brain, but on the fact that the the brain is a living organ that is itself part of an organism. It is corporeality, with its needs, that makes human intelligence develop in a social context and in constant relation with technology. In this framework, we can understand the essential role of responsiveness. Insofar as it is embedded in a body, intelligence has functional principles that are not found in machines. But insofar as the human body is embedded in a technological and social context, the bodily root is enhanced through the technological and social supplement.

The second argument is about anthropophilia. People wonder – and, of course, answer negatively – whether robots will take over and, turning to a more tender aspect, whether sex robots will ever be capable of genuine feelings, and whether interaction with them is permissible or disrespectful of humanity. Yet an aristocratic marriage, or a peasant marriage in the 18th century, had not a single one of the characteristics of internal purpose, respect, and idealised sentimentality that are typical of bourgeois marriage and the romantic sexuality that surrounds it, and which are put forward as ideals valid for every human being in every age. People also debate the reliability of self-driving cars, perhaps while lamenting the disappearance of another job – presumably lawyers specialising in defending people driving under the influence. More seriously, people complain that there can be no future or projectuality, because algorithms are made to work in the very short term, reflecting the momentary instincts of an anonymous multitude; but, on the other hand, they fail to take into account that throughout history tyrants had a projectuality of their own that most people would gladly have done without. In all these cases, there is a perspective error, which consists in contrasting real automata with ideal humans.

The third argument is that of rationality. Reason is the emergence of a conceptual purpose above the natural purpose, which in any case is its foundation. And if there is, has been and will be artificial intelligence, there can never be an artificial *reason*, because reason arises from the encounter between the urgencies of an organism, which sets the time and objectives, and the resources of a system of mechanisms, which strengthen the urgencies and implement them. Reason thus appears to be characterised by a mixture of calculation, which is our mechanical component, and incalculability, which is our organic component, structurally open towards the future, and therefore endowed with a direction. Just like consciousness and spirit, free will is not an illusion, just as colours or time are not an illusion.

My fourth argument is about freedom. Obviously, we are subject to efficient causes just like any physical object, but in addition to that, as organisms, we have final causes directing our behaviour. Our conduct is free because it does not take place in a closed system (the rare times when this happens, we say things like "I had no choice"), but in an open and therefore incalculable system, determined by an immensely large number of causes. Born at a time when explaining meant identifying the causes of something, the debate on free will necessarily leads to the denial of freedom, since it is obvious that an effect, once it has occurred, could not but follow from the cause that produced it. If we see the scene a second time, it will become clear that Sam *had* to play *As Time Goes By* again and

Blücher could only precede Ney on the fields of Waterloo, and that, for the same reason, we couldn't help but do whatever we did.

But let us look at the scene from the other side, *before* it took place, and bearing in mind that it may well have gone differently. In other words, let us look at it from the point of view of the end (goal) and the end (finish point): nothing is decided and nothing is determined because nothing is yet. Interruption, especially if it is irreversible, gives a direction to the process and outlines its meaning, in the theory of meaning as well as in the philosophy of history, in the granting of loans as well as in the life of each of us. And the fact that, in front of each of us, there is a nothingness or a not-yet that might not be actualised following a causal order is precisely what makes us free, what gives us an end (goal) precisely because there is an end (death). Teleology, the fact that we find ends for ourselves, our neighbours and the world, originates precisely from this encounter between the organic and the mechanical. This can help us find meaning in what we do, and what others do, but at the same time it is the cause of the perverse tendency (of humans and not, as far as we know, of animals and automata) to hypothesise super-individual, intentional and evil or benign entities intervening in their path.

By definition, free will is not manifested in experiments where subjects are asked to press two buttons quickly and without emotional involvement,[54] with the entirely predictable consequence that the action performed without thinking precedes the neural discharges in the brain. Humankind is overestimated as the bearer of absolute and innate freedom. But we forget that Moses came down from Mount Sinai carrying a tablet of ten algorithms that had been dictated to him by the Supreme Programmer. And that, coming from myth to history, the steps that led to the outbreak of the First World War, or determined the collapse of the New York Stock Exchange, are to a very large extent algorithmic in character ("if X, do Y") even though there was not a single computer on the face of the Earth. The difference does not lie in how intelligence works, but precisely in the fact that artificial intelligence is found in a machine, and natural intelligence in an organism. And this changes everything, because an organism has objectives, desires, aims and fears that are completely absent in a machine, and which define an opening towards the future that transcends the order of causes, precisely because they refer to something that is yet to be. This is why a computer can do astronomical calculations without getting bored or tired, but also why it is futile to fear that the computer will take over: it will not, the machine will not do anything at all if humans do not ask it to. It is therefore humans that we must

---

[54] Like Libet's famous experiment. Cf. B. Libet, *Mind Time. The Temporal Factor in Consciousness*, Harvard University Press, Cambridge (MA) 2004.

watch out for: it is humans that might become too full of themselves and choose to become great dictators.

3.3. Epistemology: Alteration

The third phase of hysteresis is alteration: iteration can result in a qualitative change, for example – as noted in 1.1.3 – *praxis* can become *poiesis*. This alteration is a passage not only from event to object, but also from the state of affairs to the proposition that refers to it, and that is why the paradigmatic sphere of alteration is epistemology, what we know or believe we know. Indeed, epistemology is a reorganisation of the known that leads us to the unknown, a doing that becomes a knowing and, more profoundly, a passage from automaton to soul, from what is simply mechanical to what is human, that is to say, born of the interaction between organism and mechanism. *Praxis* has, so to speak, hysteresis behind it, as the iteration of gestures learned and perfected through exercise; *poiesis* also has it in front of it, with all the effects deriving from imitation and exemplarity, that is, from following the example of previous actions.

Here, too, I would like to emphasise a substantial point. The relationship between ontology and epistemology guaranteed by technology is not one of construction, from top to bottom, but one of emergence, from bottom to top, whereby epistemology emerges from ontology through technology, i.e., competence without understanding. At a certain point, iteration can turn into alteration, into the reconfiguration of a given field that is independent of knowledge. This alteration, following a technically determined method, may generate a meaning that was not previously visible: the butler did it; the Earth is round; Madame Bovary is a fictional character. In this framework, epistemology is the (not necessary, but always desirable) result of the encounter between ontology and technology, between recording and iteration, that can give rise to something qualitatively different, namely alteration. In fact, iteration is not only the possibility of guaranteeing the persistence of objects in space and time, but also the possibility of differentiation, i.e., of introducing variation in repetition. Think of the genetic changes in species on the basis of random mutations, made possible thanks to an enormous quantity of time and an even greater availability of matter. Moving from biology to epistemology, thanks to this same process, something qualitatively new entered the universe: truth, which, just like taxes, debts and jokes, would not exist without humans – even if that does not mean that it is relative (I point this out as a warning to relativists and tax evaders). This emergence is such precisely because it rests on a foundation of iteration, without which alteration would be strictly inconceivable.

Let's have a closer look at this last point. *Iter*, "again", and *alter*, "otherwise", derive from the same Sanskrit root, *itara*.[55] Reciprocally, it is worth highliting the intrinsic duplicity of the notion of *inventio*, which indicates the retrieval of the old, the archival search for precedents (to make an inventory) and the construction of the new (inventing in the proper sense).[56] In ancient rhetoric, this is the first thing an orator must practice: searching for precedents, common places suitable for persuasion. However, it is precisely the search for the old that allows the invention of the new, not only because the old is not necessarily known to the listener, but above all because a different arrangement or a changed context makes the old appear new.[57] The modern rhetoric of creativity, or *creatio ex nihilo*, even in the field of ideas, is blind to the fact that nothing is born from nothing, whether in physics or in the world of ideas, and what appears to be a creation (absolute genesis) is in fact a rediscovery or re-evocation. This should come as no surprise, since this aspect is logical and ontological and not only linguistic. I am aware that this may seem like a futile speculation, but bear with me. Think of a Hopper-like bar, late in the evening: a client sits in front of his whiskey, drinks it, raises his glass and says to the bartender: "Give me another one". This does not mean that he wants a different drink (let alone a different glass). He wants the same whisky: another glass (alteration) of the same drink (iteration).

Natura non facit saltus, and that doesn't just apply to bars. There's no real leap from nature to culture: a memory can become corrupted and complicated to the point of becoming an original thought; a long series of data can be simplified to the point of becoming a theory; a series of regulated actions can deliberately treat a part of the world in such a way as to obtain true statements about that part of the world (I get on the scale, look at the display, utter or think the phrase "I weigh 86 kilos, I need to lose weight"). The leap, rather, takes place between ontology and epistemology. On the ontological level, in the case of the whisky glass, there is iteration (it is still a whisky glass), but on the epistemological level there is alteration (it is another, one more, a second glass). Indeed, this is the beginning of a process of capitalisation and, in this case, intoxication: "first a man takes a drink; then a drink takes a drink; then the drink takes a man".

---

[55] *Itara* (adjective); Vedic *itara* (= lat. *iterum*, "a second time"): other, second, next, different.

[56] G. Simondon, *Imagination et invention (1965–1966)*, Éditions de la Transparence, Chatou 2008; C. Westbury, D.C. Dennett, *Mining the Past to Construct the Future: Memory and Belief as Forms of Knowledge*, in *Memory, Brain, and Belief*, ed. by D.L. Schacter and E. Scarry, Harvard University Press, Cambridge (MA) 2000, pp. 11–32.

[57] "*Inventio* refers not so much to invention (of arguments) but rather to discovery: everything already exists, you just have to find it again: it is more of an 'abstractive' than a 'creative' notion". (R. Barthes, *L'ancienne rhétorique [Aide-mémoire]*. In *Communications*, 16, 1970. *Recherches rhétoriques*, pp. 172–223).

The technological practice that secures the link between ontology and epistemology is *verification*. "To verify" comes from the Latin *veritas facere*: to make the truth. This has two sides to it: that of invalidating the proposition "snow is white" (if snow is not white) and that of convalidating it (if snow *is* white). The example of snow, though canonical, has the defect of concealing the technological element of truth-making, which is why I'd propose another one: there are 22 beans in a can (ontology); I count them (technology); I utter the sentence: "there are 22 beans in this can" (epistemology). The sentence is true for any rational being (teleology).

In this framework, instead of understanding the ontological foundation as a "truth maker",[58] given that ontology is what provides the matter, I would suggest indicating the ontological layer as a "truth bearer". The function of "truth maker" instead, given what I said earlier, goes to technology, which indeed is responsible for making the truth. Lastly, epistemology has the function of "truth teller" and teleology (the end for which truth is made, which is not a natural fact) is the "truth user". Truth bearers are the states of things with respect to which a proposition is true (the beans in the can); they therefore constitute the ontological element of the process. Truth makers are the operations necessary to produce true propositions about states of affairs (in this case, counting the beans)[59]; they therefore constitute the technological element of the process. Truth telles are the propositions necessary to communicate the results of the operations carried out by the truth makers with reference to the truth bearers (in this case, the proposition "there are 22 beans in this can"); they therefore constitute the epistemological element of the process. Finally, truth users are the recipients of the message, because without subjects the truth would make no sense, and have no "address" to be sent to.[60]

---

[58] According to the canonical version, K. Mulligan, P. Simons, B. Smith, *Truth-Makers*, in "Philosophy and Phenomenological Research", 44, 1984, pp. 287–321.

[59] That the truth maker should be interpreted in a technical and practical sense, rather than as an ontological foundation, is an insight found in Derrida, *Le Facteur de la Vérité* (*Poetique*, 21, 1975, pp. 96–147) (which plays on the fact that "facteur" in French also means "postman").

[60] In this framework, in addition to differentiating the elements that make up verification, I propose three theories of truth: *hypotruth*, which corresponds to mainstream hermeneutics; *hypertruth*, which is mainstream analytic theory, and *mesotruth*, which is the one I argue for in this book.

Hermeneutic thinkers have developed an epistemic theory of truth which is in fact a hypotruth, i.e., a subordinate truth, since it is detached from ontology and consists rather in the conceptual schemes that mediate, and in fact constitute, our relationship with the world. In this version, with varying degrees of radicality, "true" becomes synonymous with "conforming to a shared belief". Thus, hermeneutic thinkers rightly note that truth does not go without saying but requires context and actions, but they go too far when they claim that the truth consists only in the verification procedures, and that the idea of a world "out there", independent of our concep-

3.3.1. Truth bearers

The first element in the process of verification is found in truth bearers: the state of things, which exists independently of all knowledge – viruses, beans, galaxies ... As said, ontology does not depend on how much we know or believe we know. And epistemology is true if it corresponds to something that is there. Now, what

---

tual schemes, is a pre-Kantian naivety. Not only do they provide a theoretical bailout to post-truth (which doesn't know what to do with it), but, above all, they miss the opportunity to give hermeneutics its proper dimension, which is technological and not ideological.

Most analytics have developed a very strong notion of truth. I'll call it hypertruth, because it postulates a necessary correlation between ontology and epistemology, in which the proposition "snow is white" is true (epistemology) if and only if snow is white (ontology) implies: if snow is white, then it is true that snow is white, so that it would be true that snow is white even if there was (now or ever) no human on the face of the Earth. For the supporters of hypertruth, if it is true that salt is sodium chloride, then this proposition was also true for Greeks in the Homeric age, although they did not have the tools to access that truth. To describe the truth as the relationship between the proposition "snow is white" and the fact that snow is white is an easy thesis to share. From this, however, supporters of hypertruth draw the conclusion that that proposition would be true even if there had never been a human being on Earth capable of formulating it. And this is far from obvious.

This second part of the thesis of hypertruth appears motivated by the concern that, otherwise, hermeneutical and hypotruthist drifts would be given too much room. But this is by no means inevitable. For example, Heidegger's thesis according to which, before Newton, the theories he enunciated were not true is not in itself relativistic: Newton's laws (epistemology) did not exist, but the reality to which they referred existed (ontology). Newton's work consisted in revealing something that was already there. To say this does not mean – regardless of Heidegger's conclusions – to claim that the motion of the planets was created by Newton, but that the true conception of the motion of the planets depends on the system thanks to which Newton was able to elaborate his laws (in this case, mathematics, unthinkable without pen and paper). Likewise, the true conception of the Medicean Planets depends on the technical device (in this case, a telescope) with which Galilei discovered them, and this conception is not so despotic as to turn into planets what in fact are the four main satellites of Jupiter.

The fact that salt is sodium chloride or that there once were dinosaurs on Earth does not depend in any way on us and our conceptual schemes. However, it does depend on us that there is such a thing as chemistry (which might very well never have been developed), that we have studied bones and fossils, and that we have come up with classifications and interpretations. So, the planets were there before Newton's laws and of course they were exactly what they were without the intervention of any conceptual scheme. To claim instead, with the proponents of hypertruth, that these laws were true even before they were discovered is tantamount to either formulating a meaningless assertion, or – *involuntarily agreeing with the supporters of hypotruth* – to making interactions between planets depend on conceptual schemes.

Both hyper- and hypotruthists live under the constraint of Kant's dictum that "intuitions without concepts are blind": being a philosopher means dealing with concepts, either to analyse them like hypertruthists or to deconstruct them like hypotruthists, neglecting the other side of the argument, namely that concepts without intuitions are empty. Above all, both overlook the fact that intuitions often have the tendency to *disprove* concepts, expectations, rules. But if this

does it mean for a proposition to correspond to a thing or a state of affairs? It is this difficulty that usually generates alternatives to correspondentism. The fact that no satisfactory alternative has yet been found, however, suggests that the defect does not lie in correspondence, but in conceiving it as a kind of contemplation, which of course it is not. If this were the case, the truth would be within everyone's reach, and error would be inexplicable, or explainable purely as a kind of eye defect: after all, that is how Plato puts it. If, on the other hand, we recognise that there is an intrinsically technological component in truth, i.e., competence without understanding, the problem is easily solved. This explains why the correspondence theory of truth is both intuitive and obscure. It is intuitively clear that the sentence "snow is white" is true if and only if the snow is white, but the processes that led to this observation transcend our immediate understanding, and do not require it. Even a machine is capable of making the truth. We can see it very well in the many tools that humanity has been using throughout its history – scales, abaci, calculators – which are generally capable of making the truth, even if for them the truth means nothing (and from this point of view the most sophisticated computer is no different from a scale). The indispensable condition for being a truth bearer is not to depend on the subjects that make the truth, on the selves that want to know something. Yet the risk of this happening, if you give it a moment's thought,[61] is minimal, as we have seen when discussing ontology.

---

is the case, how does one move from epistemology to ontology? Only by a negative route, when reality resists us? No, there is a positive way. Hypotruth and hypertruth are opposed by what I call *mesotruth*. This term is due not so much to the fact that it is halfway between the two, but rather to the fact that it insists on the role of *technical mediation* between ontology and epistemology. In mesotruth, truth is neither the epistemology that shapes ontology (as claimed by hypotruthists), nor the ontology that is reflected in epistemology (as argued by hypertruthists), but a three-term structure, which includes ontology, epistemology, and *technology*. The latter is to be considered as the element, so far largely underestimated by philosophers, that ensures the transition from ontology to epistemology, and that allows the truth to be *made*. For mesotruth, truth is the technological result of the relationship between ontology (what there is) and epistemology (what we know). This relationship does not require prior knowledge, since, as we saw in the previous chapter, there can be – indeed, most often there is – competence without understanding.

[61] The appeal to realism, in the non-speculative sphere, consists simply – though this is no small matter, given the spread of contrary beliefs – in emphasising the independent character of reality with respect to the knowledge we have of it. For negative realism it is primarily a matter of accounting for a salient feature of objectivity, namely invariance (R. Nozick, *Invariances: The Structure of the Objective World*, Harvard University Press, Cambridge (MA) 2001). Invariance is the circumstance whereby an objective fact must have three characteristics: accessibility from different points of view, the possibility of intersubjective agreement, and independence with respect to the beliefs, hopes, and interests of the observers (it is worth noting that the concept of "invariance" also plays a central role in one of the first and most relevant theorists

When we came into the world, the world was already there, and not as some inert residue, but as a structure from which living beings emerge with their social and ideal worlds. Indeed, coming into the world means coming *from* the world, a world that surrounds us and comes forward with a complex reality, made up of obstacles and resources. So, this being-in-the-world makes it implicitly senseless to ask how we access the world. If I were to wake up in an unknown room, the right question for me to ask wouldn't be "How can I know this room?" (of course

---

of quantum physics, P.A.M. Dirac: "The important things in the world appear as the invariants [...] of [...] transformations"; cit. in Nozick, *Invariances* cit., p. 76). These three elements are also distinctive criteria of realism as opposed to correlationism, the thesis according to which: "In general there is no subject as subject or object as object, but [...] knower and known are one and the same, and therefore, as such, neither subjective nor objective" (F.W.J. Schelling, *System der gesammten Philosophie und der Naturphilosophie insbesondere* [1804]). Cf. G. Harman, *Fear of Reality: On Realism and Infra-Realism*, in "The Monist", 98, 2, 2015, pp. 126–144. Cf. L. Braver, *A Thing of This World: A History of Continental Anti-Realism*, Northwestern University Press, Evanston 2007. Accessibility from different points of view is one of the most historically attested characteristics of objectivity, and also constitutes the (in no way relativistic) origin of the notion of "perspectivism": something is objective insofar as it can be examined from different points of view. Herein lies the difference between a real bed and its pictorial representation (Plato, *Republic*, X, 596a), or between a real city and its representation in a print (Leibniz, *New Essays on Human Understanding* cit., I, I). The possibility of intersubjective agreement, once again, constitutes a salient feature of objectivity as opposed to opinion, and underlies the privilege traditionally accorded to observation over evaluation, where this agreement appears more difficult. This criterion alone is certainly not the ultimate measure of objectivity: if everyone in Salem believes that a woman is a witch, that does not mean that witches exist. And this is obviously true for every object of scientific knowledge, where disagreements can occur even over very long periods of time (this has been the case with physics over the 20th century), just as millenary agreements (geocentrism) can turn out to be unfounded. Independence from our beliefs is, finally, the contestation of the notion of "dependence" that plays a crucial role in correlationism and relationalism. This is why, for example, it is hard to agree with the hypothesis that the existence of objects is related to fields of meaning (there is a field of meaning in which unicorns exist, that of literature, and a field of meaning in which they do not exist, zoology) because in this case existence would depend on belief. This last point suggests, however, that the true ontological measure of objectivity is independence, whereas multiple accessibility and subjective agreement are epistemological features of objectivity, as they concern how things are known by us, and not how things are. From this point of view, *cogito ergo sum* is a principle that oscillates between tautology (it is obvious that I am there: I am thinking) and irrelevance: I could, for example, be anything, for all I know, in which case existing would mean very little. Vice versa, *est quod ego non cogito* seems partially wiser, at least on an ontological level: if the existence of things depended on my thinking about them – and I do mean me, because obviously the other "mes", in this hypothesis, depend on my stressful thinking exercise –, the world would be uninteresting and very tiring. If, on the other hand, I assume that, for example, one of the salient features of my friends and my bank account is the fact that they exist even when I am not thinking about them, then not only do I feel less lonely, but I also assert something that seems to conform to the experience of the world that you and I, who fortunately exist independently of me, agree upon without too much trouble.

I can know it, I'm in it), but rather, like Jonah and Pinocchio in the whale's belly, "How did I get here? Why am I here?" The problem is not how the mind refers to the world, but rather how the mind emerges *from* the world, and the answer to that question lies in the process of capitalisation (iteration and alteration of what is recorded) that we are examining here. In this framework, the process of ideal-isation as an outcome of iteration and alteration is decisive for the passage from technology to epistemology, from becoming to knowledge (or presumed knowl-edge), that is, from iteration (the same is repeated) to alteration (the same is known and becomes the object of understanding). As soon as we have objects, we are also able to formulate statements about something, carrying out processes in which epistemology is revealed as the result of a cooperation between ontolo-gy and technology: the former offers the matter, the latter the process, and epis-temology the form (the proposition "S is p", where p claims to be true of S).

In this framework, idealisation and objectification constitute two parallel pro-cesses. In the first, a trace, insofar as it is indefinitely iterable, escapes the spa-tio-temporal contingency of its genesis and becomes an idea. In the second, an act takes the shape of a material object (the past is repeated by matter) or a social object (the past is remembered by memory). In both cases, we are dealing with two performances made possible by hysteresis. And this applies to all levels: a natural process, such as the transmission of the genetic code; an artifactual pro-cess, such as the chipping of a flint; and a cultural process, such as the formation of a language or a ritual. In each of these cases we are dealing with a succession of repeated acts: the copying of the code, the repeated striking of the flint, the iteration of certain gestures or sounds. Little by little, each of these iterations gives rise to an object. And this occurs both in the form of codified *praxis*, i.e., a meaningful and complete action, for example a ritual dance or a conversational exchange; and in that of *poiesis* or production: an arrowhead, a literary work or a law. This alteration may take various forms: an unconscious and accidental phenomenon, an error in copying the genetic code, the mispronunciation of a name leading to a linguistic transformation, the straw that breaks the camel's back. It can also be the birth of consciousness, of meaning, of understanding, or of the illusion of understanding. Suddenly, an illumination interrupts a cycle of iterations, and tells us – say – "Tat Tvam Asi" or, more prosaically and more ef-fectively, that puerperal fever reveals the existence of hitherto unknown entities called bacteria, or, again, that the poverty of the working class does not depend on their bad moral dispositions, but on the existence of surplus value.

In the strictly epistemological domain, when *praxis* turns into *poiesis* it gener-ates a peculiar product: truth. Unlike an artefact or a social object, this adds nothing to the ontological furniture of the world (there is little point in ordering a cold, frothless, but *true* cappuccino) but much to the meaning that the world has for us. To illustrate this point I would like to mention an image used by Aristotle, which compares sensible experience to a fleeting army.[62] In the flow of experience, the sensations scatter around like a panicked troop, until a soldier stops and reassures his companions, so that the phalanx can get back together. In the same way, the flight of sensations stops when a belief is formed. This then becomes knowledge when we are able to transmit to others the beliefs we (rightly or wrongly) consider to be true and justified.

The Aristotelian image is interesting in many respects: it explains the gradual passage from feeling to thought; it suggests that, just as sensations are fixed, so categories are formed, precisely through the interaction of experiences – that is, with a "rhapsodic" process that seems so reasonable and that yet Kant disliked so much. Most of all, it compares a natural process to an intentional process – the soldier who chooses to stop, and the others who follow him. Between the fixation of sensation and the soldier's decision there are many differences, but there is no real gap, for Aristotle – there is no magical property that differentiates the soldier's decision from the way in which water, looking for its way down, turns right or left. To put it in my terms, we first have the flight of sensations, which does not depend on the chaotic nature of the world, but on its complexity: this is ontology. Then we have the creation, through iteration and its correlates (idealisation and objectification) of a system open to understanding: this is technology. Then, as a fortuitous circumstance (the soldier might not have stopped, or might not have been followed by the others), there is alteration, the formation of something comprehensible and linguistically expressible. Finally, leaving the military metaphor but sticking to this Aristotelian passage, we have interruption and teleology. Truth becomes knowledge: it accesses its full epistemological dimension only when the person who possesses it is capable of communicating it to others. That is to say, once again, that there is no truth without truth-users.

3.3.2. Truth makers

Let me get back to what I was saying. The second element is technology, the truth maker. The bean count is an analytical operation (it is necessary to count the beans), but of course one can also perform synthetic operations. Indeed, the latter offer something that is usually, and for good reasons, more interesting than analytical knowledge: there is little interest in knowing, analytically, that no bachelor is married – it is much more interesting, synthetically, to know that a certain bachelor is a good friend of Emma Bovary's. Now, when in primary school we are asked to calculate how long it takes for a tub of water to fill up, we are asked

---

[62] Aristotle, *Posterior Analytics*, 100a 4–14.

to formulate a synthetic a priori judgement, since the answer cannot be derived from the analysis of the flow of water and the capacity of the tub, but from the synthesis of the two. In other words, Kant has allowed himself to be distracted by the a priori requirement and has neglected the most interesting aspect, namely the synthesis, the operation, the truth maker.

What did Kant say? He argued that an operation such as $7 + 5 = 12$ is a synthetic a priori judgement: 12 is not thought of in either 5 or 7, because 12 can also be obtained by adding $6 + 6$ or $11 + 1$. Well, good for the a priori. The most important thing, however, is that this operation is a technology, it is doing and not knowing, following a rule that does not require any understanding, but only a form of training. The real keystone lies hidden in a place that Kant did not foresee and ultimately did not even understand: synthetic a priori judgements are not to be found at the beginning of the process, in the table of judgements that, for Kant, precede the categories. Instead, they must be found further down, in the schemes: the methods of construction that for Kant mediate between concepts and objects. In other words, synthetic a priori judgments are found in technology, which can, though does not have to, become epistemology. If we think about it for a moment, the law $O = RA$ represents, in the case of social objects, the equivalent of $7 + 5 = 12$ in the case of ideal objects; and like these, it can be iterated many times with capitalisation effects.

In the case of social objects (promises, bets, holidays, money ...), just as in the interpretation of force in both Kant and contemporary physics, there is a force which, manifesting itself through a form (which need not necessarily be understood: how much do we understand about the terms and conditions we mindlessly accept on the Internet?), produces an object. This object is literally a synthesis in the Kantian sense: a social object, say, "marriage", cannot be deduced from the analysis of its components – which, if you think about it, explains a lot. This synthesis lies at the basis of natural objects no less than social objects, because the concept of water cannot be derived from the analysis of hydrogen and oxygen.

I have illustrated the ontological foundation of truth with a passage from Aristotle; now I would like to specify its technological dimension by referring to an equally famous passage from Augustine. At some point, in the *Confessions*,[63] Augustine poses an elementary and almost comical question: why should I confess to God, who already knows everything? What is the point of pouring one's heart out to someone who knows more about me than myself? The answer is enlightening: Augustine says he wants to "make the truth", not only in his heart, but also in writing, before many witnesses. Does he mean that one makes the truth just as one makes post-truth? Of course not: one can hardly pass off some

---

[63] Augustine, *Confessions*, X, 1.1.

fashionable nonsense to an omniscient being. He rather means that truth is not only an inner process, but also a testimony that is made publicly and has a social value, and is above all something that entails an effort, an activity, a technical skill. The truth is the production of propositions) about a state of things. It is therefore a *poiesis*, the making of a product. This circumstance has three implications that deserve to be discussed.

The first, and probably most obvious, is the document dimension of truth, its link to the capitalisation processes examined in 2.3 and 3.2.1. Even the greatest truth, if discovered by an amnesiac, would fall into oblivion, and if we do not take care to communicate and record our truth, it would perhaps continue to exist in some hyper-uranium but would disappear from the world of history and humans, the many witnesses to whom Augustine addressed his *Confessions*. This is why truth needs texts, witnesses and documents. And this is why the documedia revolution is imposing so many transformations and raising so many new problems, precisely because, as I said in 1.2, the docusphere is rarely a transparent infosphere.

The second dimension of truth-making relates to the competence without understanding that is characteristic of technology. I can ride a bicycle without knowing the laws of physics, and I can perform body-weight exercises with extraordinary dexterity without having the slightest notion of physiology, just as Homer, much to Plato's annoyance, could describe battles with great pathos without ever having led or even fought in one. Conversely, reading all the cookery books in the universe does not guarantee that I will become a chef worthy of the name. The notion of competence without understanding allows us to solve a long-standing problem. The intuitive definition of truth is the correspondence of a proposition with a thing: the sentence "snow is white" is true if and only if the snow is white. So far so good. The trouble begins when it comes to explaining what we mean by "correspondence": does it mean that words resemble things? This is an impasse much like the concept of time was for Augustine, which is resolved precisely by emphasising the fact that truth has a technological dimension, and like any technology it can function perfectly well even if we do not know why it works.

Hence a third dimension, illustrated by the factual character of truth. Rather than the harmonious development of a hypothetical "human nature", knowledge should be considered as an alteration, much more beneficial than neoplasms, albeit of the same kind. Where there could have been only competence, at a certain point understanding arose, with an alteration that has been maintained and traditionalised in the various forms assumed by knowledge. Two crucial circumstances have played a role in this: the development of *competence* and the outcome of a function that is in itself mechanical, but which in humans constitutes a response to organic needs – *understanding*. In this framework, the concept of truth as a true opinion accompanied by reason[64] is an ideal that is seldom found in reality, not only in the domain of artificial intelligence (it would seem futile to ask a computer to account for the results of algorithms applied to some endless field of documents), but in the world in general. This has been the case since the dawn of time, or better, since humans – as humans – invented tools: it would be strange indeed to demand that a thermometer accounts for its measurements in order for it to work.

3.3.3. Truth tellers

"To say that what is is, and what is not is not, is true". That is how Aristotle put it,[65] and we agree. Truth is something that is first made and then told, so it depends on language (truth does not exist in nature), but of course this does not mean that what is talked about is identical to the language that talks about it, because then our words would be empty.

So, truth tellers always come third. This domain is neither an inert sphere, which only mirrors what is there, nor an overactive sphere, which fabricates everything there is. Rather, it is the field in which concepts take shape. The concepts of "number" or "disabled" do not exist in nature. They are gradually sedimented and culturally selected: concepts like "calories" and "phlogiston" survive, but as historical and non-physical notions; with time, new concepts are formed. In the case of social objects, concepts have not only a descriptive, but also a performative value, as they can highlight ontological unseen layers: workers will discover that they are the victims of the surplus value, Black people will find out they are discriminated against, and hopefully the day will come when humans become aware of the capital they hand out to tech platforms. Nevertheless, even in this case, truth refers to a different ontological stratum that is independent of epistemology, which merely gives it a linguistic guise, taking advantage of the mediation introduced by technology.

There is the same difference between ontology and epistemology as there is between a land and its map. And this difference also implies the dependence of knowledge on being: something must be there for something to be known. Going back to the arguments presented in 3.1.2, the supposed epistemological dependence – the supposed dependence of being on knowledge, which is summed up in postmodern slogans like "being that can be understood is language", or "there is nothing outside the text" – arises from the confusion between the axiological

---

[64] Plato, *Theaetetus,* 201–210a.
[65] Aristotle, *Metaphysics*, 1011b.

relevance of something (language is *important*, history and the subject are *important*, and what's even more important is having a roof over one's head) and ontological relevance. Language, thought and history matter with respect to reality (who would ever deny it?), *therefore they constitute reality* – and this is simply absurd. Now, if by "epistemological dependence" we mean that the existence of, say, the Tyrannosaurus Rex depends on our conceptual schemes, then it follows that when the Tyrannosaurus Rex existed, paradoxically the Tyrannosaurus Rex didn't exist, as we humans didn't exist yet. However, if we mean that the *word* Tyrannosaurus Rex depends on our conceptual schemes, then this is no dependence in any serious sense of the term. The Tyrannosaurus Rex never knew it was called that, which didn't stop it from being exactly what it was.

It seems equally strange to claim that the proposition "salt is sodium chloride" was true *in* Homer's world (that is, to be more precise, at Homer's time). Many philosophers[66] argue that it was rather true *of* Homer's world (that is, to be more precise, for us today considering Homer's world). Now, if by "Homer's world" one means planet Earth, that world is neither Homer's nor ours nor anybody else's, so saying that "salt is sodium chloride" is true of Homer's world" is meaningless. If, however, one means Homer's age, then saying that the proposition "salt is sodium chloride" is true of Homer's era is simply false. Being true today of the world as we now know it – and which probably has many characteristics in common with the world in which Homer lived, although we cannot be absolutely certain – does not equal being true (consequently) of an epoch in which, among other things, it was not known that salt is sodium chloride.

It seems also strange that a proposition like "salt is sodium chloride" (not to mention "stalking is a crime", "observing Christians do not eat meat on Friday", "Nelson won the Battle of Trafalgar") can have meaning regardless of language, and therefore of humankind. So, those same philosophers insist on the non-linguistic but logical nature of propositions. However, I do not think that this changes much, not only because often philosophers themselves speak of propositions as linguistic expressions (and one could hardly do otherwise), but especially because logic, like mathematics and language, is a technology whose existence depends on the existence of humans. Unless one is worryingly anthropocentric, the proposition "there were once dinosaurs on Earth" has only appeared at a certain point in time: before, dinosaurs had still existed at one point, but nobody knew it. And saying that for Homer it was true that salt is sodium chloride is like saying that "dinosaurs exist" (in the present) is true for us.

We have to conjecture a time when knowledge became conceivable: that was the birth of epistemology and the emergence of truth. In fact, what seems to have

---

[66] Cf. D. Marconi, *Per la verità*, Einaudi, Turin 2007.

taken place in a process that began 120,000 years ago and culminated around 5500 years ago is not the birth of truth, but rather its emergence. In general, the first meanings, as invitations found in the environment[67] and as the ancestors of truth, are to be sought in the sphere of pre-adaptation: land, size, hand, fire, bivouac, division of labour. This procedure, on the level of knowledge and reflection, mirrors the process of trace inscription underlying perception and remembrance. It is the intellect that adapts to the thing, in a movement that goes from the world to the mind. This circumstance establishes the link between truth, memory, inscription and documentality in a combined system of ontological correspondence and epistemological coherence mediated by technology. For the former, truth is the correspondence of the proposition to the thing; for the latter, correspondence takes place within a conceptual scheme, a vocabulary, and measurement systems, but these do not override ontology. The sentence "Turin is 140 kilometres from Milan" requires more than a conceptual scheme, and in particular the kilometric measurement of distance, but the latter conceptual scheme does not determine the state of affairs; if, by adopting a different measurement unit, I said "Turin is 86.991966913226 miles from Milan" I would still be uttering a true sentence, albeit a verbose one, which refers to the same state of things.

Beyond the infinity and variety of examples one could offer,[68] the philosophical question is: could these alterations have taken place without iteration? Or, in other words, could an epistemology ever arise from an ontology without the mediation of a technology? Obviously not. It is only repetition – perhaps long to the point of boredom, and above all full of mistakes, mixing up with barbarisms and solecisms – that has determined the passage from the Latin *sermo rusticus* to the Neo-Latin languages. It is only the creation of a system of analogies made possible by iteration, and therefore a regular code, that allowed the Indo-European voiceless plosive **p, *ph$_2$tér* ('father') to evolve in Germanic as a voiceless fricative, *fáðar*, and in Latin, instead, as "p", "*pater*". It is not surprising that memory recreates the past – rather, it would be surprising if this recreation took place in the absence of memory. This is the fundamental difference between

---

[67] Gibson, *The Ecological Approach to Visual Perception* cit.

[68] Consider a simple fact that we are often annoyed by: it is very common, in nature and in culture, that iteration, due to a transcription defect or to a desire for variation, generates an alteration. This is how an incorrect copy of the genetic code produces evolution in the species or a tumour in the individual; how a mispronunciation or an error in the parish register generates, from the same source, "Ferraris" and "Ferrario", or "Sean" and "Jean". Likewise, this is why the southern part of the German languages underwent a consonant rotation which did not take place in the northern ones; or why the diphthong *oi*, traditionally pronounced *weh* in French, took on the pronunciation *ua*, under the influence of the Parisian *argot*, with the result that in France *troi* and *roi* are pronounced "trwah" and "rwah", while in Québec they are pronounced "trweh" and "rweh".

hysteresis and mere recording: recording is not enough, just as perfect memory is not absolute knowledge, but Ireneo Funes' idiocy. And it is hard to disagree with the idea that too much memory, a long repetition and rumination, paralyses life. Above all, we know from experience as well as from infinite psychology books – many of which we have rightly forgotten – that memory recomposes, modifies, and recreates the past, instead of simply accumulating facts as if it were the dumpster of our experiences. All this is not only obvious; what is more, it is right: when we say that memory does not merely record, but iterates in obsession, alters in invention and interrupts in oblivion, we are simply bringing back to experience a metaphysical phenomenon that involves all there is.

3.3.4. Truth users

The interruption brings us back to the soul's distinctive element: its finitude. Truth does not exist in nature, we said. In fact, Mont Blanc does not care that it is called Mont Blanc, nor that it is 4810 metres high, nor that "4810" is the name of the Mont Blanc nib. But, at different levels, all this can be of interest to humans, who therefore, as "truth-users" or end-users, constitute the *telos* of a process that in the absence of humans would make no sense. One of the reasons why truth does not exist in nature is that without humans and the practical constructs that make them so, no one would be interested in truth. As I said many times, artificial intelligence thinks better than humans – and this does not only apply to now, but has been the case all along. Conversely, within humans, as responsive beings, there is a great deal that is artificial, starting from intelligence. But there is one thing that does not exist in nature, souls, or automata, and only exists in a responsive being: truth. This is because a responsive being is the only one that needs truth: an automaton capable of comparing billions of documents in order to find correlations certainly does not need the shortcut that is true opinion accompanied by reason.

In other words, an essential ingredient of truth is humanity as the set of subjective users of objective truths. Making the truth also means finding out whether Albertine is unfaithful (if you are jealous like Proust), reporting the butler to the police, or discovering a vaccine: truth has a dimension that is not only epistemological (this is completely obvious) and not only technological (this is a little less obvious), but also teleological. The truth does not depend on someone's wishes – as posited by nihilism which thus renders the truth useless by decreeing the triumph of interpretations – but is made in view of something or someone. Since the truth is not found in nature like sodium chloride, it is addressed *to* someone, who is the aim of the research: the patient who has asked for a check up, the curious user who looks something up on Wikipedia, the child who asks the reason for something and the jealous boyfriend who secretly reads his partner's WhatsApp messages are all part of the truth, as its final cause. In other words, they are the truth users, and are no less important than the truth bearers (Albertine) the truth makers (her careless revelations), or the truth tellers ("Albertine is unfaithful"): all these elements receive meaning only in light of an end, which is precisely the user or, more prosaically, the cuckold.

This example obviously indicates the practical genesis of truth, and does not affect its value in the least. The reasons that Rousseau accumulates to discredit the sciences, and which are all attributable to the fact that they respond to human needs, do not at all constitute a genealogical argument against it. It is natural and right that the truth initially responds to individual needs and practices. But it is equally true that the truth can detach itself from this practical genesis (and writing is one of the most foolproof methods in this sense, as it can address completely different recipients from the original user). This is how truth becomes a (virtual) heritage of all humanity, and history presents itself as the tale of humanity's infinite tendency towards truth.

I would like to illustrate this point with the example of a philosopher who is much closer to us, at least in time, than Aristotle or Augustine, namely Richard Rorty. According to him,[69] the truth is of no use: it is perhaps a nice thing, but it remains useless, a kind of compliment or pat on the back. All this was posited with the best of intentions, in agreement with Rorty's idea that human solidarity is more important than objectivity. It is not clear, however, how there can be solidarity without objectivity, and how there can be justice without truth. Going back to the origins of the pragmatism Rorty refers to, reconsidering William James' *The Meaning of Truth*, can help us reflect on the role of truth in human affairs. Indeed, James's thesis, which in this respect is at odds with Rorty's, is that truth is the most useful thing there is. As we read in *Pragmatism*, "*true ideas are those that we can assimilate, validate, corroborate and verify. False ideas are those that we cannot*".[70] When defending his position in *The Meaning of Truth*, he went on to say: "Good consequences are [...] a sure sign, mark, or criterion, by which truth's presence is habitually ascertained."[71]

Let us not forget, in fact, that in the absence of truth and objectivity, trust would disappear and, with it, the foundation of the social bond. Let me conclude this section with a trivial example. Having mushrooms at a restaurant implies having much faith in truth: those who picked the mushrooms knew (or, in an

---

[69] P. Engel, R. Rorty, *What's the Use of Truth?*, Columbia University Press, New York 2007.

[70] W. James, *Pragmatism: A New Name for Some Old Ways of Thinking*, Longmans, Green & Co, New York-London 1907, p. 97.

[71] W. James, "Two English Critics", in *The Meaning of Truth*, Longman Green and Co, New York 1911, p. 273.

unfortunate scenario, thought they knew) that said mushrooms were not poisonous, and this knowledge corresponded to a property of the world: the fact that the mushrooms were not poisonous. It also implies equally great faith in humankind: people that we have never seen and will probably never see again are serving us mushrooms that could be poisonous, but are not. It's unclear why one should posit an antithesis between the chef's solidarity and the mushroom picker's objectivity. Why should the chef's solidarity be more important and true if, regardless of objectivity, he gave us poisonous mushrooms, in line with the humanitarian ideal of sparing us the pain of existence? In a way, truth does pop up like a mushroom: it emerges from the world and goes towards other parts of the world like us. This happens regardless of any convention, because a poisonous mushroom is poisonous (if it is) even if the UN decreed that it isn't.

3.4. Teleology: Interruption

This brings us to the fourth stage of the hysteresis, which might well derive from fatal overconfidence in a mushroom dish, but which, mushroom or no mushroom, concerns us all as mortal souls: interruption. Nabokov used to say: "To leave is to die a little, but to die is to leave a little too much", and he was right. It means leaving *for ever*.[72] And this "forever" is what distinguishes means and ends and gives life its unique and ultimate meaning. On 21 March 1955 Einstein, who was to die the following month, wrote a letter to Vero and Bice, the children of his close friend Michele Besso. In it, he said that for those who believe in physics "the distinction between past, present and future is but an illusion, however persistent it may be".[73] Well, it would be an even more persistent illusion to think that Michele Besso, or Einstein, or anyone else, could be resurrected.

Our ordinary experience is made up of memory, which restores the past and gives meaning to the present and the future: today, it is also made up of machines, which greatly enhance this possibility of retention. Our experience is made up of iterations that give rise to the most common facts: linguistic codes

---

[72] This is what makes the exchange between Brutus and Cassius at the battle of Philippi less dramatic, when imminent death (and therefore "forever") is still and only a possibility. "Brutus: Forever and forever farewell, Cassius. / If we do meet again, why, we shall smile. / If not, why then this parting was well made. / Cassius: Forever and forever farewell, Brutus. / If we do meet again, we'll smile indeed. / If not, 'tis true this parting was well made" (W. Shakespeare, *Julius Caesar*, Act 5, scene 1, vv. 118–122).

[73] "Für uns gläubige Physiker hat die Scheidung zwischen Vergangenheit, Gegenwart und Zukunft nur die Bedeutung einer, wenn auch hartnäckigen, Illusion" (Einstein Archiv Dokument 7–245).

with which we express ourselves, traffic lights that turn on and off, obsessive rituals, religious beliefs and inveterate habits. It is made up of alterations that bring about variations against the background of continuous retention and iteration: tumours and disasters, strokes of genius and unprecedented stupidity, guilty distractions and revolutions after which nothing will be the same. But if all this makes sense, it is because there are interruptions. The fourth stage is therefore *interruption*, that is, the greatest alteration: the series comes to a halt, and this brings about the sense, the aim, the purpose of it all. This last function of hysteresis concerns everything that is, but is of special significance for humans in particular. Hysteresis is not an infinite process, as shown by the basic experience that everything that is, sooner or later, comes to an end: mountains eroding, stars exploding, washing machines obeying their destiny of programmed obsolescence, and organisms following the fate awaiting all flesh. This may seem depressing, and indeed it is, yet if we think about it for a moment the opposite would be worse. Not only would we find ourselves faced with eternal boredom, but above all everything would lose its meaning. Indeed, meaning is such precisely because it relates to organisms, humans in our case, who, knowing that they have a limited time and a fixed direction, produce all the things that make us unique – at least in our eyes.

Hysteresis, as we have seen, is a principle that goes to the heart of the paradox of time, which both is and is not. This paradox is as old as philosophy itself, and is no less baffling than the quantum paradoxes, but it cannot be resolved by claiming that time does not exist: in that case perhaps little would change in the quantum world, but a great deal would change in the world, much closer to us and relevant to us, of mortgages, taxes and death. Indeed, time is and is not at the same time, and yet it contains everything that is, because it is really difficult to deny that things are in time, and that time takes and gives.

3.4.1. End

Our universe is very old, accidental, and doomed to death – if we can use that expression for something that has never been alive – but in one corner of it it has been able to give birth, and here the term is really appropriate, to increasingly complex organisms. How is this possible? Is it a miracle? There are many answers to this question, and they have to do first and foremost with our imperfection. The miracle would have occurred if we were the pinnacle of creation; in reality, we are very imperfect, and of these imperfections, the most onerous is death. On the other hand, imperfection,[74] which is also ignorance of the totality of causality and

---

[74] Cf. the already mentioned Pievani, *Finitudine*. However, Pievani implicitly errs on the

destiny, makes us free, exposed to a future that we cannot foresee but in relation to which we are called upon to make decisions. Death, like the end of all things, is therefore not radical frustration and meaninglessness, as many philosophers and non-philosophers think. Nor does it receive meaning from the succession of generations, whereby the death of the individual is partially remedied by the continuity of the species: whoever dies is an individual, and that individual really does die. Likewise, death cannot really be compensated for by the continuity of matter, since the resurrection of a few pieces of us in a pear or a fly is hardly a great consolation. Rather, there is a constitutive relationship between being destined to an irredeemable end, therefore to pure meaninglessness, and being able to confer external ends that, in a circular fashion, can restore something like a meaning to our existence. This relationship, although intuitively quite visible, is anything but linear, and to analyse it I will break it down into six stages.

The end (death). The first stage is very simple. As we saw in 2.1, the distinctive feature of the soul as opposed to the automaton is that the former, unlike the latter, is destined to an irreversible end, whereas the automaton can be (and generally is) programmed for a series of on/off positions. We thus have two fundamental differences: the soul lives only once; the automaton iterates, never tires, and even though sooner or later it breaks down, until then it can be resurrected at any time, like the street lamps that are turned on every evening. That of the soul is a defect but, seen from an anthropic perspective, it defines our superiority over machines: the latter have no purposes or needs, no ennui or ideals, simply because they do not have a limited lifespan, and no irreversible alternative between on and off. The sense of the end, which in higher organisms can assume different degrees of consciousness, manifests itself first and foremost as organic needs dictated by the demands of metabolism. It is these circumstances that, at different levels, determine the sense of temporality, intentionality, consciousness, and in general the sphere of meanings, which as such belong only to the sphere of organisms insofar as they are endowed with sense and direction. And it is precisely these phenomena, which emerge from the organism, that give meaning and purpose to the mechanism, which therefore has meaning and purpose only as derived from the organism.

The end (purpose). Let's move on to the second phase. Only that which has an end (death) can have an end (purpose), because only something mortal can feel the uniqueness of choices and the historicity of existence (if it is a professor), and the pressure of food and vital needs (if it is either a professor or a duck). Suppose

---

side of excessive optimism (why shouldn't we be both finite and imperfect?). Still, this attitude is preferable to the excessive optimism displayed by N. Bostrom, E. Yudkowsky, *The Ethics of Artificial Intelligence*, in *The Cambridge Handbook of Artificial Intelligence*, Cambridge University Press, Cambridge 2014, pp. 316–334.

I am fidgeting with kitchen tools to make a pasta, and the various parts of the stove, which have explicit purposes, undergo the efficient causes that derive from my action, triggered by my desire to eat. Rationality is precisely this ability to assign ends, from that of making a pasta dish to that of restoring the Roman Empire. Souls have bodies, hence ends, and humans – who are responsive souls – possess, at least in principle, reason as the faculty of purpose. When placed in the social system, need is transformed more clearly into consumption,[75] capitalisation and finalisation, and this tangle of external ends, whether good or bad, enriches the life of the soul, gives it a little more meaning (depression is merely the revelation of bare life, of the soul without automaton). This is why the soul has such a need of automata, from the wheel to fire, from society to culture. In accordance with what was said in 2.2, however, having automata is precisely what differentiates us from non-human souls. Having no purpose outside themselves, human souls have developed the capacity to produce automata, such as technological devices and social structures, which are expressly endowed with an external purpose – not the internal purpose manifested in physiology. To speak of a desiring and consuming automaton is to speak metaphorically, such as when one calls an energy transformer a "feeder" (*alimentatore*, as we say in Italian).

Responsiveness. Hence the third stage. An interesting consequence follows from what I have said: in order for human beings to have an end (purpose), there must be souls that as such do *not* have any, and these souls cannot be generic, but human, since non-defective animals do not know what to do with nails, hammers, software and search engines. It is within this horizon that we can draw the complete picture of humanity as the intersection between an irreversible organic process, which is therefore endowed with an internal purpose, and a reversible mechanical process, which is therefore endowed with an external and instrumental purpose. But – this is the point I will highlight in 4 – if we remove the soul, the automaton loses all meaning. Conversely, interruption, by breaking the order of causes and its survival in hysteresis, enables the emergence of finality, of *telos*, and this is why interruption corresponds to the sphere of teleology. And by referring to teleology it becomes possible to clarify the meaning of "intention" and "intentionality". In this framework, responsiveness provides human

---

[75] As we know, Heidegger defended an idiosyncratic thesis, to say the least, namely that only humans die, while animals simply perish, being unaware of their own death. This seems to be a problematic criterion, given the little we know about the feelings of animals about their own deaths, whereas we have evidence of the sensitivity of animals to the deaths of others. What is certain is that humans, like any other animal, are subject to metabolic urges; and that, unlike any other animal, they satisfy these needs through consumption. We will never find a non-human animal in a supermarket, unless it is on the leash of its owner; and what supermarkets offer for pets is targeted at humans who care for them.

animals with a significant difference from non-human animals. Like the latter, humans are endowed with an internal purpose, i.e. a fundamental absence of any purpose beyond the functions of nutrition, development and reproduction. Precisely because it is embedded in a living body, that is, a complex system that struggles to survive but is destined to die, the human automaton is endowed with intentions, purposes, teleologies: things that are also found in a chicken, but not in a computer.

The paradox of finalism. So here is the fourth stage. What is our end (purpose)? Simply our end (death). What a prospect! The embarrassment arising from the intrinsic lack of purpose of animals is manifested not only in the problematic notion of "internal purpose", but mainly in the idea that one of the organism's purposes is self-reproduction. The conceptual move is very clear: reproduction, in fact, seems to introduce an external purpose into organic life – that of producing other organic life. Obviously, this is not the case: the soul has no reason to exist, and yet it is capable of giving meaning to things, with a *doing* that precedes *knowing*. How is this possible? Internal purpose can also be defined as follows: a characteristic aspect of human beings is that, compared to automatons and tools, they serve no purpose. The rose has no reason to do what it does, it blooms because it blooms. Scissors, instead, do have a reason to exist: they are made to cut things, for example a rose. Sounds have no purpose, but language does, for example to give a name to a rose. In short, a tool has a clear purpose: to serve the needs of the user.

This is why the soul has no purpose, while the automaton always has one: a pen is for writing, a clock is for telling the time, and if, by chance, someone were to create a useless machine, the question would arise: It serves no purpose, is it not a soul? The crucial consequence of this double rooting, in somatic interiority as well as in technological and social exteriority, is the paradox of finalism: a soul that has no purpose imposes purposes on automata that have no soul. To those who think that one cannot give something one does not have, I would point out that here we are dealing with this very case: if there is a meaningless thing, it is life, but it is precisely within this general meaninglessness that meaning is created.

Giving an end to means. The fifth stage is that automata are produced expressly by human souls in order to remedy their inadequacies and respond to their urgencies: from internal purpose we thus pass to external purpose. The absence of external purpose typical of the soul is reflected in automata that are endowed with an explicit and, I would say, emphatic external purpose: from automata that can largely perform only one function (a flint) to automata that can in principle perform any function (the computer as a universal machine), but never in the first person and always by proxy – near or remote – of a soul, as they do not have the irreversible internal purpose that determines vital pressure and urgency. Herein lies the soul's superiority over the automaton. The automaton might seemingly retain its purpose even if the user and all of humanity were to disappear, but in fact the machine would prove useless in such a scenario, and would therefore lose its purpose. Not only hammers and nails, but also software and search engines depend on the existence of humankind.

Giving meaning to the end. The difference between the soul and the automaton, as well as between the internal finality that characterises the former and the external finality that characterises the latter, consists in a defect of the soul's finality compared to the automaton's finality. The soul alone has a short and meagre life; but the automaton, with no soul, would not exist at all, and its finality would vanish with the disappearance of all souls. The sixth and final step is that the lack of any obvious external purpose is what creates an immense spectrum of motivations in souls, in contrast to automata. The irreversibility of the processes involved in this causes souls to act in a finalistic way, under the effect of an urgency and pressure that are completely absent in automata.

3.4.2. Sense

One more point. If humans, together with all other organisms, did not die, resources would soon prove insufficient. If there are excellent reasons to be paid to live, this means that there are excellent reasons to be paid to die. This payment is all the more due because death is the only debt that must necessarily be paid, being the ultimate end of the life process. There is no difference between the functioning of the solar system and that of a rotisserie. Within the sphere of mechanisms, however, a more circumscribed sphere emerged that is dominated by irreversible processes: the sphere of organisms. Salt dissolves in water, but if you evaporate the water you get salt again; instead, when a single-celled organism has split in two, you cannot go back to the original cell. It is precisely this irreversibility that gives meaning and rhythm to life, history, economics and morality.

In this framework, giving a purpose means giving a meaning and answering the question why. The universe is headed for thermal death. This concerns us so little that we are currently preoccupied with global warming. But that destiny of thermal death is, on a cosmological level, the destiny of every organism, natural or social. Everything that is will end, and it is better that way, otherwise nothing would make sense or be comprehensible. We understand what a hammer is when it breaks; likewise, we understand what a person is when they die and, unlike the hammer, do not rise again. Death is a master from Germany, as Paul Celan said, and he might seem to be far away but is actually behind every gesture we make, every item we buy online, every photo we post. This might sound like something sad or unfortunate, but it's not: it's what makes up the world of culture and spirit – the bittersweet circus that humans have managed to put on by mixing the skills enabled by technology and the thrill and motivation brought about by death.

A cyclical world – Nietzsche was right about this – would shatter all meaning. But this very fact suggests that resistance is futile: let us accept that there is an interruption at the end, and let's try to make sense of the world and life. What would be the meaning of a meaning that was simply a sense, a direction, the arrow of time dictated by entropy? For this meaning to be more than just a direction, there must be living beings involved in the process, beings for whom there is meaning in the idea that slowly but surely, the iteration of gestures and metabolism, ever slower and slower, will come to a definitive halt. Only that which has an impending end can exist in a time which has meaning and direction, and therefore have purposes – i.e., a teleology. It is therefore no coincidence that final causes concern organisms and history, the two spheres in which time and transience are everything. Everything has an end, which manifests itself as an interruption. If alterations are things that sometimes happen and sometimes don't, there is something that, sooner or later, cannot fail to happen, and that is the interruption, the end of all things. It may seem like a bad thing, but it is the greatest of goods, because it gives meaning to every act of our lives. That is where you were going, and that is the meaning of everything you have done, *les jeux sont faits, rien ne va plus*. Thus a circle is drawn that can be described as the set of three components.

Direction. The first component is direction: only that which has an irreversible end can have a direction, hence a sense understood as meaning. *Sinn, to send, Sendung, sentiero* – all these words recall the fact, represented with extreme coherence in Indo-European languages, that the formation of meaning depends on the existence of a purpose. In short, there can be no sense, significance and scope without paths and directions. Hence meaning: what was simply iterated by a mechanism or an organism now *means* something to someone. This someone must be an organism, subject as such to the processes of interruption I mentioned earlier, and connected with the processes of capitalisation described in 2.3 and 2.4. Elephants, beavers, crustaceans or amoebas do not have a philosophy of history or a shred of storytelling, nor do they need an end (purpose) beyond the end (death) of their organisms – which is not a limitation, but a sign of the success of their life forms. *We* are the ones who can and must ask ourselves "where will we end up?"; *we* must furnish the world with the few resources we have, starting with the urgency imposed by death. This is why finalism is much more pronounced in souls than in automata. In general terms, a mechanism is subject only to causal laws: event A causes event B. An organism, on the other hand, also obeys final laws. It would be a strange way of describing the capture of a stronghold to claim that the fragments of gunpowder, in response to a spark, came out at such a speed as to push a hard, heavy body against the walls; one would rather say that the invaders wanted to take the city, and *to this end* they bombed it.[76]

Capitalisation. The second component is capitalisation. Devoid of internal purpose, automata are nevertheless predisposed to hysteresis, and thus to the capitalisation of souls' behaviour through processes of recording, iteration, alteration and interruption. Suppose souls were immortal, and Hamlet could tell himself: "I have sworn, of course, but there is no hurry, I will think about it in a million years". The whole system would collapse. Conversely, everything holds together because souls are mortal: some know they are mortal, many others do not know or do not think about it, but they are still subject to the needs of the body: hunger, desire, anxiety, boredom. The very fact of dying is what gives meaning to life, primarily because it gives life time (more or less long, but never infinite), and therefore gives meaning to promises, aspirations, desires and happiness that would have none for an immortal. Hence the genesis of meaning, which draws a circular picture – the one we have seen at work in the processes of capitalisation, in the relations between intentionality and documentality, in the genesis of value, and so on. That which has an explicit sense can only receive sense from that which is outside it, and return it to the sender as a second-degree external purpose: a mobile phone has no internal purpose, yet it can push someone to attempt to steal it. What distinguishes us lies not so much in what we have inside us, but rather in what is outside us: libraries, homes, supermarkets, universities, and of course also jails, lagers and factories. This is what allows me to write at the moment, because by nature I should have been dead for forty years now, probably at the hands of an animal stronger than myself.

Signification. The third component is signification as the result of the encounter between the soul's direction and the automaton's capitalisation. From the point of view of the universe, entropy is the end of all things. From the point of view of organisms destined for the end and therefore capable of projecting purposes outside themselves, as humans are, it is instead the beginning and the condition of possibility of signification. To be clearer, the phrase "in a hundred years we will all be dead" has the same meaning if it appears in a book, if it is repeated by a loudspeaker at the station, or else if it is said by a lecturer in the mood for jokes. But only for the lecturer and for their listeners does the phrase possess, in addition to a sense, also a meaning. And this is due to the trivial reason that in a hundred years they will indeed all be dead (this, incidentally, is also the reason why the phrase means something to me and you).

---

[76] G.W. Leibniz, *Discourse on Metaphysics* (1686), Oxford University Press, Oxford 2020, § 19.

The totality of the soul's life is meaningless, possessing only an internal purpose, that of being born, growing up and dying. However, once placed in a circle of capitalisation (society, culture, documentality), the soul acquires a huge number of intermediate external purposes, ranging from scratch cards to the Nobel Prize, from heroism to wealth (the hallmark of intermediate purposes, the most obvious in its link with capitalisation), to holiness. These are imperative purposes, because if you don't eat, you die, and at the same time they are futile: what is the point of living? Yet, as we have seen when discussing capitalisation, there is a sense in which humans have managed to create ends (purposes) that go beyond the end (death) immanent in biological life. Indeed, the human animal has the characteristic of multiplying external ends, developing an instrumental reason that is wrongly regarded as a degraded form of reason, whereas it embodies the quintessence of reason as the faculty of purpose. Humans, unlike any other animal, externalise ends through the construction of technological devices that compensate for their inadequacies and at the same time externalise finality: they transform their spurious internal finality into a pure external finality.

3.4.3. Intention

In this framework, the alleged "mark of the mental" is nothing but the relationship between a body and a purpose, which in a tick manifests itself in the attraction to heat and blood and in Albert manifests itself in writing books on intentionality. In common speech, "to intend" or "to have intentions" means to set out to do something. However, in philosophical language it also means to represent something in consciousness. In fact, the distinctive character of the mental would be the capacity to form images of objects, whereas objects are not capable of forming images of subjects. At this moment I can represent in my mind the computer screen, which thus constitutes an intentional content of my consciousness precisely because I have the image of the computer screen in my mind. Having this image is the distinctive characteristic of the mind with respect to that which is not mind; for some it is even the condition of possibility of experience.

However, suppose I switch on my computer's camera and turn it towards me. I have a representation of the screen, and the screen has a representation of me. Does this mean that the computer has intentionality? If "intentionality" refers to the ability to have representations of things outside ourselves, then not only computers and cameras, but also mirrors have the distinctive feature of the mental. And the fact that the notion of "secondary intentionality" has been used for centuries to answer this objection only adds to the embarrassment, because it is not clear why one would call the intentionality of a mirror or a computer "secondary", and that of a professor or a beaver "primary". The difference is not between primary and secondary, but between organic and mechanical, and the question remains as to why we should introduce an animistic notion such as "intentionality" instead of "representation", which does not require the use of soul supplements to be understood.

This discussion shows that, for historical reasons that I cannot retrace here, the term "intentionality" is misleading, because it does not simply suggest having an image, but clearly evokes an act of will. To have the intention of doing this or that is not simply to represent this or that – something that both the mind and the computer undoubtedly do. It means to propose to carry out an action, something that I can do and the computer cannot, not because I possess the representation and the computer does not, but because I have attitudes towards this representation (fear, desire, projects) that the computer does not. Once again, recourse to intentionality appears to be a *petitio principii*: a notion that should define the distinctive character of the mental actually presupposes this distinctive character, i.e. the fact that the mental is a sphere of representations (and, let us not forget, of operations that are not represented) that takes place in an organism rather than in a mechanism.

This is also true regardless of the circularity whereby the mental is defined based on the intentional and the intentional based on the mental. To hold that extramental representations acquire intentionality only insofar as they are used by a mind, but do not have intentionality in themselves, is at best to hold a solid truism, namely that a mirror or a computer are mechanisms while a beaver and a professor are organisms. In the worst case, which is the most frequent, it means claiming that the mind confers intentionality on representations, i.e. using the *definiendum*, intentionality as the alleged character of the mind, as *definiens*, intentionality as a property that the mind confers on representations.

The distinctive feature of the human mind with respect to a computer, as far as intentionality is concerned, does not lie in representations, but in the *purposes* connected with these representations, that is, with what we intend to do with them, and more generally with what we intend to do full stop. Having an unavoidable end (death) is what dictates the urgencies, contingencies, timeframes, anxieties, boredom, lacks, and all that philosophers call "intentionality". The latter is simply the result of the specific condition of a soul that, being able to live only once, is interested in and projected towards the world in a completely different way compared to an automaton. In other words, both a tick and Albert have intentionality, whereas Watson, the supercomputer, does not. And this is not because the tick and Albert are more intelligent than Watson, but because, when they die, they do so forever (although the tick, as Uexküll teaches us, is superior to Albert also from a vital point of view), whereas Watson, if switched off, can switch on again. That said, how do we distinguish Albert from the tick? Rationality with respect to purpose is not the appropriate criterion, because even on this point the tick beats Albert by far. One could say that the tick does not contemplate the starry heavens above and the moral law within, but we know too little about both the tick and Albert to make such a claim. The only decisive difference between the tick and Albert is that the latter, being endowed with hands, can equip himself with supplements that make up for his deficient organism with the resources of the mechanism, or rather of the multiple mechanisms endowed with purpose and resistance which, unlike the tick, he has access to.

It follows that the characteristic of the mind, if by this term we refer to the natural mind and not to the artificial one, is not intentionality in the philosophical sense, but intentionality in the legal sense, which is also in accordance with common parlance. Computers have intentions as representations, humans have intentions as representations and as intentions to act. In other words, what characterises human intentionality with respect to artificial intentionality, just as what characterises human intelligence with respect to artificial intelligence, is not a capacity to represent (in the case of intentionality) or to manipulate signs (in the case of intelligence), but the practical purpose and emotional tone that accompany representation and manipulation. Humans have an end (death) and therefore pursue ends (purposes).

3.4.4. Narration

We treat the brain as a computer and the eye as a camera because the introduction of final causes (thought, vision) helps us explain how they work. However, as such, neither the brain nor the eye have any purpose whatsoever. They are simply part of an organism whose only and often unconscious purpose is the need to stay alive. It is this need that makes organisms operate as guided by final causes – the prototype of which is the stimulus of hunger – while mechanisms operate as guided by efficient causes, usually a responsive animal pressing a button. I use scissors in order to cut, Caesar crossed the Rubicon in order to become a dictator, you bought a big bunch of roses in order to impress your date on Valentine's Day. All these aims would be inconceivable for a mechanism: and in fact scissors are made to cut, but in themselves have no inclination to do so. Speculation also consists in turning towards what is not yet by identifying final links, and, as we shall see in a moment, the quintessence of speculation is the philosophy of history, which takes up the question of the meaning of being. The laws of meaning are laws of sense, which emerge only retrospectively – that is, speculatively, as a narrative. If there is an end, an insuperable threshold, then there is a meaning and a purpose to the story. I am not defending any ontological optimism here. I am merely arguing that if history has a meaning (it may well not have one), then this meaning is good, as evil is not a lack of being, but a lack of a sense in being, of the resistance that allows the dove to fly.

This, in my opinion, is the point of honest storytelling,[77] which is not to tell a fairy tale to put children to sleep, but to construct meaning to keep adults awake: and there is nothing like catastrophism to make people sleep soundly in the belief that only a god can save us. This is where I ground my apology for the philosophy of history, and for a positive one at that. The ancients conceived of the cosmos as stable, or subject to a cyclical nature comparable to the seasons; if anything, they saw history as decadence, i.e. as a departure from the golden age. Instead, moderns see history as the great driving force of humanity. This propulsion follows an arrow that goes from creation to the end of time, and sees each phase, at least ideally, as either an advance (in the euphoric perspectives) or as a step back (in the dysphoric ones). However different or antithetical they may be, these versions share the non-obvious and highly questionable assumption that meaning is only found within a story – in other words, that the meaning that humanity can attribute to itself depends on its position within a storytelling.[78]

---

[77] C. Salmon, *Storytelling: Bewitching the Modern Mind*, Verso, London 2008; Cometa, *Perché le storie ci aiutano a vivere* cit.

[78] I propose to isolate four variants of this storytelling, two dysphoric ones (nihilism and messianism) and two euphoric ones (the Enlightenment and utopianism). The first storytelling, which has constituted the mainstream of recent decades, is *nihilism*: *humanity is getting worse and history equals decadence*. Nihilism presents itself as the dysphoric face of utopianism, and as the realisation that revolutions fail. Nihilism was formed in the 19th century and gave rise to a perspective that, since the 1980s, has been referred to as "postmodern". Postmodernism maintains the typically modern idea of overcoming, but introduces an opposite approach: history does not have a progressive destiny, but a regressive one, and in particular is characterised by a loss of the values and meaning that humanity gives itself. As well as being a dysphoric result of the failure of utopia, nihilistic storytelling is a negative adaptation of the Enlightenment storytelling, because, like the Enlightenment, it conceives of the human being as a bearer of values and historical projects, but unlike the Enlightenment, it sees the human being as originally perfect and heading for decadence. Nihilism is therefore an entropic narrative which, starting from the hypothesis of a perfect humanity corrupted by technology and society, describes modernity as a progressive annihilation of the meaning of history and life, a crisis of values exchanged for futile material well-being. The second storytelling, as dysphoric as the previous one, is *messianism*: *humanity is getting worse and salvation can only take place outside of history*. I call this storytelling "messianism" because it remotely recalls the feeling of waiting for the Messiah in the Jewish tradition. Unlike Enlightenment storytelling, messianic storytelling denies that what has been achieved so far is really the goal that humanity set itself: in messianism, the Messiah is always yet to come, and those who claim to be such are only false prophets. The problem with messianic storytelling is the fact that it continually postpones the realisation of the ideal, and therefore ignores the concrete achievements that have taken place. A particularly striking example of this is the insistence that our age is characterised by a total lack of economic and social justice, even though we are doing better in that respect than in any previous age. The third, euphoric storytelling is *utopianism*: *humanity is getting better but*

If I were to say that the desert is advancing and humanity is going down, I would not shock anyone. I would like to spend a few words on this point, though. Selling a negative philosophy of history means offering a mainstream service; proposing a positive one, as I am doing in this book, means being naive, at best, or complicit with evil, at worst. Therefore, I would like to propose an experiment similar to Descartes': let us suspend all prejudices, be they right or wrong, and, substantiating what I stated in the prologue, examine some "signs of history", clues that prove that humanity is indeed getting better. Note: not one of these signs entails resignation to, or acceptance of, the status quo. It simply means that

---

perfection, as such, is not achievable in history. Strictly speaking, utopianism is merely a hyperbolic variant of Enlightenment storytelling. It indicates an ideal of human development, which, however, as such, does not take place in our history and geography, but in a non-place. The etymon of "utopia" is indicative in this regard, since it includes both a reference to a non-existent place (*ou-topos*) and a reference to a desirable place (*eu-topos*). The form is Greek, but the word was coined in the 16th century, and, as we have seen, nothing is further removed from the thought of the ancient Greeks than the exploration of a desirable future called to redeem the present. In order for utopian thought to take on its full vigour, it requires at least the intervention of a religion such as Christianity, which places the meaning of existence not in the present, but in the future. It also requires the encounter with worlds that are indeed different (it is not by chance that *Utopia*, Thomas Moore's book that is said to have fathered the concept, was published in 1516, a quarter of a century after the discovery of America – which, moreover, suggests one of the reasons that may have prompted it). Above all, however, it requires a certain impatience with the long timescales envisaged by Christianity, and a desire to intervene in the present by anticipating the future, not by imagining a paradise, a gift from God, or the festive interruption of a dull and monotonous life, but a utopia: a human construction that aims to create a new world. The fourth storytelling, equally euphoric, is the *Enlightenment: humanity is getting better and perfection can and must be achieved in history.* This storytelling began with the Renaissance and developed into the 19th century, taking the shape of idealism and Marxism. Human history is characterised by a progressive emergence from barbarity and the acquisition of happier and more enlightened living and thinking conditions. Over time, this concept encountered increasing difficulties due to both the side effects of progress and to doubts as to whether the goals of the Enlightenment had actually been achieved. Typical examples of this dissatisfaction with the achievements of the Enlightenment storytelling are social, ecological and technological criticism. The common thread running through these attitudes is the observation that the goals humanity set itself through the Enlightenment have not been realised and have in fact given rise to antithetical results: growing social inequalities, environmental crises, technological dehumanisation. It is nevertheless worth noting that, albeit in a dysphoric form, the assumptions of the Enlightenment storytelling are preserved, in particular the one according to which the meaning of humanity is to be found in the place it occupies within a historical becoming.

As far as I am concerned, I propose a storytelling that lies somewhere between utopianism and the Enlightenment, considering nihilism and messianism as bad conceptions of history dictated by the Rousseau syndrome, i.e. by the idea that humanity is incomparably better than it actually is: in that framework, any regret about a supposed decline is an implicit praise of how perfect humans once were.

we are going in the right direction, setting the conditions for any transformation to make sense.

First sign of history. In the last fifty years, CEOs of large companies have seen their profit multiply. This is solid evidence. Should we conclude that the rich are getting richer and the poor are getting poorer, and that humanity is therefore heading for catastrophe? Let us think about it. I am struggling to find any moment in human history that justifies a nostalgic tone. And I ask myself: to which previous historical epoch would we like to go back? When one says that we have lost our freedom in the digital world, the good question would be: as opposed to when? In this same vein, during a lecture in Mexico, when a dissatisfied listener exclaimed: "Life sucks!" Ortega y Gasset appropriately replied: "Compared to what?"[79]

Given how hard it is to answer either question, it follows, I believe, that humanity is getting better. In a way, it cannot be otherwise, for if it were true that the rich are getting richer and the poor are getting poorer, our preaching would be empty, our hope would be void, all social criticism would be made in bad faith, and history would be nothing but a nightmare from which we cannot wake up. The point, however, is that this is false, both *de facto* and *de jure*. The gap between rich and poor has widened in recent decades, but only in relative terms: the rich have become richer than the poor, who are in any case less poor than before, at least according to statistics that are always difficult to verify. In addition to the empirical difficulty of measuring how rich the rich are and how poor the poor are, there is a transcendental difficulty to which, in my opinion, not enough attention is paid. If the rich were really getting richer, the current richest man on earth would also be the richest man in history: but this is not the case. The richest man in history, as far as we know, was Mansa Musa I, king of Mali in the Middle Ages. Then there have been many others, including Augustus, Caesar,

---

[79] In fact, we have no terms of comparison between life and its alternative, but – and this is the interesting bit – every life can and in a certain sense must be compared with the lives of others, in order to understand its value or dis-value, just as every event receives meaning when compared with other events. The philosophy of history is precisely this comparison turned into narrative, which gives meaning and a sense of direction to an event as correlated with other ones. Humanity is not subject to the tyranny of mediocrity and petty pleasures that de Tocqueville saw in American democracy (cf. A. de Tocqueville, *Democracy in America* [1830], Saunders & Otley, London 1840, II, 4, 6). It is better to be dominated by the smallest pleasures than by the greatest pains, shortages and deprivations. As for claiming that our society, in particular, is a performance society means forgetting the kind of performance that is required to survive in the savannah or to win in a duel. Fortunately, there are many human beings today who get by without any skills or abilities, which would be unthinkable in the state of nature – it was perhaps possible for the Roman plebs of the golden ages, but they lived off the exploitation carried out by the Roman empire.

Napoleon, and Genghis Khan. So, to our knowledge, the richest men in history are all dead, and the rich living today are much poorer than they were.

More seriously: what criteria are used to make these calculations? What authorities certify them? Mere reference to monetary parameters is misleading in any case.[80] How do you compare an absolute ruler who owns an entire state, a constitutional ruler and a captain of industry? And of course wealth is not immediately synonymous with power or great well-being. Camorrists have to stay locked up in videosurveilled compounds, and billionaires who fall under the blows of sex scandals (made possible by a harsher and more active public opinion) have objectively less power than the tycoons of the generations before them, and all of them have less wealth and less glory than Attila the Hun. Not to mention that it is hard to imagine Attila under house arrest for abuse of power, or as the victim of a #MeToo campaign. These common-sense considerations suggest not only that mankind has not ceased to progress, but that the thesis that the rich are getting richer is unreliable and, strictly speaking, nonsensical: we struggle to establish, in the same culture and even in the same neighbourhood, the real estate value of an apartment. Yet we are able to competently compare the assets of Basil II Bulgaroktonos, the 11th Duke of Marlborough and Alexander the Great and conclude that they were all *unquestionably* poorer than Elon Musk?[81]

If the thesis that the rich are getting richer is unfalsifiable, and therefore nonsensical, the correlative thesis that the poor are getting poorer is highly falsifiable, because while it is difficult to measure wealth in different eras, or even in the same era, it is very easy to measure poverty. At the beginning of the 19th century there were one billion of us, now there are seven billion, and there are more people alive today than all the dead in history. This means that the poor are getting richer and richer, because more and more human beings are crossing the threshold of absolute poverty, namely lack of means of subsistence. And, perhaps even more significantly, in the developed countries where you and I are lucky enough to have been born, the average life span and active life span of people have been extended beyond all expectations.[82] Of course I acknowledge

---

[80] On the need for a multidimensional analysis see R. Aaberge, A. Brandolini, *Multidimensional Poverty and Inequality*, in "Temi di discussione", Banca d'Italia Working Papers, 976, September 2014.

[81] Ironically, those who argue that the rich are getting richer are often the same people who have spent years splitting hairs in a doctorate devoted to some very specific thesis, such as that cockfighting on the island of Java should be interpreted in a very different way from how we conceive it in the West, where we are inclined to compare it to a gambling practice once in use among us.

[82] The increase in divorces has shown that the "till death do you part" oath was calibrated on a much shorter average life span than today. Now, if it has become normal to change partners, it is not clear why it should be abnormal to change jobs. There is nothing wrong with this;

that there are fewer opportunities for young people in developed countries compared to forty years ago, and forty years ago things were not as great as during the postwar economic boom.

This does not detract from the fact that the comparison between humanity in 1920 and humanity in 2020 favours the latter, for more than one reason. Take France and the United Kingdom, for example: together they make up 120 million people who are partly worse off than their ancestors. Partly, of course, because after all *Les Miserables* and *Oliver Twist* are about France and the UK. This is not surprising because after the war they lost their colonial empires, which were perpetuating slavery in another form. On the other hand, 3 billion Chinese and Indians, almost half the world's population, have an infinitely higher standard of living than when they were direct or indirect colonies of the West. It takes a racist bias towards the West to say that the poor are getting poorer. Perhaps the only ones who matter are our neighbours? It is estimated that the middle class will make up 61% of the world's population by 2030 (without this meaning that being middle class means being better off or happier than a petit-bourgeois of the 20th century). This is precisely because the prophecy of automation has come true, in a way that Keynes himself could not have foreseen. It has certainly proven true that industrialisation has had as much impact as the arrival of gold from Mexico in Europe. The main problem for the survival of capitalism, at that point, was to achieve a fairer redistribution of wealth.[83]

---

on the contrary, it is in principle the best a human being could expect, only we need to arrange, and first of all think about, the conditions in which this discontinuity can become a promise of happiness. Not to mention the availability of consumer goods, such as cheap clothes, food and technology: for the first time in the history of the world the number of overfed children is double that of underfed children, and there is nothing surprising about a beggar with a smartphone. Even if one were to evaluate negatively, for reasons that are difficult to understand, the greater availability of consumer goods (the point is not their availability, but the inability to dispose of them), the fact remains that in order to seriously address the issue of the "poor-growing-poorer", one cannot overlook the benefits that every human being derives from scientific and technological progress, which means: lower infant mortality, greater civil protection, more effective healthcare, greater literacy – in other words, all that accounts for the impressive demographic curve of the last two centuries.

[83] We need only consider the historical development curve of global GDP: a curve that runs parallel to and not far from the horizontal axis and then shoots vertiginously upwards from the end of the 18th century, with an unstoppable growth ever since. Basically, from the industrial revolution onwards, per capita income has progressively increased, as has the quality of life, thanks to new discoveries and inventions as well as the slow but concrete conquest of civil rights. Incidentally, although global capitalism also increased inequalities between countries in the first phase, since globalisation became convergent, these have also been reduced since. The impression of impoverishment stems from the specific situation of the Western generations who have grown up in times of globalisation, and who have a lower standard of living than their parents.

History is a hard discipline, because it confronts humankind with the facts it has produced. Nevertheless, if we look at it without prejudice, the balance is positive. The geopolitical upheavals that have decreed the end of colonial empires, the birth of new industrial powers, and what we, in Europe, perceive as a crisis – albeit accompanied by positive signs, such as the increase in the average age of the population – are merely the redistribution on a global scale of widespread prosperity. This redistribution is difficult, given the exponential growth of the population (another indication of the fact that more and more human beings are crossing the survival threshold), and is still profoundly unequal, but it is going towards the cosmopolitan progress of the whole of humanity.

Let us not forget that less than a hundred years ago, the very same Ethiopia that is now referred to as one of the most exploited countries, and with labour costs so low as to be competitive with automation, could easily have been a military target for Italian colonialism. What we are witnessing is a pro-cosmopolitan upheaval in world history. The grandparents of the Ethiopians exploited today were gassed by the Italians, and the grandparents of the exploiters, the Chinese, were massacred in Nanking by the Japanese. When people say that humanity is getting better, they do not mean that it is close to paradise, but that it is a bit further away from hell. If this is the case, history is not at all the tale of the decline of humanity, but of its undeniable progress.

What awaits us, then, is not the unjust and unhappy growth in which the rich are ever richer and the poor are ever poorer, but – if we manage to understand the present and transform it – a new welfare, in which machines are entrusted with toil, boredom and repetition, and humans with invention, consumption and education. In a history considered from a cosmopolitan point of view – the only standpoint deserving to be the subject of a philosophy of history worthy of the name – it is therefore undeniable that humanity is getting better. Countries that were previously colonised by the West, such as oil producers, have formed autonomous cartels, causing the energy crisis in the West, and accumulating wealth that in some cases has been monopolised by local elites, but in others has been redistributed, generating welfare. From a cosmopolitical point of view, it was a phenomenon of growth, redistribution and social justice.

Social immobility is a characteristic of traditional, family-based societies, and can only be maintained in small places.[84] Instead, globalisation tends much less towards the centralisation of capital and the maintenance of role-based revenue than towards redistribution of wealth on a much larger scale. It is not surprising that former colonising countries have become impoverished: they have had to

---

[84] G. Barone, S. Mocetti, *Intergenerational Mobility in the Very Long Run: Florence 1427–2011*, in "Temi di discussione", Banca d'Italia Working Papers, 1060, April 2016.

renounce large amounts of free resources and cheap labour; they have had to bear the costs of decolonisation processes; and they have had to take in the inhabitants of their former colonies, redistributing resources that were now modest, and facing all sorts of social problems. It is also worth bearing in mind that what has been lost in the West in terms of income has been gained in terms of civil rights, and that in more than one case civil rights have led to a decrease in income. Divorce was introduced in Italy in 1970, and since then the accumulation of capital, which was the primary vocation of the family unit, has been drastically reduced.[85]

Second sign of history. Today we count every single dead, whereas it used to be normal to have ten thousand fallen in any given battle – and I'm not referring to Zhukov (under whom hundreds of thousands were killed) but to Frederick the Great or Napoleon. There is much indignation about the treatment of prisoners in Guantanamo, but no one, not even their torturers, ever thought of impaling or crucifying the inmates – something that was normal practice for prisoners in Europe in the age of Raphael, Tasso and Descartes. And as for less famous wars, such as that in Rwanda, such conflicts would no longer be possible today, because they would be immediately come under public scrutiny (these unknown wars, on the other hand, remind us that the worst slaughters in history were made with machetes, not thanks to some advanced technology). And to claim that trade and financial wars are worse than military ones shows dramatic insensitivity to the victims of Gettysburg and Verdun, Hiroshima and Dresden.

In 1950, a simple man named Harry Truman – someone who was not exactly an intellectual, and who had only become president of the United States because Roosevelt had died – found himself making the decision whether to drop the atomic bomb on Korea, during the Korean War. Eventually he fired General MacArthur and rejected his proposal. Why? Because he knew what Hiroshima and Nagasaki had been like. Such a thing must never happen again, just like the Shoah and many other things – we all agree on this because humanity has become more civilised. Failure to recognise this fact, and seeing human achievements as a simple turn of power that found it more convenient to control life rather than give death, is tantamount to saying that all progressive political action is futile and all social criticism laughable.

Third sign of history. This is also why the environment has become one of humanity's greatest concerns: there is an increasing number of humans who can afford to worry about the environment, something that, say, during the Battle of Kursk in 1943, the most gigantic tank battle in history, was not at the top of any-

---

[85] If we wanted, we could abolish divorce and restore patriarchal, frugal and ecological family units. I doubt, however, that the overwhelming majority of those parading at Family Day would welcome such a restoration.

one's list of priorities. Of course, as we have seen, there is a great deal of conceit in this new sensibility, as if we were the lords of nature called upon to save it, but we forget how indifferent the planet is to the ephemeral layer of humans who for a moment in the history of the universe have affected its surface. Still, this sign should not be overlooked. Because, let us not forget, the devastation of the environment is a characteristic of humans, not of the Anthropocene[86]: Europe was once covered in forests, and ever since the Neolithic period humans have been busy deforesting it. And it is hard to imagine a worse environmental catastrophe than what happened on Easter Island: in order to transport the Mohai the people

---

[86] This is a problematic notion. The endless ages that came before us have picturesque names that fascinate the young and old alike: the Cenozoic, 65 million years old; the Mesozoic, which began 251 million years ago; and the Paleozoic, which began 542 million years ago. The recurring "zoic" suffix is a signal that deserves reflection: we study the epochs of the earth by marking them with epochs of life on earth. And this is far from obvious, since there is no mention anywhere that the task of the earth consists in hosting forms of life, or that the organic is superior to the inorganic. It is easy to see the anthropocentric nature of this division, which works by marking increasingly complex life-forms up to the most complex one, that is, the human race. At last, once dinosaurs and other childhood dreams have gone, comes the list of hominids, also defined according to their supposed intelligence, with the peak being again us – the *sapiens sapiens*. Here we have a problem, of course. As there aren't too many ostensible proofs of the *sapiens sapiens'* intelligence, we resort to time frames defined by the materials they used to make their tools: stone (Palaeolithic, Mesolithic, Neolithic), copper, bronze, iron ... So the unit of measurement of all these epochs is made up of two axiological principles: the organic as superior to the inorganic, and the human as the ultimate organism, because it is capable of producing artefacts. Now, I do not think this is a wrong choice. Why should we introduce other ways to measure time (say, the number of volcanic eruptions or environmental devastation caused by meteorites)? But if we agree on this point, then we must admit that the concept of "anthropocene" is problematic to say the least. Disputes about its dating are a sign of this difficulty. When did the anthropocene start? The oscillation, and therefore the approximation, spans across several tens of millennia, from the Flintstones to the day before yesterday. (cf. R. Meyer, *The Cataclysmic Break That (Maybe) Occurred in 1950*, in "The Atlantic", 16 April 2019). Some say it began about 40,000 years ago, when humans started to exterminate the great animals that had preceded them. Some say that it began when, through breeding and agriculture, humans literally changed the face of the earth. For some, the threshold is even closer to today. In this group, some place the beginning of the Anthropocene in the geographical discoveries that, by another convention, would mark the beginning of the modern age. Others, with a prevailing aesthetic sense, believe that the Anthropocene began with the industrial age (so that, as with cholesterol, there would be two Anthropocenes: the good one, the age of the *Eclogues*, and the bad one, the age of *Modern Times*). Finally, others, relying on the sure marker of radioactivity, make the Anthropocene coincide with the atomic bomb. This, however, is a contradiction that is difficult to ignore, because on the one hand the Anthropocene is "the period during which human activity has had the greatest influence on climate and the environment" (Oxford English Dictionary). On the other, the Anthropocene would begin at the very moment when, for the first time in the history of the world, there were the premises for the disappearance of human action as a result of a nuclear catastrophe.

had to cut down all the trees, to the point that they could not even leave the island, which they had made almost uninhabitable, because they couldn't make a boat.

On the positive side, think of how many regulations for the protection of the environment and health exist today that did not exist in the past. Half a century ago London was full of pollution and the Thames was extremely dirty. Not anymore. Half a century ago smoking was a sign of virility and intellectual flair and buildings were full of asbestos (even when it turned out to be harmful, things went on unchanged for a while). Now only those who can't quit still smoke, and everyone is discouraged from doing so, and buildings are asbestos-free. Obviously, we do this for us, and certainly not for the planet, which looks upon us (to use an anthropocentric expression) with commiseration.[87]

Fourth sign of history. Those who find today's societies, and in particular the cosmopolitan and complex world made possible by the Internet, less free than earlier ones, should really think about other kinds of life. For example, what life can be like in the country, spent from birth to death under the weight of the neighbours' judgement, marked by its origin, limited in its possibilities. Let alone life in a totalitarian state, which paradoxically offers a few more possibilities (for example, to change social status). I am not at all denying that society and consumption hetero-direct us, just as we are hetero-directed by customs, culture, family and so on, but this does not detract from two aspects that I consider crucial. The first is that the automation of production means that the alienation that comes from repeating the same gesture over and over again has been passed to machines rather than humans. To me, this is a good thing from any point of view: if this was the only transformation that took place, it would still be an enormous step forward for humanity. The second point is that, of course, there is no end to alienation because humans are never entirely their own masters. From this point of view, those who live in a traditional society that offers no choices are much more alienated than those who live in an industrial society, and those who are aware of being conditioned by Internet platforms are much more self-aware than a fundamentalist Christian is aware of being conditioned by the Bible. This awareness is, again,

---

[87] Am I saying that global warming is a lie? Of course not. The stock of greenhouse gases in the atmosphere has never stopped growing (it could only do so after many years of net zero emissions). And in fact we continue to chip away at the so-called "carbon budget", and we continued to do so even during the pandemic, albeit at a slower pace. So, the world today is less polluted than before in terms of emissions, but more polluted than before in terms of stock. Fine. But what did we expect? If one stops smoking, this does not remedy the damage already done – it only curbs the damage one would suffer if one did not quit at all. But there are two things we know for sure today: the fact that smoking is bad for you, and that global warming is a serious problem.

a step forward for humankind on a path of infinite refinement whose ultimate (ideal) goal is self-awareness and the overcoming of alienation.

Hence another thing to think about. For decades, it has been assumed that the standardisation of consumption, goods and customs is the result of globalisation. What has happened, however, is something diametrically opposed to such expectations. We have never had a humanity so diverse (in terms of tastes, ideas and behaviour) as the one that is before our eyes today. The fragmentation to which populism bears witness, whereby consensus can only be created through negation, affirmation being purely monadic, derives from globalisation which, I repeat, is not homologation but – quite the opposite – hyperdiversification. Where do we get the idea that the progress of modernity and the dictatorship of technology have homologated us – as if we were all the same inside and out? I would like to remind you that nothing is more homologating than transmitted customs, folklore and tradition. Today everyone is dressed differently[88] and everyone watches whatever they want (be it intelligent or stupid), whereas before, between the Greek tragedy and television via religious processions and storytellers, options were rather limited.[89] Sovereignism itself is the expression of a hyperdiffusion of tastes (no one has ever claimed that every taste is good), but in a cosmopolitan context. However, to point out that the creation of a sovereignist international is not conceptually different from the creation of a solipsist international is to ask too much of one's interlocutor.

Fifth sign of history. To claim that critical consciousness is declining today amounts to idealising the past or, more accurately, contrasting humans as they are with what (according to the speaker) they should be. What are these supposedly illuminated periods that we have declined from? In January 1726, at the beginning of the Age of Enlightenment, the great Voltaire, born a bourgeois (his name was François-Marie Arouet and he was the son of a notary) was addressed at the theatre by a member of the noble family of Rohan: "Monsieur Arouet, Monsieur Voltaire ... What is your name anyway?" Voltaire's reply, "Whatever my name is, I know how to honour it", earned him a beating from Rohan's henchmen, until the latter said: "Spare his head, it's still good to make people laugh". I would say

---

[88] Except for the few who have decided to follow Dolce & Gabbana's aesthetic canons.

[89] In its early days, television used to offer one or two channels, now the Web offers thousands of films – were we more homologated then or now? Did the canteen at the factory offer a variety of options? We now expect food providers to cater for vegans, Melarians, Muslims during Ramadan, and kosher diets: is this not progress? These vast choices are hardly noticed in canteens, mainly because people are working less and less, and yet the same young men who a hundred years ago would have been eating their ration during their military service (a homologation that has disappeared, and which we do not feel nostalgic about) today hang out in sushi bars. Do we see homologation here too? Sometimes it seems to me that the meaning of words is lost.

that we have come a long way on human rights, not only in comparison with Rohan but also in comparison with Voltaire, who for his part and like many others got rich from the slave trade without seeing anything deplorable in it. This is not only about people like the aristocratic and choleric Rohan, but also involves great and enlightened men who are usually associated with the revolutionary turn of thought in the 19th and 20th century. Nietzsche, for example, did not hesitate to approvingly quote the saying "Are you going to see a woman? Do not forget the whip". It is impossible not to notice how far we have come since a past that, fortunately though not without difficulty and resistance, is now behind us.

Let's come closer to us. In the 1940s, Horkheimer and Adorno deplored mass culture, except that they were in the United States because a more imperious and not-too-intelligent mass culture was in force in Germany, not to mention that there are no serious musical or ideological reasons to prefer the *Horst-Wessel-Lied* to *Tutti Frutti*. Marcuse complained about the one-dimensionality of man (which as we have just seen is anything but true) and about the "repressive desublimation", but that did not prevent him from teaching in La Jolla, on the Pacific coast, surrounded by bougainvillaea, hedonists and pacifists, when he could have gone back to teaching Hegel in Germany, perhaps in Karl-Marx-Stadt. There is no need to embrace an unreserved panglossianism to conclude that mass culture – often much more intelligent and amusing than that of the elite, especially if the latter takes on the ultimately fascist hues of the *Sociology of Music* – is still better than illiteracy or forced indoctrination, and that repressive desublimation is still preferable to repression without sublimation. All this is in support of a philosophy of history that is anything but Spenglerian. We are not declining: on the contrary, we are progressing, and the comparison with the heroes of critical theory shows this. Adorno despised jazz as "nigger music" (I quote); Horkheimer compromised the health of his listeners through the passive smoke of his cigars; Marcuse imagined a sexual liberation that today would lead one straight to court.

These attitudes are now unacceptable to us. Similarly, we are disconcerted by the fact that important and enlightened intellectuals, in 1977, signed a manifesto minimising the seriousness of paedophilia. It would be a very serious lack of historical sense, even in such a short period of time, not to recognise the change in context, whereby nothing of the kind is permissible anymore. But it would be an even more serious failure, and one that would be truly unforgivable, to ignore how far we have come in half a century and to stick to the refrain of our supposed decline. The point, in conclusion, is very simple: when we claim, quite rightly, that it is unfair to evaluate Aristotle, Fichte or Nietzsche, and often, as we have just seen, much more recent thinkers, according to our current criteria of judgement, we are affirming something of capital importance, namely that *humanity is progressing*. We should be aware of this, not out of futile self-satisfaction (which no one cares about anyway) but for two much more substantial reasons. The first is that, if humanity is progressing, then our being in the world, like everything we do and think, has a meaning and a justification. The second is that the fact of progress also imposes a right and a duty: the need, for every human being, to contribute to this progress not only with criticism, but with positive action.

4. Transvaluation: Where Are We Going?

Kant wrote that philosophy is the teleology of human reason: it gives purpose and meaning to knowledge and action, by trying to understand where we have to, and want to, go. Now, looking into the philosophy of history, as Augustine did, feeling old and worn out in a world even more worn out than himself, means asking: are we going in the right direction? I promised I wouldn't simply mimic Augustine, but here the reference to his greatest invention, the philosophy of history, is inescapable. So let us ask ourselves the question "where are we going?". From a cynical standpoint, the answer would be "Where do you think? To the grave". So, what is to be done? What can we do about this? As for the grave, not much – at most we can have our ashes scattered instead. As for the rest, instead, there is very much that we can do. This is what makes the social world interesting, as well as possible. Significantly, this world has been able, among other things, to capitalise on death – indeed, to make death the great motive of history, thought, the economy and society. Let us never underestimate this, especially when death, which is always around and within us, makes itself felt with its full power: as we have seen, it is precisely the perspective of the end that generates ends, meaning and history. It is precisely our mortality that harbours the political centrality of us humans. Herein lies the principle of a revolution that requires a *meditatio mortis*, but not an acceptance of the existing state of affairs, whether outside us or within us.

Let us never forget that. Contrary to a widespread and unfounded belief, history is not at all over, as was predicted thirty years ago,[1] echoing speeches that had already been heard in the 1930s and had been thoroughly refuted in the 1940s. According to such views, after the modest event of the fall of the Berlin wall, which dispelled the spectre of a third world war, there was nothing more to be expected: thereafter history would be dead calm, and humankind wind-still. This picture sounds a bit boring but is certainly reassuring: Events (with a capital E) always bring something catastrophic with them, be it wars of conquest, genocides and colonisation following geographical discoveries, or religious wars. But dead calm, in turn, harbours a messianic anguish: something has to come along

---

[1] F. Fukuyama, *The End of History and the Last Man*, Free Press, New York 1992.

(the spread of Christianity in the era of the Pax Romana cannot be explained otherwise), it will come from outside, to change, reshuffle, break up and rebuild everything – for better or for worse, it does not matter. This may seem like a bizarre attitude, and indeed it is, but it is human, so some philosophers have theorised about it. Others have been more down-to-earth, like Saint-Exupéry's tail gunner, who during a clash with German fighter planes said: "It's all very well and good, but these things don't happen in civilian life".

So, we were waiting for a catastrophe and a miracle. And the first one apparently came, unexpectedly, in the form of a virus. It came and stirred up everything, in life, in the economy, and in politics, just like wars always do, and in particular like the Third World War that (fortunately) did not take place. I do not at all mean to minimise Covid here: I only wish to put it in perspective. We certainly need a vaccine and the disease to go into remission, which sooner or later will come, though after too many deaths. Just as we need an economy that does not collapse under the weight of such an Event. But we also need a broader reflection that places Covid in the framework of the great problems of humanity, both those that cannot be solved (we are mortal) and those that *can* be solved (just think of the gap between the average life of a European and an African). We need to understand that anguish is such because it has no object and, however paradoxical it may be, it finds peace when it identifies one, in our case Covid – something that lent itself very well to this role, because it affected our lives much more significantly than other anxieties.

What benefits would come from such a reflection? On the one hand, a better knowledge of human nature, but this is a matter of theory. On the other hand, an understanding of the true scope of the demands for justice, fairness, redistribution, security, and above all the conviction that we are going somewhere – things that have always plagued humankind, and which are now felt much more strongly because they bounce around the Web every day, in a world that has finally become, also socially, the globe it has always physically been. That was its manifest destiny (that's for sure), fulfilled over thousands of years of human history. And within this destiny, which is progressive and cosmopolitical, we must think of the future, with the same hopeful outlook as those who have survived the war and at the same time with the awareness that, however terrible, this war between man and virus is not as deadly as the war between man and man. The world was there before us and will be there after us, and the features manifested in our history naturally reflect principles that were already in place before human history began, and that will continue to operate when humanity has disappeared, leaving room for other forms of life, or no life at all. In the meantime, the space remains open for a political reflection which, in the present case, I regard as a transvaluation of values, and more specifically of the value of labour.

Thus, history never ceases to go forward, and the revelations of the present could not be more eloquent in this respect, because they show us a human being profoundly different from the one we knew in industrial modernity. Just like the industrial revolution in Marx's time, the documedia revolution appears as the fundamental perspective through which a historical humanity (our own) can mirror, interpret, understand itself, and above all improve and transform itself. But where history ends (the ontology of our present) politics begins (prospecting towards the future). We have not always been producers, and very few of us will long be able to call themselves such. But all of us – the hunter-gatherer, the worker, the creative and the recreational – have been and will be mobilised. What lies at the heart of the ongoing mobilisation is the body as a carrier of needs, drives and consumption, as well as organs capable of manipulation. The task of this final book is therefore to provide a concrete practical proposal for the new world. Once we have understood the nature of the documedia revolution, the anthropological and sociological revelation that derives from it, and the metaphysical foundation that determines the emergence of both, it is a question of thinking about the place of humans in an automated world. This place is not at all marginal, because if machines no longer require humans as cooperators in production, they still need them as the essential pole of consumption that gives meaning to the whole process. If we leave *homo faber* in the tool shed, we can generate a webfare that means freedom from material needs as well as from ignorance and prejudice – in other words, culture. From here we can conceive of a future that rests on three bases: invention, (the autopoietic uniqueness of humans); consumption (which if properly understood is the true *animation* of the entire technical system); and education, (something that is outside the reach of any computer).

4.1. Deconstruction. The Other Side of the Hill

In 1771, Louis-Sébastien Mercier published *L'année 2440*, describing a future Paris full of scientific and civil progress, but still ruled by another King Louis, the 34th in the series. The road to utopia is full of refutations, and I would simply propose a change of perspective: let us not try to envision the next thousand years, but simply guess what is on the other side of the hill, as the Duke of Wellington said, putting ourselves in the position of possible winners instead eternal losers against some Napoleonic power.

Rather, let us relaunch the master/servant dialectic without taking the servant's point of view. It is easy and unsatisfactory to lament the exploitation of one's fellow human beings. It is much better, though more difficult, to think that one day this exploitation will no longer be economically worthwhile. And it is even better to make sure that that day comes soon, organising the conceptual scenario of a world in which consumption, as a human characteristic, takes the place of labour as toil and alienation. For consumption cannot be automated, but produces infinitely more value than what could come from using humans as machines and not as humans.

It goes without saying that complete automation, like the Messiah, is an idea that will never be fully realised, for reasons both *de facto* and *de jure*. However, there are two other things that are equally certain. First, that if we understand the true nature of artificial intelligence, which is not to organise a taxonomy from above, but to record our behaviour, it is clear that every day is a step towards full automation. Secondly, that there is no need to wait for this automation to create the conceptual scenario that makes it feasible: we can already rethink the notion of labour, trying to understand what humans really are, what they want, and above all what can be done for them – that is, (by stepping out of the dubious position of benefactors) what we can do for ourselves. To do this, we first need to get rid of four assumptions that paralyse or divert action: the idea that we are masters of nature, the idea that we are slaves to technology, the victimhood that sees any revolution as bound to fail, and the messianism that sees communism as a fading hope rather than a reality that is already present in our world.

4.1.1. Slaves to nature

On 30 August 1755, at the Délices, near Geneva, Voltaire acknowledged receipt of Rousseau's *Discourse on the Origin and Basis of Inequality Among Men*:

I have received, sir, your new book against the human race, and I thank you for it. [...] The horrors of that human society – from which in our feebleness and ignorance we expect so many consolations – have never been painted in more striking colors: no one has ever been so witty as you are in trying to turn us into brutes: to read your book makes one long to go about all fours. Since, however, it is now some sixty years since I gave up the practice, I feel that it is unfortunately impossible for me to resume it: I leave this natural habit to those more fit for it than are you and I. Nor can I set sail to discover the aborigines of Canada, in the first place because my ill-health ties me to the side of the greatest doctor in Europe.[2]

It is hard to argue with Voltaire. In fact, without culture and society, humans would have followed the destiny of their natural life, which is short, solitary and nasty, and it is very possible that our species (assuming that something like this can be established, given that we are descended not from one ape, but from many) would have become extinct hundreds of thousands of years ago. With the

---

[2] Quoted in J.-E. Joullié, R. Spillane, *Philosophy of Leadership: The Power of Authority*, Palgrave MacMillan, London 2015, p. 53.

result that we would not be here and no one would ever have even uttered the word "human".

We cannot be human without mechanisms, structures, and automatisms, i.e. technologies, because the human being – in the absence of such things – is an animal destined for a short and far too frugal life. If this is so, and if we reject the hypothesis that the human being was fashioned by God from clay and animated by a breath of life, we should investigate the conditions under which one of the many animals on earth developed the form of human life – the form that machines envy us, so to speak, to the point of recording, archiving and profiling it through documedia capital. These conditions certainly come from far away, from a capitalisation of rituals and behaviour that generated roles, goals, norms; these things emerged from a remote antiquity that only began to exist with social hysteresis (provided, of course, we discard the alternative hypothesis that they came from heaven).

If our ancestors had remained in the water, they would not have been able to light fires and set up camp; nor would they have been able to make objects with their hands, while telling stories with their mouths, which remained free for expressive rather than practical purposes; nor, consequently, would they have been able to create more or less conscious communities, develop rites and customs; and perhaps, with sticks, build fences and canopies to protect themselves from the weather and ensure relative safety from stronger predators, be they lions, members of other clans, inhabitants of the neighbouring village or (as today's easily retrodatable statistics teach us) hutmates. Anthropology is the other side of technology: the better we understand technology, the better we understand humanity. The revelation of who we are is a historical process, and if the birth of history can be traced back to the arche-technology of writing, let us not forget that the evolution of devices, from the simplest chipped flint to the smartphone in our hands, tells us better than anything else what we are and what we want.

What determines our difference compared to non-human animals is not our psychophysical endowment (we are highly imperfect animals, and above all more aggressive, fearful, impulsive and dependent than others), but the great externalisation and capitalisation device offered by technology. In all of this, there is no trace of our supposed dominion over nature, for the very good reason that no one has given us the task of saving the planet any more than God gave Adam the mandate to name animals in order to assert his supremacy over them. Nobody gave us any tasks with regard to the planet, but we have always experienced the struggle to survive, to feed ourselves, to find shelter, to fight against animals that are much stronger than us. It is our salvation that is at stake, not that of other living creatures, of nature or even the planet.

As for saving other living beings, there are billions of non-human animals ready to take our place, just as we took the place of dinosaurs. The latter were doomed not because of their doing, but because of a climate change that can be attributed, as is currently assumed, to the crash of a meteorite more powerful than all the atomic bombs that crowd our military arsenals today. As humiliating as it may seem compared to the high concept we have of ourselves, and of our powers for good and evil, from the point of view of nature (of what for us, and only for us, is "nature") the self-destruction of the human environment taking place in the Anthropocene is a great instance of self-assertion.

As for saving "nature", here too there is a strange speciesist pride in the task we have given ourselves. This is entirely based on the difference between natural and artificial: the latter would be all that is done by humans, and the former would be everything else. Rather presumptuous, I'd say! On the one hand, it is hard to see why a termite mound or a dam built by beavers would be "natural", while the same artifacts, if made by human hands, would be "artificial". On the other hand, if we think about it for a moment, what underlies the alternative between natural and artificial is actually the alternative between natural and *supernatural*. What the human being does is allegedly the absolute other compared to nature – the hand of man is in fact the *mano de Dios*, called to work miracles by reversing the order of nature. But let's not forget that the plastic island in the Pacific is also natural, its elementary components are the remains of the dinosaurs we have replaced, and our role in the genesis of said island is infinitely smaller than that of a gardener in the Borromean Islands or of a polder builder in Zeeland.

As for saving the planet, no thank you. The planet does not need our intervention, because the fate of the Earth is already marked: first a crash into the Sun, and then, eventually, the thermal death of the universe. The goal, if anything, is to preserve the environment that makes the *human* life form possible. There is no nature as such, only an interaction between nature and culture. And this interaction is by no means exclusively destructive, but rather mainly constructive. The world in its natural state is no more sensible or benevolent than ours. It is up to us to aim for something better, so as to save ourselves. It is not up to the planet, which has no need for our help whatsoever.[3] Now, one cannot simultaneously

---

[3] Marie Antoinette had a model farm built next to the Petit Trianon, with goats, cows and so on. She would spend her happiest moments there, indifferent to her subjects and anticipating today's popular passion for organic and bio farming, at least for those who can afford it. It may be entirely legitimate to see this inclination towards natural things as a cunning move of the market, but it may very well be a cunning move of reason, which by making the natural and organic a symbol of distinction leads to greater care for the environment. One could hardly have any doubts between buying a biological soap and a synthetic product that is frighteningly

combat the ecological crisis and hold on to a future of industrial labour. That is
the bad news. The good news is that we are leaving the industrial world, and that
technology has never commanded anyone, except those who voluntarily submitted to it.

4.1.2. Masters of technology

Hitler's dictatorship differed in one fundamental point from all its predecessors in history. His
was the first dictatorship in the present period of modern technical development, a dictatorship
which made complete use of all technical means in a perfect manner for the domination of its
own nation. Through technical devices such as radio and loudspeaker 80 million people were
deprived of independent thought. It was thereby possible to subject them to the will of one man.
The telephone, teletype and radio made it possible, for instance, for orders from the highest
sources to be transmitted directly to the lowest-ranking units, where, because of the high authority, they were carried out without criticism. Another result was that numerous offices and
headquarters were directly attached to the supreme leadership, from which they received their
sinister orders directly. Also, one of the results was a far reaching supervision of the citizen and
the maintenance of a high degree of secrecy for criminal events. Perhaps to the outsider this
machinery of the state may appear like the lines of a telephone exchange – apparently without
system. But like the latter, it could be served and dominated by one single will. Earlier dictators
during their work of leadership needed highly qualified assistants, even at the lowest level, men
who could think and act independently. The totalitarian system in the period of modern technical development can dispense with them; the means of communication alone make it possible
to mechanize the subordinate leadership. As a result of this there arises a new type: the uncrit-
ical recipient of orders.[4]

This was the final statement by Albert Speer, Hitler's Minister of Armaments and
War Production, at the Nuremberg trials. If you replace Hitler with the microphysics of power and the tangle of cables with the web, you get a picture of the
resigned attitude prevailing today, though often in the guise of critical radical-
ism.[5]

---

polluting, and this too is a sign of progress. But if it were specified that the organic soap is such
because it is made from pure human fat, only a monster would still go for that option, despite
it being, strictly speaking, the more ecological and less anthropocenic and anthropocentric
choice.

[4] Jacqueline George, *Goering Cross Examined*, Q-Press Publishing, Victoria, 2014,
pp. 206 ff.

[5] The Allies mocked the justification provided by former Nazi leaders ("so you were nothing but well-paid postmen, right?"). No such irony is found in Heidegger when he speaks of the
dictatorship of technology, comparing the death camps to mechanised agriculture, and invoking
salvation by some god. We need to stop indulging in the ultimately unaccountable fantasies of
a *Gestell*, an "imposition" that would rule humanity, or of machines seizing power under the
guidance of some demonic controller. Heidegger first mentioned this "imposition" in that very
same 1946, in an otherwise eccentric passage from *Anaximander's Saying* (in *Off the Beaten*

In *The Betrothed*, Don Ferrante, after catching the plague that in his opinion did not exist, being neither substance nor accident, died cursing the stars like a character from Metastasio. Today he would have died cursing power (or capital, or neoliberalism, or biopolitics, or the microphysics of power). This is not a socio-economic reality, but a deity that is malignant for most and providential for few. Such a notion lends itself to being the primary cause of good and evil, and exempts us from any analysis or conceptual invention, as it is enough to name the infernal deity and curse it to obtain the consent of one's tribe. It is an anthropomorphism, one of the many with which we explain the world, and which in this case has the face of a super being dedicated to ensuring the greatest evil for the many and the greatest good for the few. Take the image of Stalin as a great octopus taking over the world: could it not be replaced by Power? Power, however, is faceless, precisely because it is microphysical, some would say, thus excluding the hypothesis that it is simply chimerical. As well as failing to explain social phenomena, this attitude, which I call "fatalism", leads to a practical outcome that is indeed demonic: the absolution of humankind, even in a society that claims to be secularised. For fatalism, there is some supreme entity endowed with great intellectual, scientific and economic resources. This cognitive increase determines a radical discontinuity in the exercise of power. The latter, which until then had been exercised as the right to give death, is transformed into the capillary control of life by the power of technology. This thesis combines an error in metaphysics, an error in the philosophy of history and a political error.

As for metaphysics. It takes a good deal of bad abstraction to set life against death. As I have repeated many times, if life has a meaning, if organisms are different from mechanisms, it is precisely because, unlike the latter, they can die, and they fight a structurally losing war against their natural destiny by respond-

---

Track, Cambridge University Press, Cambridge 2002, p. 242): "The court of justice is complete: non even injustice is missing, though admittedly no one is properly able to say in what it consists". What the court has got to do with it, only Heidegger knows, and we can only make a guess. If my conjecture is correct, this is further proof of how easy it is to blame external agents. In Speer, as in so many critics of the dominance of technology, we see an inversion of cause and effect, whose assumption is that man is naturally good but is corrupted by technology and the market. Yet it is not the market that produces pornography, it is perverse and polymorphous sexuality, as rooted in our biological past (who induced the bonobos to debauchery? techno-capital?) and repressed by civilisation and its discontent, which generates the pornography market. This is true of all the other human weaknesses that do not seem to us to live up to the ideal we have of ourselves, and which we impute to technology and capital. When we do so, we effectively end up absolving humanity. Among these weaknesses, the worst is the voluntary servitude that is postulated by those who claim to be under the rule of technology. Don't believe them, they are lying, and they do so to justify themselves, just like Speer did in 1946 – though hopefully for lesser misdeeds.

ing to the needs of metabolism and trying to technically perfect the conditions for their survival. To disregard this means instituting an inconsistent discontinuity both in the heaven of ideas and on the earth of humans. From a metaphysical point of view, Tamerlane and a ruler of a contemporary state, be it liberal or totalitarian, exercise the same power, which is precisely, let us remember, the "power of life and death". This is the essence of politics – the power of life being the field of medicine, and the power of death being the field of the executioner.

But it is obvious that there is a stark difference between a mediaeval despot or Hitler and a contemporary politician. This brings me to the error in philosophy of history committed by fatalism. It consists in not recognising that, for example, there is a difference between Damiens being drawn and quartered at the beginning of *Discipline and Punish* and the fact that today more than half of the world's states have abolished the death penalty. Speer had some true atrocities to justify: the Holocaust, the fact that conspirators of 20 July 1944 were hanged on butcher's hooks, and the Nero order of March 1945, which decreed the destruction of Germany's entire production system, and which Speer himself, together with German industrialists, tacitly opposed. But, as it was rightly said, Hitler's madness turned history backwards, in a tragic regression from the manifest destiny of humankind.

Based on a metaphysically insubstantial opposition and a mistaken philosophy of history, which insists on seeing regression and barbarity or will to power where there is progress and emancipation, fatalism meets its worst defeat in the field it cares most about: politics. Here it turns into victimhood. As I mentioned in 1.1.4, the traditional view of the struggle between master and servant led to the victory of the latter. This was already the case in Hegel and even more so in Marx. The argument, as we know, is that at some point the servant stops being afraid of death and asserts his power over the world. Now, neither of these options are contemplated by fatalism, which sees humanity as a slave to technology (and worse still: a voluntary slave). In this view, humanity is so afraid of death that it will sell its freedom in exchange for an anti-smoking campaign or a cancer prevention programme. Humanity has discovered that it is worth more than cannon fodder. And we must emphasise this human value instead of raising altars to the unknown god.

Against fatalism, it is necessary to recognise that humans are metaphysically the masters of technology, even though they may, of course, be slaves to other humans; and this, as we have seen, is due to an undeniable fact: we cannot think of technology without humans, while we can think of humans (albeit in a purely animal state) without technology. There is no imposition of technology, we have no excuse, because technology does nothing without the vital drive of mortal, needy, fearful and curious organisms such as we are. Let us keep this in mind instead of negotiating downwards relationships between humans and technology, which are actually negotiated with other humans who have understood the essence of technology better than we do. Capital is not satanic, but Faustian[6]: it takes the good and the bad of what we are, it is not an omnipotent entity come from outside. Capital does not plot against us, except insofar as it is itself a network, a plot, a web, a system. It neither spies nor monitors, for the simple reason that it does not think, but rather notes, records, keeps track. As we have seen, capital is the outcome of any recording of traces on an external medium by human agents. Capitalisation starts with experience: we get familiar with things, and eventually we become experts. For this capitalisation to be of value, however, there must be a common external object. A potter makes a vase, a hunter a trap, a writer a poem. In all these cases, *praxis* is externalised as *poiesis*, as the production of objects that stand outside the mind and result from actions. Objects, which bear within themselves, in their form, the traces of their composition, are the visual asset, the archive of acts that represents the essence of capital. Starting from here, it is necessary to prevent Power and other generic divinities from being used as a distraction, a *tete de turc*, or rather, a Mechanical Turk. Instead of focusing on such things, we should try to understand the ongoing process, and the first rule to do this is not to be guided by prejudice and catastrophism.

Do you remember Sénécal, the resentful tutor from *Sentimental Education*? "Senecal went on to say that the workman, owing to the insufficiency of wages, was more unfortunate than the helot, the negro, and the pariah." Do we not fall into the Sénécal syndrome when we speak of slaves, miners and colonised people in reference to today? This is a lack of historical sense that translates into a lack of human respect and, what is worse, into a fatalism that is content with big statements and, after denouncing chimerical forms of slavery, actually promotes voluntary servitude. One things is certain. We are not colonised countries, and there is no new slavery, because the old one is enough for us. Perhaps we are silicolonised,[7] but it remains to be understood what this means, and whether it is a good or a bad thing; surely those who complain of minatorialisation do not work in mines. Besides being false, these metaphors are disrespectful to the historical memory of real mining, real colonialism and real alienation. Classical alienation meant working ten hours a day every day, repeating the same gesture over and over again; colonisation was the exploitation of a large part of the world by European countries; working in a mine meant being underground in the dark, risking death in cave-ins and explosions. To call the profiling of our behaviour "aliena-

---

[6] This term was used in reference to money by O. Spengler, *The Decline of the West* (1918), Oxford University Press, Oxford 1932, p. 408.

[7] E. Sadin, *La silicolonisation du monde. L'irrésistible expansion du libéralisme numérique*, L'échappée, Paris 2016.

tion", to speak of "colonialism" at a time when technology has allowed decolonised countries to overturn traditional power relations, and to define as "miner" someone who is anywhere but in a mine, means making dishonest use of words.

Above all, it means covering up the lack of political action with an indignation that is within anyone's reach. This is a double blindness, here, which does not take into account the specific, and therefore singular, characteristics of the present, relying on a past that is easy to regret as a golden age or to deprecate as an iron and steel age that would continue, against all evidence, into the present. The other face of this fully-blind Janus is the neglect of the past, which allows comparing the present alienation, if there is such a thing, with a past that is gradually disappearing thanks to automation. At a time when, in many parts of the world, need is giving way to consumption, thus revealing aspects of the human that were previously the prerogative of the elites, let us look at the actual reality, which has another face, less pathetic and yet as harsh as reality always is. And instead of lamenting our chimerical voluntary servitude, let us try to understand our involuntary but very real mastery.

How is it that we want to interact with technical devices, producing value, without gaining any obvious economic benefit from it? These are facts that outline the profound structure of humanity and the social world. The latter relates to technology out of an essential need, devoid of economic interest in the strict sense – even if, as we have seen, in a broad sense, the economy and capital are always on the horizon for humanity, although we tend to forget this. When production is automated, humans do not idle, as would appear logical from the standpoint of *homo œconomicus*, but get mobilised, engaging in an activity that is unpaid and therefore inexplicable in the light of economic rationality in the narrow sense. Faced with this circumstance, it is easy to speak of voluntary servitude, with a self-victimising *ethos*; it is more difficult to lay the foundations for transforming servitude into mastery, first by putting victimhood out of play, then by understanding that we are much closer to realised communism than the resigned may believe.

4.1.3. Political victimhood

An elite of families has managed finance and money since the dawn of time. This financial "cartel" is also the same that keeps hidden principles of physics that would free us from economic slavery, like antigravity. The elite creates ad hoc financial crises to control the flow of money in the world and direct it into their hands. This immense amount of money is then used by the elite to fund research in alternative physics. The knowledge of this "secret physics" allows the elite to create, influence and manoeuvre earthly events in their favour.

This is the back cover of the Italian edition of Joseph P. Farrell's *Babylon's Banksters*.[8] As Descartes would say, "[he is] mad, and we should not be any the less insane were we to follow examples so extravagant".[9] Still, there is a lesson to be learnt from all this: the louder the protest and the more apparently radical the criticism, the easier it is to miss the essence of things, i.e., actual reality. Of all demonologies, conspiracy theories are the most dangerous, because they justify every form of human weakness, while at the same time clouding the analysis of the situation by constructing a myth (the too rich, the too poor, capital ...) that dispenses with knowledge and reflection. If conspiracy makes its own justice, fatalism shows a more thoughtful and reflective face, yet it tells a similar story of defeat without redemption, turning into victimhood.

Unfortunately, I am familiar with this one. When I was a child, more than half a century ago, people kept talking about the forthcoming collapse of capitalism and the imminent advent of communism. This tale was messianic and a bit boring, and what's more, it won votes for right-wingers who were supposed to save us from the Bolsheviks. After 1989, the refrain changed: capitalism has won and is racking up one success after another – a frustrating position, because it presupposes a highly intelligent capital and a very stupid communism. This situation was made possible, paradoxical as this may seem, by a philosophical error: out of deference to Hegel, Marx considered contradiction to be the engine of history, so he expected capital to collapse under the weight of its contradictions, leaving room for communism. The supporters of capitalism were quick to object that these contradictions did not exist, but they did not see that the new capital would actually bring about communism. They can be forgiven: after all, not even the supporters of communism seem to have noticed this fact.

Still, history does not proceed by contradiction, but by revelation: it progressively brings to light a human essence that as such is perfectly contingent, but acquires necessity through an emergent process that, in hindsight, is called "natural selection". If we were to believe the philosophical literature that has characterised the last quarter of a century,[10] it is still difficult to avoid the impression of an irrecoverable defeat. Marxism is behind us, reduced to a messianic promise that is invoked without much conviction. Conversely, the undisputed winner of the battle between capital and labour is capital, which has emerged triumphant

---

[8] J.P. Farrell, *Babylon's Banksters: The Alchemy of Deep Physics, High Finance and Ancient Religion*, Feral House, Port Townsend (WA) 2010. Italian translation (where this description comes from): *Bankster. Come un gruppo di banchieri ha conquistato il mondo*, Uno editori, Turin 2017.

[9] R. Descartes, *Meditations and Other Metaphysical Writings* cit., p. 19.

[10] Let's say from *Specters of Marx* onwards. Cf. J. Derrida, *Specters of Marx*, Routledge, London 1993.

from all its clashes and proved immune to all those insuperable internal contradictions that were supposed to lead to its collapse. This position is halfway between two historically contemporary phenomena: the messianic socialism of the
German neo-Kantians and the dance of the spirits, the melancholic peyote-driven
circles of the last Native Americans, now confined to reservations.

Apart from being melancholic, this perspective is limited to messianism, the
most seductive but problematic of standpoints, which eternally postpones the
achievement of the goal. As for the dance of the spirits, it is politically useless
and socially harmful. It generates depression in the reader: capital has never lost
a battle, its antagonists have lost them all. It generates futile pride in the writer:
everyone has given up the fight, except me, even though my denunciation is
pointless.[11] And this is precisely the problem. The dance of the spirits is about
regretting a world that not only no longer exists and cannot be brought back, but
which is not even worthy of regret, since it was full of alienation and toil, which
no one can seriously claim to miss. The appearance is discouraging enough, and
only ensures that we smile as we go down: it is not our fault if the world is going
wrong, it is the fault of capital and its actors – not us. Indeed, if there is one principle that reigns supreme in victimhood, it is this: *le capital c'est les autres*.

Let us turn the page. In 1997, Richard Rorty gave three lectures, which were
collected and expanded the following year in a short book that is arguably one of
the most eloquent, expressive, revealing and unacknowledged texts in Rorty's
oeuvre: *Achieving Our Country*.[12] The basic assumption was that the new American left, unlike the old or very old left (Rorty's parents were Trotskyites, and
abandoned their beliefs when they had evidence that the organisation was fi-
nanced by Moscow), had moved from the position of political actor to that of
political spectator. And the same could be said of philosophy, which had moved
from civil commitment to nihilist critique, from Whitman and Dewey to Foucault and Heidegger. It is significant that analytical philosophy is completely
absent from this picture, as in Rorty's eyes it probably appeared as a hopeless
case (one can hardly blame him).[13]

---

[11] To refer to Adorno's categorisation (*Introduction to the Sociology of Music*, Continuum,
London 1989), this attitude is reminiscent of the resentful listener who only listens to baroque
music.

[12] R. Rorty, *Achieving Our Country. Leftist Thought in Twentieth-Century America*, Harvard University Press, Cambridge (MA) 1998.

[13] Bertrand Russell described the analytical philosophy of his time, which had been reduced
to scholasticism and mannerism, by means of a personal anecdote. He once got lost on a bicycle
ride and asked a passer-by if he could show him the way to Oxford. The man asked him: "You
mean the way?". "Yes." "To Oxford?" "Yes." To which the man replied, "No." All in all, it was
a good thing for Russell, because if he had met a continental instead of an analytic, the continental thinker would have entertained him with a long conversation about the impossibility of

On page 79 Rorty quotes Stefan Collini's observation that in the US (but not in the UK) "cultural studies" means "victim studies" – which applies, *a fortiori*, to post-colonial studies. It might be better to look at the world differently, with Rorty's sympathetic but not condescending outlook – just the opposite of the pitying and pharisaical *ethos* that Rorty stigmatised in the American intellectual left with brave, non-conformist, and surprisingly still relevant words. There are in fact at least four major revolutions that Rorty could not take into account in his argument, three of which he could not even have known about: 9/11, with all that followed; the documedia revolution; the 2008 financial crisis; and the health crisis we are currently experiencing. How can we abandon the rather comfortable position of compassionate spectator and take on a different one? Since everyone has to do what they know how, or at least think they know how, I wish to end this analysis with a positive political proposal.

At the end of the 18th century, Kant spoke of a magniloquent tone being adopted in philosophy, referring to the exaltation and pretentiousness of the nascent idealism. At the end of the 20th century, Derrida went even further by criticising the apocalyptic tone then prevalent in philosophy. Today, the main tone is a victimistic one. I could analyse the endless literature on this tone,[14] a rhetorical

---

finding the right way, complete with supporting texts about the ability to get lost according to Benjamin, as well as the Heideggerian praise of going off the beaten track.

[14] I'll only give three examples (there are many). First: D. Byler and C. Sanchez Boe, *Tech-Enabled 'Terror Capitalism' Is Spreading Worldwide. The Surveillance Regimes Must Be Stopped*, in "The Guardian", 24 July 2020. "There are concentration camps in China where facial recognition is used as a permanent control system." Objection: in Auschwitz, they used dogs; in the American penal baths, even a century ago, they chained prisoners by the foot, which solved the surveillance problem. "Terror capitalism, an 'American invention', puts millions of people under control by calling them terrorists." Objection: Communist terror, a Stalinist invention, exterminated millions of people by calling them deviationists, revisionists, etc. The same goes for Mao's Cultural Revolution. Terror fascism, a Nazi invention, exterminated millions of people because they were Jewish, communist, Roma, or homosexual. Second: N. Klein, *How Big Tech Plans To Profit from the Pandemic*, in "The Guardian", 13 May 2020. There is no doubt that platforms want to profit from the situation, the question is whether they have caused it or not, and especially whether this profit does not, as in the case of electricity and gas, also benefit the community. Health and distance learning, broadband ... Where is the harm, if it responds to a catastrophe? And how can we deny that this ongoing process is also an advancement of civilization? Here's the punchline: this, of course, only applies for the privileged. But, one could object, if the fact is negative in itself, that it only affects the privileged should be a rare and admirable case of social justice. Third, older but still relevant: L. Winner, *The Whale and the Reactor: A Search for Limits in an Age of High Technology* (Chicago University Press, Chicago 1986, republished in 2020), argues that with technology we sign a social contract of which we are not aware (one could argue that today we are much more aware, we are constantly being asked to sign, which we do without really caring). The consequences can be horrific, as when the highway from New York to Long Island was built with bridges too low for

exercise that would keep everybody happy, because it combines penitence with reluctance. Victimhood paints the picture of a humanity that is crushed and powerless, and frankly a bit stupid, the victim of a lying capital – which is strange, if we consider that it includes radical sceptics like the Flatearthers. This humanity is secretly conditioned by the media and the elite, as if it were not already conditioned by religion, political parties, and what is worse, by friends and relatives. This humanity is blackmailed by hidden puppeteers, all under a cloak of political correctness and preventive censorship that confines dissidents to a transcendental Siberia. The opposite is also true: any victim can become a political subject. This is how the Anthropocene took the place of the 20th-century revolutionary hope: with the workers gone, we now fight for animals, then plants, and finally the planet.

My counter-proposal is that philosophy should stop taking the winners' side, and instead at least consider the possibility that human rights *may* be achieved, and no longer in opposition to labour: the latter will have disappeared and been replaced by something better and more worthy. There is little point in sugarcoating shackles, making them more flexible, and exercising one's ingenuity in criticising exploitation. My suspicion is that these attitudes, not devoid of *Schadenfreude*, perpetuate the evil they want to fix, as it constitutes their *raison d'être*. Indeed, whereas theology was able to overcome the death of God by developing a theology of the death of God, a critique of capitalist exploitation, being a critique and not a celebration, would not survive the end of its favourite enemy. The political victim is merely a secularised version of the ascetic priest who called for renunciation and submission to the evils of the world. With the aggravating circumstance that while the ascetic priest, like the 20th-century communist, submitted to the renunciations that came from his principles and paid for his choices, nothing of the sort happens to the political victimist, who risks appearing as the military chaplain of the situation, helping people relieve their conscience and self-consciousness. It is highly probable that SS Sonderkommandos also had their military chaplains, who in all probability felt less responsible, or even redeemable, in their role as critical conscience.

It is nice and easy to appeal to principles. It is much more difficult to make these principles match our actions, be they good or bad (yes, even a bad action requires courage and energy, if it is more than just inaction). And here the real threat does not lie in imbecility, but in Pharisaism. By "Pharisaism" and "Pharisees" (and their equivalents: bleeding heart, complacency, *angélisme, moral*

---

buses, so as to prevent the poor from crowding the beaches. Who could be so foolish as to make such a choice? Indeed, it is economically disastrous in addition to being morally reprehensible: *c'est plus qu'un crime, c'est une faute*.

cant, *Gutmenschen*, prigs ...) I mean the ethical and psychological attitude according to which our moral value depends on the ideas we profess and not on the actions we perform.[15] We do what we can, for better or for worse, and the measure of what is possible is given first and foremost by experience, notwithstanding the fact that we can all bring something new into the world (if we are particularly lucky or unlucky). This is why exemplarity, i.e. a *form of doing*, is the real remedy to Pharisaism, in the threefold sense of giving an example, taking as an example and setting an example.

As for *giving an example*. No one is obliged to do something that is impossible, as we are taught from an early age. But if an action *is* possible (i.e. if it is possible to find an example of it), then it can also become the subject of evaluation and, above all, of moral action. In other words, if no one can be rewarded or blamed for fighting against the Martians, the fact that one acted where the others didn't is an essential criterion of moral judgement. This is why the actions of figures such as Schindler, Perlasca or Peshev[16] are so important: great or small, they are real, and make the actions of those who wanted the Shoah and the inactions of those who let it happen all the more appalling.

It is then a question of *taking* things *as an example*, finding exemplary actions in the great repertoire of culture, which indeed is nothing but an inventory of examples and good or bad models. In childhood, we were all recorders and parrots, storing words and gestures that we did not understand. And in traditional education, a very high dose of machine learning was, and in many cases still is, a prerequisite for understanding. A Chinese mandarin learning thousands of ideograms, a cellist practising to become one with her instrument, an athlete, an acrobat and a baker are all cases of mechanical learning leading to freedom. The mimetic nature of desire, imitation as an essential tool of social behaviour, and repetition as the foundation of motivation ("pray, faith will follow")[17] are all intuitive as well as experiential proofs that motivation is not an innate and human-specific endowment, but emerges through a process in which following examples plays an essential role.

At the peak of this process, we might find the ability to *set an example*, i.e. to perform a coherent action that can be a source of inspiration for others. Think of

---

[15] Cf. M. Scheler, *Formalism in Ethics and Non-formal Ethics of Values* (1913–1916), Northwestern University Press, Evanston 1973.

[16] He is the least known of the three, because there have been no books or films about him to date. He does have a page on Wikipedia, though, and I have spoken about him in *Emergenza* cit., pp. 104 ff.

[17] Georges Brassens, *Le Mécréant*: "Mon voisin du dessus, un certain Blais' Pascal / M'a gentiment donné ce conseil amical: / 'Mettez-vous à genoux, priez et implorez / Faites semblant de croire, et bientôt vous croirez'".

the ending of the *Buddenbrooks* when, after Hanno's death, firm convictions falter, just as the pious Buddenbrooks' faith in the Resurrection and my own faith in Emancipation.

But now Sesemi Weichbrodt stood up, as tall as ever she could. She stood on tip-toe, rapped on the table; the cap shook on her old head. "It is so!" she said, with her whole strength; and looked at them all with a challenge in her eyes.

She stood there, a victor in the good fight which all her life she had waged against the assaults of Reason: hump-backed, tiny, quivering with the strength of her convictions, a little prophet-ess, admonishing and inspired.[18]

I believe that there is a Sesemi Weichbrodt in every professor, and I am no exception, but as far as I am concerned, my fight has been a different one. I have always been torn between the pessimism of the will (the consideration of humanity as it is, starting with me of course) and the optimism of reason. Unlike Sesemi, reason does not lead me to despair, but is the very foundation of a hope which, if I were to abandon myself to feelings and the instinct of the will, would not even touch me. As I mentioned when discussing the philosophy of history, if humanity were not getting better, all our actions would be meaningless. But in fact humankind is getting better, and we can only come to a different conclusion by imagining humanity to be much better than it is, making up evil creatures, human or superhuman, as the cause of our misfortunes. And for those who view this principle as an unproven panglossianism, I wish to reiterate Candide's conclusion: "One must cultivate one's own garden". With one important variation: when, as will hopefully happen in the foreseeable future, and as is often already the case, the garden will be cultivated by robots, we will have time to cultivate ourselves, and above all to monitor our actions, so that it never happens again that someone feels entitled by the ideas they profess rather than by the deeds they perform.

4.1.4. Communism realised

In my view, Pharisaism is the morally unacceptable aspect of victimhood, which on the one hand claims to be critical (in the world of ideas) but in the real world preaches resignation. We are losers against some omnipresent power-knowledge, and there is nothing we can do about it. Except complain and feel sorry for ourselves, or accept the pity of philosophers wiser than us. I can imagine an objection: what would be the alternative? Pick up a gun, take to the streets, start a revolution? Nothing could be further from my mind, nor, I imagine, from the thoughts of the victimists. Here too, fortunately, we have moved on from Molo-

---

[18] T. Mann, *Buddenbrooks* (1902), Martin Secker, London 1922, p. 358.

tov cocktails and the storming of the Winter Palace. To answer "what is to be done?" – and I am talking to philosophers here, as that should be their job – we must not pity the last (this is the duty of every educated and civilised, as well as good-hearted, being), but elaborate concepts, for example by updating Hegel's and Marx's emancipatory paradigm. In their case, the servants had to become aware of the fact that without them no one would pick cotton on the Haitian plantations and the factories in Manchester would not run. And in our case? For starters, we have to recognise that we are no longer on a plantation or in a 19th-century factory.

"Fancy living in one of these streets – never seeing anything beautiful – never eating anything savoury – never saying anything clever!"[19] That's what Winston Churchill told Eddie Marsh in 1906, on a visit to a poor neighborhood in Manchester. Is this still the case today? Let us try to look at the world without prejudice. If communism is the realisation of the principle "from each according to his abilities, to each according to his needs", today we are in the best conditions to implement it; and as strange as it may seem, the second part of the slogan – the most traditionally difficult to achieve – is now, if not within reach, at least achievable in principle. Failure to recognise this fact today constitutes the most serious paralysis of conscience and ideas that our history has known, and is therefore also the greatest stumbling block to understanding and transforming the present.

In order to do so, one must look at actual reality and abandon the unfounded conviction that communism is dead: rather, it has evolved, making ours the closest epoch to realised communism ever – more than Thomas Müntzer's revolts, Lenin's programmes, the Great Patriotic War, and the institutions of the Warsaw Pact.[20] It is understood that communism is an ideal, an asymptote towards which the development of humanity must tend, aiming at minimising injustice as far as possible. China, the hegemonic power of the new century, is communist,[21] and is

---

[19] E. Marsh, *A Number of People: A Book of Reminiscences*, Harper & Brothers, New York-London 1939, p. 150.

[20] Cf. A. Bastani, *Fully Automated Luxury Communism: A Manifesto*, Verso, London 2019, who, like me, argues that the current era is the most propitious for the realisation of communism, although he does not (and neither do I) address the problem of reconciling communism and individual rights.

[21] With an outlook that could not be more Eurocentric, it was assumed that communism ended in 1989, with the collapse of the Warsaw Pact and the Soviet Union. Not true. Communism is alive and well in China, as it is in Vietnam, Laos, Cuba, and North Korea. Also, if we consider that in 1989 there were one billion Chinese, while today there are almost one billion four hundred million of them, there are more communists today than ever before, because the Warsaw Pact has never exceeded three hundred million inhabitants. If we give up the provincialism according to which communism would have fallen along with the Berlin Wall, in the heart of Europe, we discover that in China communism has not only not fallen, but has made it

essentially based on the documedia revolution, of which it is the most coherent and clear-cut manifestation, anticipating – albeit in an illiberal form – the webfare I advocate in this book.

---

the political and economic leader of the future: in the 21st century it will play the same rolle that America played in the 20th and England in the 19th. Now, at the end of an ongoing struggle, of which it will most likely be the winner, the new international will be led by China. What makes China the cradle of documedia capital is first and foremost the demographic factor. The number of Chinese is ten times the number of Russians, four times the number of Americans, three times the number of Europeans and twenty-three times the number of Italians. Numbers have always been a military resource, but for a while they have also become a commercial resource, often in combination with the military aspect, which made it possible to expand markets through war. Today, this massive demographic size has a very special value, linked to the data collection: of 1.4 billion people in China, one billion have a smartphone. This invisible, or at least pocket-sized, army is China's greatest asset. So, after having been, and largely remaining, the world's largest producers of goods, the Chinese are also the largest producers and collectors of documents. And this, when documents are the fundamental commodity, translates into a gigantic competitive advantage, which explains the rapidity of China's economic rise. Ownership of big data also enables analytical market knowledge, so that planning ceases to look like a failed Stalinist dream and becomes a truly great economic strategy. The veil of ignorance is of no benefit to anyone, just as the invisible hand is of no benefit to anyone. This does not contradict Marx's idea of communism, which is not an ideal or a doctrine, but the actual process that abolishes the present state of affairs: in other words, deconstruction. Now, there has been no movement more deconstructive than capitalism, and the documedia revolution is itself an enterprise directed towards the overturning of the present state of affairs. This reversal *has* taken place, and must be understood and governed – not simply exorcised. This is true for any platform, but also in this case the Chinese have an advantage worth reflecting on. Western platforms are private, the Chinese ones are public, and while nationalisation is certainly synonymous with inefficiency when it comes to work, when it comes to mobilisation it makes no difference if the owner of the platform is the state, which is motivated to be efficient because it gains not only wealth, but power. Every Western patent has its Chinese double. Better than anyone else, the empire of documents has understood how important these are – much more than money, which is now marginal. This imperial nationalism has resulted in the rejection of foreign credit cards and the great leap towards handling all financial transactions by smartphone. Once you've checked the essentials, you can open up to anything, including Starbucks: indeed, there are plenty of Starbucks coffee shops in China, without this causing much of a stir at all (unlike what happened in Italy). I can imagine an objection: all this has a cost, the loss of individual freedoms. I agree, but was communism ever concerned about individual freedoms? And as we are not wrong to be fond of individual freedoms, why not hold on to them while trying to learn something from communism for once? Of course, what we should learn is not the repression that rightly outraged Western communists in the age of Stalinism, but the socialisation of surplus value. This, however, requires a preliminary gesture: renouncing victimhood and renewing our outlook by replacing the feudal master/servant dialectic, as well as the industrial capital/workers dialectic, with the humans/automaton dialectic, where the former are destined to triumph not because they work but because they *consume*, giving meaning to a whole system that would otherwise have no reason to exist.

Moreover, even in the West, despite what we say and think, we are witnessing the closest society to communism that history has ever known. The communism that is now being realised is neither the workers' paradise envisaged by the left in the first half of the 20th century, nor the totalitarian hell denounced by Russell or Gide. Let us examine the terms of this realisation in more detail, with one point in mind. If, almost two centuries after its theoretical formulation, communism had not been realised, we would have reason to believe that it was just another messianism. But this is not the case: today we are much closer to the realisation of communism than we are to the second coming of Elijah, or to the resurrection of the flesh and the life of the world to come. Here, too, I will only select a few signs, as I did with the philosophy of progressive history. My intent, I repeat, is not to cultivate an attitude of satisfied quietism (the euphoric equivalent of political victimhood), but to show that we are on the right path and that it is worth continuing along it.

The smartphone with which we generate documents (i.e., wealth) belongs to us. In other words, we own the means of production, just as we own the house we rent out through one platform or the car we drive to work through another. The difference between intellectual work and manual work is blurring: an ever-increasing part of humanity in Western societies works by tapping their fingers on the keyboard; toil is disappearing, because most of the heavy work is done by machines, and we are left with mostly one job: that of living and consuming.

The society defined by Internet platforms is classless, since the occupational identity that made it possible to categorise classes has vanished. As we know, there used to be a working class (a significant denomination); below workers there were the casteless (the underclass) who were in fact jobless, casual and undefined workers; above them were the *petit bourgeois* engaged in services, then the bourgeois engaged in professions, and finally the great bourgeois and aristocrats devoted to otium and capital management. And even at the upper level there was a ubiquitous kind of casteless people: that is, the artists, devoted to creativity.

That classification is no longer in place. What is still there, of course, is a difference in wealth, but class as a social identifier, as a criterion for predictable behaviour and decisions, has disappeared. This transformation is clearly attested to by the volatility of electoral tendencies (which capture extemporaneous inclinations and not class interests) and by the fact that it is necessary to use web-based profiling in order to find out how the electorate feels. And nothing has taken the place of class: if one has good reason to think that the notion of "class" is problematic beyond the occupational identification it denotes, this applies *a fortiori* to the notion of "people". This is where the systematic and fatal confusion takes place between the sovereign people (a *fictio iuris*), the people as a class (i.e. an *obscurum per obscurius*) and the people as a nation (an anthropological idea), which allows the infernal mixture of nationalism and socialism that we have encountered so many times in the history of the 20th century.

The classless society is also a society without a state, as the latter has lost its fundamental prerogatives: the postal service was replaced by email, money by cryptocurrencies, political dialogue by social networks, and the army by contractors. This disappearance of the state is all the more pronounced in those countries – which are the majority – where the state does not control the platforms. This suggests that potentially, and in the short term, all the prerogatives of the state will be taken over by platforms, and that the only surviving states will be those that control them. For the time being, these are explicitly communist countries; everywhere else, the society emerging from the documedia revolution manifests the rhizomatic (polycentric and non-hierarchical) dimension that in the 1970s had been considered the best response to statism, which in spite of everything characterised real communism.

Moreover, what is lamented as globalisation is actually a new international, made of consumers instead of workers. Indeed, globalisation is another indication of communism and has taken place not through the generic circulation of goods, but through the documedia revolution. This is to say that globalisation does not depend on jets, but on writing: a technology older than the pyramids, capable of transferring not physical objects, but social objects, such as money, laws, policies and identities. Not to mention that, in accordance with the $O = RA$ rule, writing (and hysteresis in general) perform a miracle: they construct social objects. And note this: the international of consumption has full credentials to become an international of education, ideas and solidarity – in other words, to become a cosmopolitan community that is much closer to the ideals of communism than the politics of communism in any one country.

We are also witnessing the end of private property. Marx predicted that the growth of information would destroy capitalism by spreading knowledge among workers. This has not been exactly the case, but a significant phenomenon has occurred. Firstly, a large part of what was traditionally protected by the legal constraint of intellectual property is entering the public domain: you can disseminate your intellectual works as much as you want, with a previously unimaginable scope and effectiveness, provided, however, that you renounce any copyright, or make do with almost irrelevant royalties.

Secondly, much of what was protected by the factual constraint of secrecy, by the inaccessibility of our private lives, is now, if not in the public domain, at least in the sphere of what can be easily capitalised by platforms. We willingly relinquish ownership of the documents that concern us, and what would once have required the intervention of detectives and lengthy investigations, we now learn with a simple browse on the net. This is the consequence of freeconomics, seen from the consumer's perspective. You can access any service for free, provided that you give up your *habeas data*, i.e. you accept the storage and processing of the documents that result from your consumption and mobilisation. On a general level, therefore, private property is reduced in individuals precisely because it is concentrated in very few intermediary platforms.

Thirdly, the prevalence of services reduces the convenience of private owner-ship: there is little point in becoming the exclusive owner of a self-driving car, taking on the task of parking and maintenance, and only getting rid of the rather minor one of driving. Finally, consumerism leads us to keep purchased goods in our possession for as little time as possible, and consumption becomes the real means of production. We therefore acquire new goods and things, but – since we will soon replace them – it is as if we never really owned them. The production of waste, a huge amount of *res nullius*, is a huge destruction of private property (in the past, even dunghills were private property). As a result, in the world we live in, private property has not disappeared, but it is less present than in the days of estovers.

A final point. Not only is the Panopticon the least of capitalism's concerns, as it is interested not in ideas but in consumption. More importantly, and politically speaking, capitalist countries are achieving the exact opposite of a surveillance state. To put it simply, the Panopticon is there, but it is turned upside down, and translates into the control of rulers, reduced to obedient executors of the will of the people – something that, though conceptually elusive, is no less pragmatical-ly effective. *What is so obscurely called "populism" is in fact the realisation of the dictatorship of the proletariat,* which dictates its agenda to politicians who are not influencers but influenced, constantly under scrutiny and busy measuring their approval rates.[22]

---

[22] That this dictatorship does not translate into clear benefits for the people does not prove much – not even the Paris Commune, the example that Marx himself referred to, actually ben-efited the proletariat. Above all, there is a flaw that Marx did not see but Lenin saw very well. A ruling class cannot be such without the possibility of making unpopular decisions. If Chur-chill or Roosevelt had followed the mood of their voters, the former would have made peace with Hitler in July 1940 and the latter would have fought Japan but not Germany. If de Gaulle had listened to the French in Algeria, the civil war would never have ended. From this point of view, there is nothing more misleading than the comparison between media populism and fas-cism. The latter was an authoritarian government, just as Stalinism was. It carried out a political project regardless of the ideas of the governed, and this, in the short term, was its strength compared to liberal democracies, which had to deal with public opinion. But democracy also requires a double veil of ignorance and a not-too-pervasive information system. Whoever for-mulates an electoral proposal imagines the expectations of the electorate only in broad strokes, not in detail, and synthesises them by harmonising them in an ideological framework; for their

Hence the most important sign, in my view. As we know, Marx saw the dictatorship of the proletariat as an intermediate and far-from-happy stage on the path from bourgeois society to communism. If what I have said about a possible humanity to come, grounded on a cosmopolitan basis, has any chance of being realised (and undoubtedly the world of the documedia revolution is the best premise for such a humanity), I think we have reasonable grounds for hope. Instead of mourning the supposed death of communism, we must strive to develop alternatives to its major limitations, in particular its total disregard for personal freedoms, wrongly regarded as a mere legacy of the bourgeois world. In the perspective I am proposing, it is necessary to carry out a threefold action. Firstly, an analysis of documedia surplus value, which has been largely neglected until now. Second, a struggle for the recognition of the enormous amount of implicit labour that humanity is now providing for free. Thirdly, a webfare system capable of redistributing this surplus value, creating the conditions for a cosmopolitan humanity, redeemed from the curse of Adam and able to do something better than mimic machines for productive or distributive purposes.

4.2. Analysis: Documedia Surplus Value

The obvious and rightful objection that, despite the philosophy of progressive history I have outlined, there is so much (real or imaginary) unhappiness and so much social discontent, has two answers. The first, which is very simple and indicates an insuperable condition, is that human beings are not necessarily made to be happy. Saint-Just said, "Let us teach Europe that you do not intend there to be one unhappy man or one oppressor on French soil." Except he didn't mention the unhappiness of at least one person in France: Louis XVI. Was the latter happy before Varenne? Not necessarily. Abd al-Rahman III, Emir of Cordoba in the 10th century, counted fourteen happy days in his life. The Prince of Salina's estimate in *The Leopard* is greater, but only slightly: two weeks before his marriage, six weeks afterwards, half an hour at the birth of his first-born son, a few conversations with his second-born son before he died, and many hours in the observatory looking at the stars, which were "some sort of anticipatory gift of the beatitudes after death".[23]

The second answer, which is complex, but which points to a superable condition, is that we are working without being paid. Indeed, claiming that humanity

---

part, voters do not fully know what their own expectations are; and the government formula can be carried forward without continuous corrections dictated by polls and public opinion.

[23] G. Di Lampedusa, *The Leopard* (1911), Collins and Harvill Press, London 1960, p. 229.

is happy with its own exploitation undoubtedly means underestimating it,[24] since there isn't that much happiness to be seen, however imaginary. There is undoubtedly a sense in which those who are angry are angry for a Socratic reason: because we are ignorant. The docusphere is not an infosphere, total mobilisation is not collective intelligence. Our ignorance does not consist so much in knowledge gaps about Egyptian dynasties or the GDP of Belgium; what we ignore is something very important to our lives, namely that by acting we produce value, and if work is the production of value, then we are working for free. The dissonance between actual prosperity and perceived malaise should therefore be seen as the sign of a mystery that demands to be brought to light. People complain about the difficulty of finding a job; and yet prosperity is widespread and clearly higher than half a century ago, if only because globalisation and automation have lowered the prices of consumer goods. At the same time, people are reluctant to admit that they have nothing to do all day, and indeed claim to be as busy and stressed as ever. There is no reason to think that they are lying. What this means, simply, is that there is a huge ongoing unpaid mobilisation of the uniquely human capacity to produce value.

Why do we not see this? Because of an involuntary ignorance that has nothing to do with voluntary servitude. We are so surrounded by the infosphere that we do not even know our own social and employment status. We are so much a part of the knowledge economy that we produce most of the knowledge, but it does not come back to us. And the light of collective intelligence is not enough to make us realise that we are producing value for free on the Internet. Therefore, being unable to find the reason for this exploitation, which is unknown to both the exploited and the exploiters, we engage in indignation at the outward signs of privilege, or even misfortune: we are outraged at those who are too rich, because we envy them, and at those who are too poor, because we are scared of them, even though we feel that they are the victims of an even greater injustice perpetrated by us, just as much as the rich and powerful. Inventing the enemy to create common resentment is a childish idea, but, like any demonisation, it is very easy to implement and always finds followers. In all this, however, the documedia surplus value goes unnoticed.

Bringing it to light is a necessary act of enlightenment if we want to come out from our self-incurred immaturity. The current representation of the give-and-take between users and platforms is that of an equal exchange. Users give their data for free, and the platform gives them other data for free. It would seem that exploitation does not take place here, but rather wherever imperfect automation

---

[24] J. Lanier, *You Are Not a Gadget*, Allen Lane, New York 2010; Formenti, *Felici e sfruttati* cit.

still needs human labour. And yet exploitation happens precisely here, in an unfair exchange that does not generate toil or alienation, but mobilisation: the platform receives much more data than the users, and above all, unlike the users, it knows how to use it. In fact, the documents that platforms acquire are much more valuable than those provided to users: they are more numerous (on average, for every piece of information you take, you give four); they are proprietary, i.e. they belong by contract to the platform; they are capitalisable, both in the form of primary accumulation and in that of profiling through comparison with millions of other documents; and they are monetisable, because they can be bought and sold.

Last but not least, these documents have the miraculous property of transforming consumption into labour, if by "labour" we mean the production of value, thereby making it invisible. The serf knew perfectly well when he was working for himself and when he was working for the master, because there were two fields: his own and the feudal lord's. The industrial worker was less clear, because he did not know where the hours in which he earned his wages ended and when the hours in which he worked for free for the master began. The web-mobilised, whatever their age and formally recognised occupation, work all the time, as they are constantly producing value, and have no awareness of working. This, it should be noted, is not a bad thing in itself, as this job is infinitely preferable and less tiring than that of the factory worker or the serf, but it is still a production of surplus value that introduces social injustice. The lack of awareness of this process is indicated by a circumstance that merits reflection: people often show prejudicial hostility towards banks, to which they lend their money, and at the same time show equally prejudicial positivity towards Internet platforms, which in some cases are even conceived as instruments of direct democracy, failing to consider that in fact the documents we give to the platforms we give for free.

As soon as the acts are recorded, and can be capitalised, everything that is given (the documents that the platform offers to the mobilised) coincides with something taken (the documents that the mobilised users surrender more or less consciously to the platform), and the other way around. But there's more: what is given by the mobilised, the documents on their acts and behaviour, does not pre-exist its giving. Hence the paradox of giving something that one does not possess: the documents given are created at the precise moment in which they are recorded, that is, produced and transferred. On the one hand, there is an injustice, because this is indeed unpaid work. On the other hand, however, because mobilisation is an expression of the world of life in all its forms, it becomes difficult to speak of alienation in general and often moralistic terms.

In short, and anticipating the crux of my argument, when I sleep, check the weather forecast, Google when Sitting Bull died, look up the train schedule or view the online menu of a restaurant, I am not engaged in a "social activity", as was thought at the time of the theorisation of cognitive capitalism: in fact, the beneficiaries of my work are the platforms, and then of course society, but only once the surplus value has been deducted. In other words, if it is necessary to recognise the documedia surplus value and then redistribute it through a webfare policy, it is precisely because at present we are dealing only to a very small extent with online social work (which is concentrated mostly in the infosphere, for example in the cooperative work required by Wikipedia) and to a very large extent with the production of surplus value for the benefit of the platforms. This means that the crucial area to focus on is not, once again, the infosphere, but the docusphere in its relations with the biosphere.

4.2.1. Profit and surplus value

It could be argued that what I am describing as documedia surplus value is only documedia profit, without surplus value, where "surplus value" means wealth accumulated unfairly during the production process. And, if this is the case, it should be treated as an exclusively distributive problem, not as a productive one. In other words, the wealth accumulated by the owner of a platform making a profit through documents should be treated in the same way as the wealth accumulated by the owner of a supermarket chain, and both could obviously be said to have positive redistributive duties. This is true for several reasons: because inequality is unfair from a contractual perspective; because it prevents the maximisation of general utility; because it forces many into a condition of deprivation that could be easily avoided; because in the long run the market will implode, as it will become impossible to consume, and so forth. Therefore, the same taxation regime should be applied to both in order to finance a webfare system.

Yet my proposal is different, precisely because it entails a transvaluation of the concept of "labour", shifting it to the pole of consumption (bodily work, although consumption also has a spiritual dimension), invention and education (work of the spirit in the proper sense). All the life that souls record on automata is production of value, hence labour, even if this labour bears little resemblance to the kind we have learned to loathe in the industrial age. There is therefore something profoundly unfair and radically misunderstood about the way automata relate to souls. The analyses I carried out so far allow us to lay the foundations for repairing the injustice hidden in the mystery of labour denounced in 1.4 – an injustice that is all the more radical the less it is perceived.

The first move is to recognise this ongoing form of exploitation and surplus value, which would remain undetected and undetectable if we considered the Web as an infosphere. As I was saying, the exchange between users and platforms appears to be a fair one: the platform gives information to users for free, and users give information to platforms for free. However, I would note that this kind of reasoning could also be applied to the feudal production mode (the feudal lord gives the serf a field for free, and the serf works for free in the feudal lord's field) and the industrial production mode (the capitalist gives the means of production and sustenance to the workers for free, and the workers give him their labour for free, the wage being the amount allocated by the capitalist to ensure the reproduction of the workforce). If we find this description of the feudal production mode and the industrial production mode grotesque and caricatured, then we must recognise that the relationship between users and platforms generates a documedia surplus value. It could be argued, however, that what differentiates the feudal lord/serf and capitalist/proletarian exchange on the one hand, and the user/platform exchange on the other, is that the first two involve an asymmetrical starting position between the contracting parties that is seemingly missing in the third.

So let's change perspective and notice the asymmetry between labour and capital, which is much greater in the documedia production mode than in the industrial and feudal modes of production, even though, unlike the latter two, the documedia production mode is rarely characterised by toil and alienation, for reasons I have amply explained above. To use Kant's categorisation, in terms of quantity, platform documents are huge, while user documents are comparatively poor; in terms of quality, platform documents are valuable, because they can be capitalised and related to countless other documents; from the point of view of relationship, the documents of the mobilisers are secret, while those destined for the users are open and in the public domain; finally, from the point of view of modality, the documents made by the users are real, in the sense that they record actual behaviour on the net, while those accessible to the mobilised are a mixture of real information and fake news. Here capitalisation describes a real process, pointing to the immense, and in itself legitimate, power of hysteresis, and this is precisely the case with the documedia production mode. Ultimately, all the advantages of platforms can be traced back to capitalisation, be it the primary accumulation of data, the computational capacity, the profiling, or monetisation that follows. At the beginning of the 20th century, many workers' houses in Turin were lit by light bulbs bearing the inscription "stolen from Lancia" (where the modest theft was, after all, a supplement to their wages), but nothing similar is conceivable in the documedia production mode.

In short, the exchange between users and platforms is an unfair trade. The documents that platforms receive from users are more numerous and valuable (they are metadata about our behaviour, not information about the weather or celebrities); they can be compared, by platforms and not by users, with those of millions of other users, with calculation tools that we cannot use, generating profiling and automations that are inaccessible to us (and, I would add, unmanageable by a generic user). Finally, they can be sold like any other good, while at most we can illegally sell one of our organs, maybe on a platform, but not the biometric documents accumulated by our smartwatch; to make a more cheerful example, none of us could pay for a coffee with the documents we get from our mobile phone: at most we could be deterred from paying an expensive bill by watching its repercussions on our heart rate monitor.

Documedia capital is thus divided into two parts. One is immediately readable and semantic, and is what is consciously produced on social networks. The other is produced through traces that are not necessarily semantic (let us call them "semiotic"), but can become so through interpretation. Obviously, platforms also have access to the semantic part of the archive, but the real great capital is the semiotic one, precisely because it does not capture ideas, but behaviour. The most decisive aspects of this process involve the economy (production of value), labour (unpaid mobilisation taking the place of paid labour) and politics (not so much ideological control over the mobilised, which is of little interest, but knowledge of their orientations for electoral purposes). Secrets traditionally used to be expressed in natural language, but today we are dealing with secrets that are often not even recognised as such, and which only constitute knowledge if one has the tools to interpret them. Here too, rather than an infringement of the private sphere, we are dealing with a new form of capitalisation and labour, the coordinates of which have not yet been determined. Similarly, if it is true that the universal blackboard is an ideal that is probably destined to remain such, the vision that the mobilisers have of it is certainly much more clear and detailed than that of the mobilised. However – and this must never be forgotten – it is also true that without the mobilised, platforms could do nothing and would be perfectly worthless.

In this framework, webfare would simply consist in redistribution: not of information, but of the surplus value produced by users for platforms, in such a way as to benefit both the platforms, which would have customers, and the users, whose labour would be recognised. This would also circumvent the obstacle of user remuneration. It is not a question of giving a small sum to the workers, because the real gain lies in aggregation: we must intervene with taxation and redistribution so as to support and educate the weakest strata of society, letting the surplus produced, for example, by the social networking of two professors discussing the revolution go directly to the unemployed and the exploited. One might object that this perspective only holds up if one espouses the labour-value theory, according to which value is the result of the labour employed to produce goods, in this case documents. And this is indeed the case. Labour-value is right and ontologically founded, otherwise we would have the most radical constructivism in place, a view like "there are no prices, only interpretations" that would be the economic equivalent of the philosophical motto "there are no facts, only interpretations". Believing that the economy is based on arbitrariness and fantasy – which is what happens if one abandons the labour-value theory – means introducing a nihilism that in no way reflects the very concrete, and far from fantastic, reality of the economy and the social world as a whole. In this case, it is very clear that value arises from the encounter between the mobilisation producing the documents and the investments made to collect them, to make them analysable, to build models that are able to match the users' preferences and the offers of the market, and vice versa.

Obviously, consumers benefit from services: competition lowers prices, thus leaving people more money to spend otherwise, such as on their children's education. Moreover, there is no reason why documents should not be used in the sense of distributive justice and improved social welfare. If there are profiling systems used by banks to identify bad payers, these same systems can well be used by the state to support those bad payers without placing the burden and risk on the shoulders of savers or taxpayers. In short, profiling based on previously volatile documents is not, in itself, a bad thing; on the contrary, it is one of the most powerful prerequisites for progress and human civilisation, both on a practical and on a political level. If medicine is more effective today, it is because it can rely on statistics and documents that were once not only inaccessible, but unthinkable. And a simple phenomenon such as the inclusion of serfs from the countryside in city baptismal registers, which of course was also intended for tax and military purposes, was one of the decisive causes of the birth of modern democracy, as well as the saying "Die Stadtluft macht frei". Furthermore, thanks to profiling, economic planning is now possible to an unprecedented scale, realising what previously appeared to be an ever-frustrated socialist utopia. In principle, all the inefficiencies of the traditional planned economy, as well as all the wastefulness of the liberal economy, can be avoided through the extensive collection of documents that is now possible,[25] and this is something worth paying more attention to than is currently the case.

---

[25] On the subject of collection, I wish to quote an interesting objection by Fausto Corvino: "I get my hair cut by a gentleman who (forgive the example) collects the hair he cuts and sells it to a company that makes insulation material out of it (or something like that, I am not sure). Hair, like data, is a resource from which I couldn't get a penny, unlike the resources that are the object of classical capitalist accumulation. And this makes it quite complicated to claim that my barber gets surplus value from my hair. In fact, he doesn't ask customers if they want to take their hair home, he just sells it. A similar argument could perhaps be applied to data. When I click on 'accept', I am depriving myself of something that has no economic value for me." Here's my response: it indeed wouldn't have any value if the barber didn't collect it and sell it. Since it has value to him, it has value to you: change barber or ask for a discount.

And it is also true that large platforms employ tens of thousands of employees, but mainly in the US, a country that is therefore reluctant to tax the web, although the profit of the platforms in relation to their employees' work is unusually high.[26] Banks and tech companies today are not very far from these standards.[27] However, the crucial point is that what platforms do is make a profit out of something that otherwise – from coughs to speech impediments – would be lost, thus minimising the weight of human input. The fact that what we give up has no obvious value to us, that we are unaware of its existence or do not suspect its value, could also be applied to bartering between gold and beads, and this is enough, in my view, to settle the matter, and to authorise a deeper analysis of something is not profit, but surplus value.

4.2.2. Proprietary documents

The documents received by users are in the public domain, while those received by the platforms become the latter's property. Therefore, we are dealing with a primary accumulation of a capitalist kind. The fact that this accumulation is made possible by platforms (no one could have capitalised on, say, the number of steps we take every day before the introduction of pedometers) does not change the fact that a primary accumulation takes place, and is profitable.[28] We are headed towards a hyper-documented economy, which guarantees enormous economic advantages and the reduction of human input in production and distribution. Soon, documents will be able to fulfil all the functions traditionally performed by money: store of value, unit of account, medium of exchange. But, it should be noted, this is only true for the platforms, not for the users, who do not have the benefits set out in the previous two points.

We are much like the natives who were once analysed by anthropologists: the latter wrote books to clarify the natives' customs and produce academically useful knowledge. However, back then it was the anthropologists who had to go and

---

[26] Facebook makes 1.6 million dollars per employee per year.

[27] Cf. Meta Platforms Inc Sales per Employee https://csimarket.com/stocks/FB-Reve nue-per-Employee.html; M. van Romburgh, *Facebook Makes More Profit per Employee than Other Fortune 500 Tech Companies*, in "Bizwomen", 8 July 2019, https://www.bizjournals. com/bizwomen/news/latest-news/2019/07/facebook-makes-more-profit-per-employee-than-other.html?page=all [31/05/2021].

[28] The importance of hereditary accumulation in the genesis of capital is underlined in Piketty, *Capital in the Twenty-First Century* cit. Piketty captures part of the truth, namely that documentality, the accumulation of records, is the constitutive element of capitalisation. De Soto, in *The Mystery of Capital* cit., insists on similar circumstances and claims that capital depends on documentality, further confirming that modern capital is not primitive, but derivative, and that what we must focus on is the production of documents.

look for the natives. Here, instead, the situation is much more convenient, because, to continue with the comparison, it is the natives (us) who look for the anthropologists and who pay for the tools through which the latter acquire apparently free information (from this point of view, the most appropriate comparison might be to say that platforms are not anthropologists, but psychoanalysts).

The mobilised tend to systematically underestimate the benefits that come from the mobilisers' constant, centralised and ever-active accumulation.[29] Yet it is on this primary accumulation that the greatest asset of Internet platforms rests: they capitalise on documents that users are not even aware of, such as the brightness of the place they are in, not to mention all the bodily documents that are recorded by smartwatches that, again, are bought by the mobilised and provide a huge amount of free documents to the mobilisers. These documents are secret in many senses, but in a different form than those involved in the explicit transfer of data on social networks. In fact, we know what we post on Facebook, but we do not know our heart rate, how many steps we have taken in a day, or all the places we have visited last week, in the banal and decisive sense that we do not remember them.

Hence the need to radically rethink the question of secrecy. While we voluntarily give up privacy, which clearly does not have the value we officially attribute to it, we produce wealth. This, in my opinion, is the essential matter and the precondition to focus on. How much do we think an unemployed person cares about their privacy? Would they be happier if their privacy was ensured, or if their mobilisation was recognised as labour and paid for, by recognising its documedia surplus value? Rather than seeing ourselves as the heroes of some espionage story, it is worth noting that the relationship between the mobilised and the platforms resembles the classic relationship between capital and labour, with one very important variant: namely that in this case labour is not paid for, and what's worse, is not even recognised as such. This recognition is what we need to achieve progress and social justice. Instead (contradicting their origins and vocation), our policies currently defend the abstract individual, focusing on the discourse on human rights and on privacy – something that is essential but insufficient, not only because it is electorally unsuccessful (the abstract individual does not vote), but above all because it does not help and support real people.

---

[29] An Austrian law student, resorting to a European law, asked Facebook for all the material collected about him and received a CD with 1200 pages in PDF format, including the comments he had deleted.

4.2.3. Capitalisable documents

From accumulation follows capitalisation. Platforms give users information worth 1 (where the nearest restaurant is) and receive information worth 10 (who is searching for the restaurant, when, what they searched for first ...), then 1,000 (by cross-referencing their search documents with data of all those who did a similar search), then 100,000 (by cross-referencing documents on restaurants with those on football matches and church services), then 1,000,000, reselling the documents like any other asset to other platforms, agencies, or individuals aspiring to become the president of the US or Luxembourg (I think there is a price difference).

None of this arouses the same protests as privacy infringement complaints or Big Brother fantasies, a paradox that has two partial explanations. The first is that surplus value takes a long time to be recognised, although it is desirable, for the sake of us all, that this time should be made as short as possible. The fact that there was surplus value in slavery and serfdom took millennia to be recognised, hidden as it was by the naturalisation of slavery, and remained under the surface until the middle of the industrial age: suffice it to say that it was the cause of the American Civil War. The presence of an industrial surplus value was also ignored for a long time by both social actors and the very economists who were interested in the phenomenon; indeed, it was only recognised by Marx, at about the same time as one of the most advanced countries in the world accepted, *bon gré mal gré*, the end of the surplus value generated by slavery.

The second is that, as theorists of cognitive capitalism know, there has been much rhetoric about content sharing on the web. However, this sharing is purely superficial, not only because what is shared could be, and often is, fake news, but above all because the real knowledge is not generated by the explicit contents that are shared, but by the mobilisation that derives from them, and by the documents collected by platforms. In other words, the sharing of often apparent knowledge produces real and effective knowledge at another level, which is not accessible to those who are mobilised, creating the greatest surplus value there is: documedia surplus value. In this framework, the great misunderstanding of collective intelligence was therefore based on a doubly optimistic assumption. First of all, that the distribution of information is a symmetrical process, whereas it is clearly asymmetrical. What we have as document consumers is almost nothing compared to what we give to the platforms as document producers. Secondly, this assumption does not take into account that nature is unfair and individual fates are not all equal. As a result, instead of a ubiquitous academia and free debate of disinterested ideas we have post-truth: the belief that the Earth is concave and not convex, that Americans did not land on the moon, and that vaccines cause autism, just to name a few.

4.2.4. Monetisable documents

Then there is a third stage. Documents capitalised by platforms can be sold like any other asset. On the one hand, documents already constitute, as such, a super-currency. In fact, there is no match between traditional currency and the documedia currency – documoney, so to speak. Traditional money only informs us about its own value; electronic money only informs us about a range of transactions and purchases; documedia capital offers profiling, and can also fulfil the traditional functions of money. In fact, it is a unit of account, where price is a much more approximate synthesis than a set of documents; it is a medium of exchange, since I can trade documents for goods or services, or, if I wish, for money; and it is a store of value, since there is no capital more powerful than the documedia capital. On the other hand, recent history has revealed how easy it is to monetise documents among knowledgeable parties (e.g. a politician and a data collection agency). Finally, one should not forget the growth of virtual currencies, which ensure an easy and secret conversion of data into money. By bringing the economy back to its origin, the blockchain ensures an immediate conversion of document assets into purchasing power.

There is little awareness of this at the moment. Humanity is strange, this is certainly no big discovery. We have witnessed the Occupy Wall Street movements and, with a resurgence of the 1930s, the condemnation of banksters, i.e. bankers likened to gangsters,[30] even though we lend the banks our money and

---

[30] This *mot-valise* helps us to recognise continuity beneath the apparent differences in time and place. The Sicilian-born American judge Ferdinand Pecora is said to have coined this word in 1933, but it had already appeared in the American press a year and a half earlier. Not before 1929, we have to imagine. And in any case, it found real popularity in Europe, as it became the key word in the speeches of the Belgian populist Léon Degrelle, leader of the Rexist party and editor of the newspaper "Le Pays réel". In the run-up to the local elections of 24 May 1936, Degrelle radicalised Rexism by lashing out at the "aristobankers", the "traitors of devaluation"; he denounced the moral corruption of the government by contrasting it with the purity of the "real country", and used a broom as a symbol to sweep away the banksters and the corrupt. "All corrupt parties are equal. They have all robbed you, ruined you, betrayed you [...]. If you want new scandals to plague the country, if you want to be crushed by the dictatorship of the banksters [...] keep following, like sheep, the profiteering politicians", wrote Degrelle in "Pays réel". In a period of six months, a non-existent party managed to obtain 11.5% of the votes, 21 deputies and 12 senators on 24 May 1936. However, after its initial success, the movement took a dead end. Fortunately for Degrelle and his men, and to the misfortune of everyone else, Belgium was invaded by the Nazis in 1940 and Rexism started an intense collaboration with the regime. When the Nazi attack on Russia began in 1941, Degrelle went to the front, first with the

nobody forces us to trade on the stock exchange. Instead, we give our documents to the Web for free, and some even see the Internet as the main instrument of direct democracy.[31] But little can be done by complaining. It is enough to type "bankster" on a search engine to find dozens of books, files and documentaries, with subtitles such as "much worse than Al Capone", "the vampires of Wall Street" and so on and so forth. Besides the difference between loan and gift, there is a crucial point here. To the banks we give the fruit of our labour, while to the platforms we give our labour *per se*, thus producing an enormously greater surplus value. In other words, we must be careful not to think of documents as the new oil,[32] thereby making it very easy to object that this is not the case at all, and that therefore there is no reason to demand any kind of payment from platforms. Here, in fact, we are dealing with both a conceptual error and a political error.

Wehrmacht, then with the Waffen-SS, forming the 28th armoured volunteer division "Wall-onien". Decorated with distinction by Hitler, who called him the son he never had, Degrelle undauntedly continued his fight against the banksters. On 27 February 1944 at the Palais des Sports in Brussels, with the Russian front broken through and just a few months before the Normandy landings, he shouted: "We are against all the banksters in the world". And again in the 1960s, as a refugee in Franco's Spain after the extermination of millions of Jews, gypsies and homosexuals, he would preach the merits of Nazism. And he was already picking on Europe, which at the time was blamed as the European Common Market. His speech can be found on YouTube by typing in "L'incroyable discours de Léon Degrelle". For Degrelle, the bankster is the perfect other, the enemy to be killed, so even a Red Army soldier can be a stateless Jewish bankster (remember that the Jewish component, according to the Nazis, built a bridge between Bolshevism and Americanism: too bad Stalin was an anti-Semite). In a remarkable book, *Le Sec et L'Humide*, Jonathan Littell analysed *Campaign in Russia*, Degrelle's memoirs from the eastern front, noting how the whole book proceeds by dichotomies: I/other, dry (the fascist)/wet (the Bolshevik), hard (man, westerner)/soft (woman, easterner). Oppositions that hold together a shaky identity, which lives only on contrasts and antagonists. Bankster or Bolshevik, woman or Jew: it doesn't matter, the dichotomy goes on mechanically generating a series of oppositions that we find as much in Degrelle's speeches as in Heidegger's *Black Notebooks*: people vs. stateless, mass vs. elite, poor vs. rich, exploited vs. exploiter. These are eternal categories, which do not relate to the power of thought, but rather to the paucity of the human being, and it is therefore not surprising that, with the 2008 crisis, the category of bankster has returned to the fore. We found it in the election posters of Marine Le Pen's Front National for the European elections of 7 June 2009: "Against the Europe of banksters", in the Occupy the Banks protests ("Jail the bankster$!"), in the condemnations of cosmopolitan, i.e. Jewish, capital, as well as in endless works of literature.

[31] While we curse the banks, we forget that the real powers are found elsewhere: Google, Apple, Amazon (and so far we all know this); Tencent, Alibaba, Batx, WeChat in China (and these names are less well known). The strongest of these powers, the companies that dig up documents and interpret them, are unknown to most: Acxiom, Criteo, Equifax, Experian, Quantcast, Tapad – who has ever heard of them? And who has heard of Privacy International, the organisation that investigates and exposes their activities?

[32] A. García Martínez, *No, Data Is Not the New Oil*, in "Wired", 26 February 2019, https://www.wired.com/story/no-data-is-not-the-new-oil/ [06/03/2021].

As for the conceptual fallacy: Alaska paid for its basic income with oil profits, and documents are said to be the oil of the 21st century. However, the analogy between documents and oil is misleading, because dinosaurs (the real oil producers) have disappeared and there is nothing we can do about it, whereas the producers of documents are here, they are human beings, they have spiritual as well as material needs, and they must share the same world in a non-confrontational way. Thus, comparing documents to oil fails to grasp the difference between a thing (oil) and an action (mobilisation) that depends on living human organisms, without which the platform system would collapse. In short, it is not a question of resources formed in antediluvian times, but of *the instantaneous capitalisation of human mobilisation.*

Secondly, coming to the political error, documents are not monetizable by users, but they are so by platforms, which have in fact been the most profitable sector for a number of years. By relying on 19th-century conceptions of work, we lose sight of the fact that mobilisation, whatever its form, produces value,[33] and must therefore be considered labour. If it is labour, it should be remunerated: not with a fee distributed to individual platform users – everyone of us, I repeat, including those who, like me and many others, are not on social networks – but with a European-level taxation that can guarantee a basic income[34] and, above all, reconvert endangered productive jobs into activities where human contribution is indispensable: consumption, education, invention, research, recreation, and care. The essential condition for this transformation, which sees consumption as the other face, and indeed both the aim and the condition of possibility, of production, is an understanding of the true nature of capital, usually confused with its most ungrateful historical hypostases, and misunderstood in its profound meaning, which I have tried to outline in 2.3.

---

[33] K. Goldsmith, *Wasting Time on the Internet*, HarperCollins, New York 2016.

[34] Here's a list of books on UBI (Universal Basic Income): G. Allegri, R. Ciccarelli, *Il Quinto Stato*, Ponte alle Grazie, Milan 2013; A.B. Atkinson, *Public Economics in Action: The Basic Income/Flat Tax Proposal*, Clarendon Press, Oxford 1995; R. Bregman, *Utopia for Realists. The Case for Universal Basic Income, Open Borders, and a 15-Hours Workweek*, The Correspondent, Amsterdam 2016; G. Bronzini, *Il diritto a un reddito di base. Il welfare nell'era dell'innovazione*, Abele Edizioni, Turin 2017; E. Granaglia, M. Bolzoni, *Il reddito di base*, Ediesse, Rome 2016; D. Raventós, *Basic Income: The Material Conditions of Freedom*, Pluto Press, London 2007; P. Van Parijs, Y. Vanderborght, *Basic Income: A Radical Proposal for a Free Society and a Sane Economy*, Harvard University Press, Cambridge (MA) 2017; K. Widerquist, M.A. Lewis, S. Pressman (ed. by), *The Ethics and Economics of the Basic Income Guarantee*, Routledge, New York 2005.

4.3. Action: The Struggle for Recognition

Lady Bracknell: Do you smoke?
Jack: Well, yes, I must admit I smoke.
Lady Bracknell: I am glad to hear it. A man should always have an occupation of some kind. There are far too many idle men in London as it is.[35]

The problem with the struggle for recognition lies here: unless you have the nerve of a Victorian lady, how can you justify that if a smoker gets a smartwatch measuring his reactions, he is actually working? Let's turn the question around. Who would be willing to work for free, uploading documedia contributions which only generate wealth for the management companies, and buying their own means of production? The answer is simple: over half the world, i.e., all those who are on social networks. And I am no exception: at this very moment, I am writing a book to assert myself and hoping to be recognised by others. The vanity of professors is a well-known fact – what's new is that this vanity has turned out to be the most widely distributed thing in the world.

This circumstance confronts us with a question that has long been silenced by the paid mobilisation of *homo faber*. Obviously, much of this mobilisation goes hand in hand with traditional, paid work activities, in accordance with the dissemination processes I analysed in 1.4.2; however, there remains a surplus that is neither recognised nor paid for. This surplus is found in activities that have nothing to do with work and that most often go back to the struggle for recognition (if not simply leisure, assuming one can draw a clear line in the matter). And recognition *per se* does not yield an economic return (although it obviously can do so, in terms of reputation, influence, etc.), but in any case it produces value, i.e. documents. Rather than psychologising the struggle for recognition, minimising it as narcissism[36] or dismissing it as a constitutive element of the human being[37] (it is, but what does it matter once this has been recognised, apart from bringing a nationalisation of surplus value?), would it not be worth trying to remunerate it, since it produces value?

---

[35] O. Wilde, *The Importance of Being Earnest*, Act I, Scene 14.

[36] Again with a moralistic reading that misrepresents reality. Narcissus mirrored himself and was satisfied with this. Those who take a selfie, on the other hand, do so in order to publish it, and their aim is not just self-satisfaction, but recognition by as many other human beings as possible.

[37] Axel Honneth (in the wake of Hegel) has insisted on the central role of recognition, even beyond its economic usefulness. See A. Honneth, *The Struggle for Recognition*, Wiley, Hoboken (NJ) 1995; Id., *Axel Honneth* (2011), Wiley, Hoboken (NJ) 2014. Specifically on the link between labour and recognition in the modern world, cf. M. Bigi, O. Cousin, D. Méda, L. Sibaud, M. Wieviorka, *Travailler au XXIe siècle. Des salariés en quête de reconnaissance*, Laffont, Paris 2015.

Let us remember: it is not true, and never has been, that we are slaves to machines – we have always been their masters, because we can imagine humans without the Internet (it was the case until a few decades ago) but not the Internet without humans. If one sees it otherwise, one should be consistent with one's own assumptions and, instead of teaching, writing books and quoting articles and authors, one should take the field and fight against the system, or at least not be complicit in any way. I realise that it is unchivalrous to invite one's opponents to go into hiding in Bolivia, and what's more, without a mobile phone, because that would make them too easy to catch. A good compromise could be to actively contribute to the progress of humankind. Apart from the fact that they are already partly implemented in practice, as we will see, many revolutionary demands, if formulated as they are presented by their theorists, would be unfeasible. For who would have the power to implement them? How could we achieve the revolution of a class that no longer exists, or a conversion of the economic beliefs of the platform owners? Should we rely on a change of heart? Or the use of force? Perhaps newly conceived rhetorical gimmicks? And if you don't want to involve the state, and more precisely, I repeat, a union of states, what kind of bargaining power do you have to create a webfare system? Let us not forget that the platforms that matter are all American or Chinese, and I struggle to imagine a soviet of European workers, students and essayists negotiating a citizenship wage with the president of the United States or the president of China, who are well settled at home and insensitive to any threat other than that of a state powerful enough to challenge them.

Let's put it another way. We are offered efficient and free services, which improve the more we use them. The documents collected by these services can be used to improve other services, to increase the self-learning capacity of machines,[38] and thus to improve production, reduce waste, make distribution more efficient, and consequently lower the prices of products, which, combined with the fact that many services are free, means that much less money is needed today to provide quality services compared to the 20th century. Incidentally, humankind was in better conditions in the 20th century than the previous one, which seems to suggest that the history of humanity is largely one of happy growth, interrupted by moments of unhappy degrowth. Stopping growth seems neither possible nor desirable, for the same reason that going back to queueing at train station ticket offices and leafing through weighty paper encyclopaedias seems like a bad idea. What is necessary, instead, is to maximise the advantages of the

---

[38] P. Domingos, *The Master Algorithm: How the Quest for the Ultimate Learning Machine Will Remake Our World*, Penguin, New York 2015. Note that this learning responds to productive rather than cognitive needs.

technological transformation that has made this increasing automation possible, and to minimise the disadvantages of the anthropological inertia that makes us insincerely nostalgic for fields, workshops and patriarchy. This is the only real goal of any biopolitics that aims at the good of humankind instead of describing prisons, hospitals, asylums, hospices and torture. This is the decisive point, and yet it was not grasped when, half a century ago, we found it plausible that industrial wages were independent of profit. If there is an independent variable – independent of wages, that is – it is mobilisation, and the same children who thirty years ago would lie for hours watching television now spend their time with a smartphone in their hands; but in doing so, and unlike the case of television, they fidget, they move their fingers, and they produce value. So much so that people hardly ever fall asleep in front of a mobile phone, a practice that was and still is very common with television – unless, of course, they use their mobile phone as a television set, but even in this case, the simple choice of entertainment creates value, so it is not just passivity, but allows producers to recognise the tastes of the public with an otherwise inconceivable analytical precision.

4.3.1. Valorising mobilisation

If this is the case, and given that machines without humans amount to nothing, we are in a position to assert ourselves, provided we know it and want to. We can recognise the true value of human beings, which does not consist in being a shoddy and needy surrogate for machines in production processes, soon to be replaced by robots. Our value lies in constituting the other, unavoidable pole of the production process, which confers meaning, need and value to the whole thing. What would a market be without consumers? And, more generally, what purpose would automata serve in a world without souls? The problem of teleology I addressed in 3.4 is not merely an academic question, but opens up a crucial economic issue: consumption as the ultimate meaning of production. I can imagine an objection: who will pay for consumption? And here the answer is simple: we are already paying for it, in every single act we perform on the web, which is a production of value that should be recognised and remunerated as such. I propose this is done through a webfare system that supports those who have lost their jobs due to automation, and at the same time gives humanity new aims and purposes as opposed to the traditional ones of toil, alienation, or even war.

Here's a very persuasive narrative[39]: automation is not the manifest destiny of technological progress; indeed, just as in Poe's *Maelzel's Chess Player* (a fa-

---

[39] In agreement with Antonio Casilli, *Schiavi del clic* (an unfortunate translation of the original French *En attendant les robots*), which is an analysis of the Golem, or "platform capi-

mous novel based on a true story), automaton does not free humans from work, but makes them work harder than before, only in more uncomfortable and disqualifying postures. I have no objection to the tale, except for the literary genre in which Poe places it, namely that of horror stories. Nonetheless, Poe classified his Mechanical Turk among fantastic tales, and I follow him not for literary reasons, but for political and philosophical ones.

First of all, if total automation were really a tale that we tell ourselves (or, worse still, that others tell us and that we foolishly believe) in order to make human exploitation bearable, then the Mechanical Turk is really an ill-chosen emblem. Indeed, chess is one of the fields in which automation has developed most easily, and those who teach the rules of the game to the computer are not squeezed into a box. Not to mention that total automation is not a cynical and barbaric destiny guided by who-knows-what dystopia: rather, it is the trivial consequence of the incontrovertible fact that a machine does not die. And in this respect, social work on the net is by far the most useful, not because it is peculiarly social in the sense of class consciousness or similar (unless comments on reality TV and hate speech are considered examples of social work), but because it provides the foundations of the mimesis of human life forms and actions we call automation.

Secondly, we are well aware that some have proposed not to oppose automation, but to bring it to its maximum power and profitability, so as to provide a basic wage that would in turn increase consumption.[40] However, from this perspective it is not clear who would pay for technological access and what consumers would pay for. Without clarifying this point, these are just empty words, and often dangerous words when one considers the populist and right-wing outcome that accelerationism has largely achieved. There is nothing desirable, and fortunately nothing immediately feasible, about accelerating total automation, the implosion of capital or its overcoming.[41] It makes no sense at all – especially if the end of capital is supposed to make way for an absence of capital, and perhaps a government of accelerationists. Rather, the capitalist process must be understood by creating webfare and social justice.

---

talism". A sociologist of online labour with a solid philosophical background, Casilli enlightens his readers by warning against misplaced comparisons between current and past kinds of exploitation.

[40] N. Srnicek, A. Williams, *ACCELERATE MANIFESTO. For an Accelerationist Politics* (2013), https://criticallegalthinking.com/2013/05/14/accelerate-manifesto-for-an-acceleration ist-politics/; Idd., *Inventing the Future: Postcapitalism and a World Without Work*, Verso, London 2015.

[41] P. Mason, *PostCapitalism: A Guide to Our Future*, Allen Lane, London 2015.

Thirdly, to conceive of humans as slaves to the click means to conceal the enormous power that humans have within themselves: in other words, it means to lay the foundations for voluntary servitude, despite the best intentions. Let us never forget: in charge of machines there will never be other machines, but humans. Now, these humans may not be us, so we must not become their docile servants; but we will certainly never be the machines' docile servants, unless we like to think so. The insistence on voluntary servitude that has characterised the last few years may well have been motivated by the best of intentions, that of raising awareness. And it may even have represented a step forward with respect to the mythologies of the revolutionary subject as a good person (and especially of the good person as a revolutionary subject) that have traditionally clouded progressive thought. But it ultimately translated into an unconditional surrender of progressivism to a capital that, in accordance with the tried and tested strategy of mediocre generals, was portrayed as invincible.

Fourth, to focus on the parcelling out and fragmentation of disseminated labour is to avert one's eyes from the real exploitation, namely the documedia surplus value. It is from this surplus value that platform managers have profited over document producers. This is not because the former are necessarily more far-sighted than the latter, but because, by controlling the production processes, they have gradually become aware of something they were previously oblivious to. The insurrogability of consumption, together with the emergence of surplus value, nevertheless enjoins platforms to redistribute their profits. But the injunction to redistribute must come from states and citizens, and requires a preliminary conceptual and political clarification: as soon as consumption is recorded, it becomes document production; the latter, as such, is more important than any production of non-document goods, which is now moving towards automation.

Fifthly, we need to rid ourselves of Eurocentrism and look at China, which represents a conceptual as well as a political problem. As I pointed out in 4.1, we have decided – with a hastiness worthy of a better cause – that China is a capitalist country, but of course it is not: it professes to be and is communist. A testament to this is that it nationalised its Internet platforms, creating a webfare system that in the West is still not only unthinkable, but unthought of. Suffice it to say that the Chinese redistribution of surplus value and economic growth have meant that – according to the International Monetary Fund – in seven years, from 2012 to 2019, i.e. since Xi Jinping took office, the number of Chinese people below the absolute poverty line, 325 dollars a year, has gone from 115 million to 10 million.

My sixth and final point is this. Far be it from me to consider the demands underlying cognitive capitalism as irrelevant. But there is a caveat. Precisely because the Web is marginally an infosphere and mostly a docusphere and a biosphere, it is misleading and unhelpful to conceive of the web user as a hyper-skilled worker, or some sort of knowledge worker. On the one hand, the web, in the great implicit labour it derives from human mobilisation, is less interested in ideas and intelligence than in our forms of life, our needs and our drives. On the other hand, the organisation of explicit labour on the Web is such as to favour, with the exception of top-level functions, a complete fragmentation and disqualification of labour.

What is needed, therefore, is a transvaluation of all labours, starting from a fundamental observation: the documedia revolution entails an axiological revolution, which overturns the traditional relations between activity and passivity. In documediality, in fact, giving is tantamount to taking, because the platform gives as soon as it receives; taking is tantamount to giving because the mobilised person takes and at the same time offers; and it is even possible to give what one does not have, for example, documents that the mobilised person never knew they had. This transformation entails a radical reassessment of the relations between production and consumption and between economy and waste, which characterised yesterday's world, and are so close and yet so remote.

Let us take a step further and link up to the considerations made in 2.4. How sure are we that the distinction between exchange and gift holds? And how useful and sensible do we think it is to explain everything by a single principle of reason, causality in physics and utility in the social world? This is not to engage in a futile defence of freeconomics, but to raise a crucial philosophical problem. If we admit, as I argued in 2.4 with considerations that I hope are not (forgive the pun) gratuitous, that the concept of "value" is subordinate to the distinction between morality and economics, then the other way around might also be true. If we are prepared to recognise that in morality, just as in society and the economy, value is an effect of globalisation, then there is no reason to omit a decisive circumstance. What for the platforms is economic value, for the mobilised is moral value (recognition). However, instead of calling for a moralisation of economic value, as is the case with the current exclusive (and exclusively formal) protection of the abstract individual, it would be more appropriate to call for an economisation of moral value.

4.3.2. If it is recorded, it has value

In this framework, the struggle for recognition that awaits humanity is the explicitation of the implicit labour we all do for free to the benefit of documedia capital. We work all day long, whether we are adults, children, or pensioners, and none of us (except for a few influencers and porn stars) is paid for this mobilisation. One could argue that you cannot complain about being deprived of something you didn't know you had, but it does happen, namely in the form of social envy. In this case, we are not talking about a class struggle, because this labour is done by everyone. Rather, it is an effort of solidarity to find resources for those who have been excluded from production, but who continue to contribute in terms of consumption and mobilisation. I believe that the discussion so far can provide the necessary philosophical and political arguments for this recognition. In accordance with what I have said, particularly in 1.1.2, 1.1.3, 2.3.1 and 3.2, there is an analytical link between recording and capitalisation (all that is recorded can be capitalised, which is not the case for what is not recorded) and between capitalisation and the production of value, such that – as we have seen in 2.3 –, contrary to the Pentecostal perspective, in the emergent perspective it is value that derives from capitalisation and not the other way around. Hence the axiom that all that is recorded has value, be it actual or potential, so that today's documedia mobilisation is the greatest production of value in human history – certainly greater than the gold of Mexico, because, unlike in that and similar cases, the growth of mobilisation does not determine inflationary processes. I would like to develop this argument through four moves.

The first move is a factual observation, namely that it is as easy to expel humans from the production cycle as it is difficult to dispense with them in the economic cycle. Platforms justify their surplus value with the argument that documents are only valuable if they are interpreted. This is true, as is the fact that platforms are not found in nature and require investment in development and research. However, this justification is insufficient, because interpretation is only valuable if there are humans at the beginning and end of the value-generating process. The fact that those humans are valuable is proven precisely by the platforms' investments in development and research. Surplus value would not exist without souls, which enable automata to record their life forms, and provide, given their organic nature, the purpose of the whole process. Certainly, without automata and their immense capacity for hysteresis, most of the documents produced by souls would have vanished into thin air. Still, it is not clear who would buy products, for themselves, their pets or their garden, if not a human. So, without humans, the whole system would collapse for metaphysical rather than economic reasons: what is the point of production without consumption? Unlike in Vonnegut's dystopia,[42] a new class of useless individuals is not conceivable, precisely because from the point of view of consumption there is nothing more essentially useful than the individual, even if the latter is radically unproductive and reduced to a vegetative state.

---

[42] K. Vonnegut, *Player Piano*, Schribner & Sons, New York 1952.

The fear that humans will end up like 19th-century horses (coachmen at least became chauffeurs) is undoubtedly strong, but it has no foundation and is a kind of cognitive error. The horse fulfilled human needs as a tool, whereas the human horse is also an end in itself, not because it has a sentient soul that is inclined to arouse Nietzsche's brotherly emotion, but because it consumes. Indeed, it is human consumers, whom we should rather represent as Swift's Houyhnhnms, that give purpose to the whole system. What used to be interpreted as dematerialisation and the end of production[43] is simply an increase in automation, i.e. in dead labour, which, however, does not exclude living labour in the least; on the contrary, it emphasises it, but moves it to the opposite pole, that of consumption. In this situation, it is therefore necessary to recognise that the relationship between labour and production is not an analytical one, and that there can be production without labour (i.e. automation) and labour without production (i.e. consumption). Conceiving of labour as exclusively linked to production, i.e. as *poiesis*, is already limiting with respect to the fact that services are *praxis*, but it is even more limiting if one does not take into account the enormous *praxis* of consumption and mobilisation. Services are but a minimal part of this *praxis*, namely the tip of the iceberg that, alone, has been belatedly recognised as labour. So, economic and social analyses are tainted by a naturalism of production, by nostalgia for a lost homeland and a forgotten way of being, where the only labour was production. Now, as we well know, such a situation has actually never arisen, and it is strange that it should be recalled at a time when the whole of production has become automatable.

The second move, on the other hand, is a legal consideration, and concerns the transvaluation of the relations between give and take, and between activity and passivity, which, as I mentioned earlier, constitutes one of the most relevant features of the documedia revolution. Television audiences and traditional media users in general have usually been assumed to be the paradigms of passivity.[44] Now, there was already a degree of rhetorical emphasis and semantic vagueness in that assessment. Indeed, what does it mean to be passive? To sit back and relax or to accept everything that is offered without question? In the first case, even a sovereign on the throne would be passive; in the second, the remote control is there to change the channel. These questions are all the more valid for the behaviour of humans on the web, which *prima facie* is much more active than the

---

[43] J. Baudrillard, *Symbolic Exchange and Death* (1976), Sage, London 2017.

[44] This is typically Debord's thesis, according to which "there can be no liberty outside activity, and in the context of the spectacle all activity is negated" (G. Debord, *The Society of the Spectacle* [1967], Black&Red, Detroit 1970, § 27). This thesis allows him to consider liberation from labour as illusory, and to disregard the fact that activity is a necessary but not sufficient condition to have freedom.

classic television setting suggests, since web-surfing can take place on the move and involves the possibility of creating content, not just enjoying it. Above all, and more profoundly, it transforms every online behaviour into capital, in a process that goes beyond the difference between activity and passivity. The shift from production to consumption is thus also, more broadly, a shift from activity to passivity, or more precisely the circumstance whereby, in an environment capable of storage, passivity (i.e., need, which can be capitalised) is transformed into activity.

The fact that potentially every action can be recorded does not only dramatically increase the production of objects, but also introduces a new logic, the one underlying the relations between platforms and users: every give is also a take, and vice versa. One might say that this is a generalisation of the exchange of ideas, but in reality there is much more to it than that. For, on the one hand, what is given is not ideas, but actions and behaviour, which therefore undergo a transformation in the course of hysteresis: I perform an act, but what remains is a document. And, on the other hand – and even more interestingly from a philosophical and economic point of view – I can give something I am not aware of, for instance my heartbeat, the brightness of the environment I am in, my exact geo-location ...

At first glance, this process takes the form of a gift of something one does not know one has. In a second sense, however, this is properly the gift of something one does not have – of something that properly is nothing, from the social point of view, but which, as soon as it undergoes hysteresis, starts being something: a social object. This something is then no longer only ours, because it is recorded by an agent other than us; indeed, most of the time, it is no longer ours at all, because prior to its hysteresis it did not exist, and now that it does exist it exists for the platform, not for us. We have no clue about this process, nor do we have any means of recognizing the object of which we are the actors but not the constructors: the constructor is the one who performs the hysteresis, i.e. the platform.

In other cases, we have an actual instance of freeconomics[45]: music or knowledge, maps, encyclopaedias, dictionaries, literary classics, old films, are today free and accessible everywhere. In still other cases, the consumer is also a producer, resulting in a sort of barter. In general, it should be kept in mind that the increased possibilities of matching supply and demand on platforms make bartering more and more easy, through an alternative to mechanisation made possible by the growth of memory and communication. Finally, we have consumer needs and the double benefit of a growth in consumption. On the one hand, consumption

---

[45] A. Romele, M. Severo, *The Economy of the Digital Gift: From Socialism to Sociality Online*, in "Theory, Culture & Society", 33, 5, 2016, pp. 43–63.

is the prime mover of the entire productive process; on the other, this prime mover is no longer unmoved when hysteresis makes consumption productive. The mobilised provide information they do not own, and which only emerges when it is given: I do not know how many steps I take in a day, but the moment I find out, I have already given this information to a platform. It is therefore difficult to speak of a gift (one does not give what one does not have) or, strictly speaking, of exploitation. It is rather the production of value, without forgetting that the condition of possibility and the ultimate meaning of the mechanism is death as the destiny that awaits every organism, which drives the mobilised to produce documents and the platforms to accumulate wealth by capitalising on them.

The third move is a transcendental consideration that follows from what I said in 2.4 about the co-belonging of market value and moral value in the process of capitalisation. The argument that commodity value and moral value, i.e. also self-interested value and disinterested value, have their common origin in capitalisation as value production does not simply lend itself to deconstructive considerations that see morality as a hidden economic interest, but also to reconstructive considerations that run diametrically counter to the former. Just as, on a transcendental level, the useful is only the other face of the useless and the gratuitous *par excellence*, so the useless and the gratuitous that unfold in human mobilisation can legitimately be recognised as generators of economic value (indeed, this is the case, as we have seen in the analysis of documedia surplus value).

This circularity is what makes goods desirable, and goods are defined as useful precisely insofar as they are desirable.[46] This circularity indicates a sequence of dichotomies – I am not going to call them "metaphysical" because they could simply be traditional –, namely the dichotomy between utility and desire, the dichotomy between use value and exchange value, the dichotomy between the symbolic and the semiotic,[47] and of course the dichotomy between labour and mobilisation. When classical economics defines exchange as a relationship between demands,[48] it becomes evident that exchange does not have a utilitarian foundation, or, conversely, that "utilitarian" does not coincide with a rational concept of "utility". Why then oppose the magical, the religious, the aesthetic, on the one hand, and the economic as rationalisation and consumption as degradation of authenticity, on the other? The real problem is that, as things stand, the provision of free goods is done by the user much more than by the platforms, as demonstrated by the documedia surplus value. Redressing this inequity is the

---

[46] J. Robinson, *Economic Philosophy*, Watts, London 1962.

[47] This distinction is of great importance, though problematic to say the least, in J. Baudrillard, *The Consumer Society: Myths and Structures* (1970), Sage, London 1998.

[48] T.R. Malthus, *Principles of Political Economy Considered with a View to their Applications* (1820), Pickering, London 1836.

starting point of the struggle for recognition and for the transvaluation of all values and labour that comes with it.

The fourth move is the political consequence of the three previous ones and consists in the thesis that every recording is actual or potential capitalisation, and is therefore value. This circumstance means that activities which once left no trace, and thus did not create value, or which even fell within the sphere of pure expenditure, are now part of value creation, and thus of labour. The greatest production of value therefore takes place during what is anachronistically still called "free time,"[49] which is expanding and yet always occupied.

One thing is certain: labour has nothing to do with our individual intentions, with what we have in mind. Although the prevailing intuition is that in order to work, one must be aware that one is working, there is nothing strange in considering the work of a mule, a cart horse or a sledge dog as "labour". If, however, we consider these activities as labour, because of the value they produce, then – this is the result of my redefinition – *any production of value is labour*. Acts that would once have been valueless, because their hysteresis was not centralised (e.g. shopping in a supermarket in pre-documedia times), as well as acts that would not even have had access to the status of social objects, because they were not recorded (what are now posts were once spoken words, not to mention the infinity of acts recorded on camera or produced by our interaction with the web), acquire a crucial role precisely because they can generate automation, perfect distribution, and enter into a circuit of knowledge and capitalisation. All this enormous labour is unseen. Bringing it out is essential for the modern age. In doing so, we could show that the progressive philosophy of history is right, that humanity is advancing, and that there is no reason to give in to unnecessary nihilism that would only lead to more exploitation.

Just as some jobs are explicitly useless from an economic point of view, but are nevertheless formally considered and recognised as such (think of the activities of the Ateliers Nationaux designed by Louis Blanc in 1848, or many of the jobs created by the New Deal), we have the possibility, in the documedia context, of enforcing the opposite principle. Moreover, the latter principle is metaphorically much more interesting, because in the former case we were simply dealing with support for consumption, whereas in this case we would be dealing with an active support for production. Since every act, even if not economically motivated, produces value, then the basically circular argument that only what is explicitly intended for that purpose is value-producing does not apply. The prime ex-

---

[49] Consider the case of TikTok, an entertainment app that generates so much value that it was banned by Trump in the US (https://www.npr.org/2020/08/06/900019185/trump-signs-executive-order-that-will-effectively-ban-use-of-tiktok-in-the-u-s?t=1638793984166 [07/08/2020]).

ample of this transvaluation is the relationship between production and consumption. Consumption beyond subsistence was traditionally only of value as a sense of ostentation, otherwise it was the satisfaction of a need, or at best a compulsive tendency. In all these cases, the consequence of consumption was squandering, dissipation. But when every act is recorded, consumption becomes a source of value, and nothing is wasted. A meal at a restaurant, a browse on the web, a walk, even a nap: everything can be used to draw documents, correlations, statistics, useful in themselves and able to train artificial intelligence for free. Those who see alienation in a hypothetical job as a mattress tester are therefore wrong. It would be more politically useful to recognise that there is no alienation in such toil-free occupations, but there is still the production of value, i.e. labour.

It is not the mobilised people who are alienated, but the documents they generate. If they were a form of intelligence, as cognitive capitalism claims, documents would pay for themselves, so to speak, and follow the general fate of ideas, which are not proprietary insofar as the idea remains the property of its inventor even when it passes to a third party. That this is not the case for documents is the crucial clue to recognize them as value: precisely because they are alienated, they can be sold and bought, and generate genuine profit for the buyer. The situation is currently disadvantageous for document producers only because they are not aware that they produce documents through mobilisation and consumption, and thus not only do they not receive the chimerical spiritual benefits produced by the knowledge economy, but neither do they receive the real economic benefits that could accrue to them from the recognition of documedia surplus value and its redistribution. A hyper-ecological situation is thus created, at least to the extent that there is no waste that cannot be recycled. The recycling and valorisation of waste, in the field of production, has a clear counterpart in the field of mobilisation and consumption.

4.3.3. If it has value, it is labour

The second step is therefore to realise that labour is all production of value, which in turn is value in view of consumption. Plato saw this very well, recalling that the judgement on the validity of production must be made by the consumer: the quality of a saddle is judged by the rider, not the saddler. Smith and Marx considered labour as the measure of value, and those same economists, like Mises, who insisted on the impossibility of giving an ultimate foundation to the measure of value,[50] have never doubted that value, as such, consists in satisfying

---

[50] L. von Mises, *The Theory of Money and Credit* (1912), Jonathan Cape Ltd., London 1934.

human needs – not the needs of galaxies or robots. Galaxies and robots do not need anything at all, and even if a robot needs electricity, it is an induced need, because it would do just as well without a power supply, and the only one who would regret this would be the human user who needs the robot's services.

Recall what was said in 2.1 about the relationship between soul and automaton. It makes no sense to speak of consumption with regard to mechanisms, because the actual energy use of a machine is not subject to the vital urgency typical of an organism. While goods are produced and distributed automatically, with a dizzying increase in dead labour, human agents are left with an enormous amount of living labour, which is not recognised as such, even though it consists of the most living labour there is: life itself. All of us humans, together with the other organisms within us and outside us, are alive: we are all subject to irreversible processes that dictate the urgency of consumption. Now, as mentioned, everything that produces value is labour. If we are interesting, it is for our flesh rather than for our spirit, and the platforms know this more than anyone else, not so much because they are primarily concerned with feeding the flesh, but because they focus on an urgency that arises in the flesh, namely the speed of delivery: the spirit can wait for centuries, the body cannot.

In this sense, consumption and mobilisation are labour to all intents and purposes, as they constitute the highest form of value production there is. A manager following the advice of *The Wolf of Wall Street* (a Martini cocktail, then another after seven and a half minutes, and then others at regular five-minute intervals) produces more value than they would by following Zarathustra or Sun Tzu – supposedly their favourite readings, judging by the management section at airport libraries. This does not only apply to Wall Street brokers, but to the whole of humanity. Conceiving of consumption as the equivalent of production, because it is the production of value, may seem paradoxical, yet it is not only necessary, but also obscurely present in all of us – we are perfectly aware that the condition of possibility of production is consumption.

There's a popular saying these days: if it is free, it means you are the product. That's not true. Rather, if it is free, it means you are the producer. You are not in a slave economy, you are a worker whose labour should be recognised, and who therefore has no interest in representing yourself as a slave or worse, as a commodity for other humans.[51] Documentalisation succeeds in transforming all that

---

[51] M. Bentivogli, in *Contrordine compagni. Manuale di resistenza alla tecnofobia per la riscossa del lavoro e dell'Italia*, Rizzoli, Milan 2019, rightly notes that the human factor is becoming more and more central to work, so any form of Luddism is unmotivated. Moving the argument one step further, I would like to propose a conceptualisation of consumption as labour. For a development of this, cf. M. Ferraris, *If It Is Free, You Are the Producer*, 24° World Congress of Philosophy, Beijing, 14 August 2018.

was previously passive and useless into something active and useful: sleep, entertainment, consumption, and *otium* – not *negotium*. *Homo faber* was but a contingency linked to a very long age of human history, which spanned from the abandonment of hunting and gathering to the perfect automation of production. As I have repeated many times, there is no essence to the human being, and therefore nothing that inexorably ties humanity to toil. If automation ensures production, then education, political deliberation, research, and care can be exercised with a breadth and focus that were unimaginable when most of human energy was absorbed by the ungrateful tasks of *homo faber* (and perhaps *homo necans*).

But human input is crucial at a much more basic level. There is an excellent reason for platforms to ensure the survival of their users, because without the latter they would grind to a halt and lose all meaning. Not to mention that even in a fully automated universe there is still an insuperable threshold for dead labour, and this threshold, unsurprisingly enough, consists in the fact that human beings die, and that the fundamental and irreplaceable living labour is precisely the labour of life, so the meaning of living labour comes from being-toward-death. It is this living labour that governs all dead labour and gives it its ultimate meaning. In short, there is no need to be enslaved or alienated in order to claim recognition. Identifying new forms of slavery in activities that share neither essence nor appearance with it is not only useless, but also generates two harmful side effects. On the one hand, we absolve ourselves of all blame and shortcomings, putting it all down to external agents. On the other, and above all, we blind ourselves to the true nature of the transformation underway.

Imagine if Marx, instead of working on industrial capital, had written (as many authors did in his day) a praise of the feudal way of life. He would have been Sir Walter Scott. This is the risk we often run: acting like a Walter Scott who regrets the good old days of the assembly line, instead of Ivanhoe and Scottish castles. What we can and must do, on the contrary, is increase cultural consumption, lifelong education, research, and *otium*, and not go looking for improbable forms of *negotium*. A humanity that can renounce both *homo faber* and *mulier mater* recognises that these characterisations – the former for a relatively short time, the latter for as long as human history – are not linked to some human essence. Human nature is never static, but is mobilised following needs, desires and consumption. Finding toil and alienation even in mattress testing, instead of working out the conceptual strategy to conceive of it as a production of value as such deserving of remuneration, is hardly a sign of friendship towards humankind – a species that is full of faults and yet thinks itself full of merits.

Even though mattress testing is not the great job that people think it is, the current situation allows each of us, and with the simple purchase of a smartwatch, to become a professional sleeper, transforming into capitalisable biometric documents statuses and acts that previously vanished into thin air, or rather were nothing at all. But how is it that we cannot sell these documents, while the platforms find it so easy to do so? This brings us back to the fundamental disparity between user and platform that I focused on in the previous chapter. What is more, the experiences of factory production in the industrial economy are the ones that lend themselves least to generating nostalgia for production, and indeed describe a situation that no reasonable being would want to recreate. The fact that this condition is instead represented as an ideal job to be restored and returned to is paradoxical, much like the vague nostalgia that prisoners feel for jail-life when they are set free.

4.3.4. If it is labour, it must be recognised

The third step follows from the previous two and goes like this: if mobilisation is labour, it should be recognised, not in terms of a cumbersome direct distribution but with the creation of a webfare system, based on the redistribution of income derived from the taxation of platforms. The idea behind 20th-century welfare that allowed the left to socialise the surplus value of industrial capital was to consider savings and investment as two sides of the same coin. If you look at capital as a whole, you have to overcome the moralistic belief that those who put money in the bank are rewarded because they save. This is not the case: they are rewarded because they make money available for investment, thereby sustaining consumption in the long term, which is the ultimate goal of any production of goods. And investment is the way to achieve what – in an age of still imperfect automation – was the fundamental objective of welfare, namely full employment. For this to happen "the gay of tomorrow are absolutely indispensable to provide a raison d'être for the grave of to-day".[52]

This is like saying that you save today only to spend tomorrow, and saving without spending makes no sense. If I hide money inside a mattress and this mattress is found centuries later, mine won't have been an act of capitalisation, but only an unconventional way of stuffing a mattress. Similarly, we must consider consumption and production as two faces of the same reality. We produce with a view to future consumption, and the only animals truly capable of consumption are human animals. On the one hand, human animals are capable of designing machines that enhance production and of interacting with them; on the other hand, they share with non-human animals the urgency of metabolism. The

---

[52] J.M. Keynes, *The General Theory of Employment, Interest and Money*, Macmillan Cambridge University Press, for Royal Economic Society, Cambridge 1936, pp. 105–106.

needs dictated by metabolism are uncomfortably peremptory, cyclical and urgent. And it is precisely this peremptoriness that brings out the radical dependence of production on consumption. We can see this easily in the housing market. There is little point in building houses if the population is decreasing, so the motive for building houses is to be found precisely in the need to take shelter from the weather.

One last point, before moving on to webfare. To those who object that no one forces us to use the platforms, and that therefore by using them we must accept their rules, I say: who, when looking for a restaurant, a health service, or simply to contact their provider, would give up, for unfathomable reasons, consulting a search engine? There is no valid alternative to this exchange, so there is not only profit, but surplus value. The proof of this is that all classical forms of intermediation are in decline. One can certainly argue that the carriage is an alternative to the train, but not that it is a competitive alternative, and in fact it disappeared along with public telephones, telephone books, paper encyclopaedias, maps, typewriters and tape recorders.

On closer inspection, the lack of an alternative is already the tacit premise of demands for privacy protection. Indeed, these demands would have no reason to exist if use of the Web were completely voluntary. After all, it would be very strange for an influencer to take action for the protection of his or her privacy with regard to content uploaded on the web, since it is assumed that he or she uploaded it in order to be disclosed. However, even those who would never think of becoming an influencer, or even just joining a social network, are forced to use the Web precisely because they have no alternative, since all essential services today are exclusively found there. It is worth noting that every time we use the Internet, we are explicitly and solemnly asked to give up our privacy, and we do so. After all, what we are giving up is not much, since the companies' interests lie not in surveillance but in profiling. The fact that this is a small matter is demonstrated by the companies' insistence on respecting privacy. There is no alternative to using the Web any more than there is an alternative to wearing clothes or using electricity, although we know very well that in both cases there were historical eras when neither was a necessity. From this point of view, the lack of a valid alternative to giving one's documents to the platforms is as clear-cut as the lack of a valid alternative for the proletarian to accepting the contract of wage labour, not to mention that, in our case, we have no form of union protection, both because the contract is iterated continuously and individually through the acceptance of cookies, and because the mobilisation provided is not recognised as labour by either the user or the platform.

Thus, the 19th-century worker and the modern-day mobilised worker share the same need, and this is the essential point. The fact that the services provided are useful and do not require physical effort only adds to the lack of alternatives. However, the differences between the two situations should not be overlooked, and it is precisely for this reason that the redistribution of surplus value should not be imputed to individuals, but centralised and allocated according to political choices. Indeed, this mobilised labour costs neither toil nor alienation, but simply produces value and does so, I repeat, without there being any alternative to this production of value. It would be grotesque if, when an entrepreneur or an intellectual produces as many, if not more, documents than a labourer or an unemployed person, they received a subsidy. However, there is nothing grotesque or paradoxical about the idea of reinvesting the surplus value generated by human mobilisation (because here too there are no alternatives: no beaver would be mobilised on a smartphone) for purposes of justice and human flourishing.

In the documedia revolution, consumers are absolutely necessary to create the *raison d'être* of those serious, thoughtful and boring individuals we call machines. The welfare objective, full employment, is much better achieved through consumption, which affects everyone, than through production, which affects an increasingly small minority. This was not possible in the past, when consumption left no trace and did not generate knowledge. Today, on the other hand, the enormous hysteresis and computational power of platforms, which in principle makes perfect automation possible, is also what makes it possible to collect information on human behaviour resulting from our mobilisation. This enormous production of value is the real human capital on which webfare is based, and which has to be paid for by the digital giants. One may well fear that this consumer world is a Brave New World, but (besides referring to the necessity of learning to live with which I will conclude this book) I would like to recall that Auschwitz was also a factory, with honest workers and impeccable 20th-century productive morals, except it was unfortunately devoted to extermination.

Fields and workshops are not noble in themselves: they are only so if they yield human flourishing. Conversely, consumption is not the equivalent of forced and idiotic happiness, but constitutes the true objective of a humanity that recognises itself as the bearer of needs, desires, and therefore ends – that is, something no machine can ever have. Once again, we must not view ourselves as victims, but as those historically responsible. It is natural and right to focus on the victims, but this only makes sense if we strive to create the conditions in which these victims will soon cease to be such, in a liberated world. First and foremost, they will be liberated from labour – a phrase that is not to be understood as the Auschwitz nightmare of "work shall make you free", but as a life that no longer needs any labour other than life itself.

4.4. Redemption: Webfare

Metternich used to say that abuses of power generate revolutions, and revolutions are worse than any abuse – the first thing should be said to monarchs, the second to the people. As for today, the first sentence should be addressed to the platforms, the second to the users and their representatives. If we do this, we may allow a redemption that does not mean otherworldly salvation, but stands first and foremost for the act of getting back, reacquiring, redeeming, and in this case returning to its authors a huge amount of unrecognised value. But how can one pay for labour that is not even recognised as such? The question is divided into three smaller ones: Who's to pay? How does one pay? What does one pay for?

4.4.1. Who's to pay? How? What for?

As for "Who's to pay?"[53] The answer is simple. The machines, of course, and more precisely the platforms by means of their owners. Souls act, and the actions recorded by the Web enable an automation that places memory rather than matter at the centre of the production process. Souls feed the automata, and the latter relieve the former of material toil. Automata, at this point, no longer need the souls' labour, but they do need their memory, the life forms that, if documented, initiate the process of perfect automation. A humanity increasingly freed from work as toil and alienation will be able to valorise the enormous implicit labour we all do on the web. This enormous creation of value can be redistributed to those who need it, with a virtuous taxation of the platforms. In addition to this, all those who – like me and I hope many of you – already have a salary can create a solidarity fund for the underprivileged and excluded with the value they produce on the web. What if platforms don't want to pay?, you may ask. Well, they *will* pay, otherwise they will be nationalised and they will lose, as far as the European Union is concerned, a market of almost half a billion mobilised people and consumers.

As for "How does one pay?", there are already many proposals being made. For example, one could seek to support consumption, through universal basic income; but at what price, with what money? It would be bizarre (and yet this is

---

[53] The problem of "Who's to pay?" is also raised by those who, like Bernard Stiegler (cf. *La société automatique* cit.), were against a redistribution of documedia surplus value, and proposed instead to invest the time gained by automation in producing a de-automation of human relations (that is, in my terms, in invention, consumption and education). The problem, however, is that it is not clear who would pay for this operation, which necessarily requires the redistribution of documedia surplus value, and in no way conflicts with the objectives envisioned by this perspective, but rather makes it possible to achieve them.

exactly what is happening with the "citizenship income" in Italy) if it were the state, a passive subject of evasion and an active subject of corruption in an impressive number of countries, that supported consumption with tax money. Instead, it is obvious that consumption should be supported by those who derive the greatest economic benefit from that consumption: the platforms. Practical and feasible programmes for the taxation of online advertising have already been drawn up.[54] But of course online advertising is archaeology, just as it is archaeology to focus on the notions of monopoly and abuse of a dominant position. As soon as we recognise the surplus value of documents, it is clear that the monopolistic interest of companies is not to take advantage of their predominant position to increase the price of services, as in the industrial economy, but to increase the number of cost-free services in order to increase the amount of documents collected.

Here we have to take things to another level: user satisfaction is also the result of increasing job cuts (which are admittedly unwanted), so it is not a matter of paying individual users. Rather, a webfare system has to be put in place to prevent the social consequences of unemployment. In other words, it is a question of launching a New Deal that starts from a negotiation between capital and labour, labour being all European citizens and capital being the platforms. The object of the negotiation would be the imposition of an excise tax on documents, and the political power would derive from the demographic mass, i.e. the document revenue, of the European Union. With the General Data Protection Regulation (GDPR), the EU has shown that it is capable of changing the rules on the use and protection of documents. To take it a step further, however, it is necessary to understand, outside of any metaphor, in what sense documents can be considered a form of capital and surplus value.

Now, when the question "Who controls the privacy and ownership of documents?" is replaced by the question "Who pays for the production of documents? And how do you make them pay for it?", a strategy for taxation and redistribution of documedia surplus value needs to be devised. China has a billion smartphone users, a mass of documents that is incomprehensibly greater than Europe and the United States; it has centralised the whole lot (everything is done by

---

[54] Cf. C. Fuchs, *A Method for Taxing Online Advertising and Digital Value*, in Id., *The Online Advertising Tax as the Foundation of a Public Service Internet: A CAMRI Extended Policy Report*, University of Westminster Press, London 2018, pp. 57–68. Cf. also M.C. Fregni, *Mercato unico digitale e tassazione: misure attuali e progetti di riforma*, in "Rivista di diritto finanziario e scienza delle finanze", LXXVI, 1, 2017, pp. 51–81 and B. Cappiello, *Cepet leges in legibus. Cryptoasset and Cryptocurrencies Private International Law and Regulatory Issues from the Perspective of EU and Its Member States*, in "Diritto del commercio internazionale", 3, 2019, pp. 561 ff.

mobile phone, and the companies belong to the state) and has nationalised the surplus value generated by the production of documents on the web, reinvesting and redistributing it with the power of a planned economy that today, thanks to the knowledge provided by big data, can be implemented successfully. That is why, in China, jobs that can be easily automated are carried out, in an openly redundant way, by employees, because the latter can thus have a job and a salary. Of course, China is a communist, and therefore authoritarian, state; we must find a solution that would work for our liberal democracies, and populism is certainly not the answer – if anything, it is the manifestation of widespread discontent.

A taxation of documedia surplus value, enforced on a European level (many people are needed to achieve this), could be the basis of a new web, and of the most solidarity-based and fair kind of capitalism history has ever known. It is important that this strategy takes place on a European, if not cosmopolitan, level, because only a massive demographic number can put enough pressure on the platforms. I am not proposing any nationalisation, not least because the promoting entity would be supranational (the EU), and would intervene in the economy not by governing industries (it is clearly incapable of doing so) but by redistributing profits not only to shareholders, but to citizens. This, I think, is all the more reason to shift from states to the European Union: those who implement webfare must not be incompetent or controlled by industries or incompetent elites, so a supranational organisation is the only solution. Of course, we must not ignore the obstacles to this project, which (apart from the fact that it might not be understood or appreciated, but that is another story) are essentially of three types: geopolitical, economic and moral.[55]

As for the geopolitical obstacle: the US and China are the two big developers of digital technologies. Taxing the platforms we use in Europe could lead to a trade conflict with the US. This is true, but it is also true that the US cannot hope to win the race against China without Europe. The latter (like South America, but with a federative structure) is a supranational power that would merely tax and redistribute as webfare the enormous value generated by its workforce, provided by its 460 million web-mobilised people. Above all, the taxation of documedia

---

[55] Currently, only two countries have introduced a web tax in Europe. One is Italy, with the 2018 budget law, which for technical reasons came into force in January 2020 and should be levied from 2021. According to it, 3 percent tax should be imposed on active online companies with an annual turnover of 750 million or more. France introduced the same tax in July 2019. The UK would like to introduce a 2% tax. However, both France, Italy and the UK are aiming for a European agreement on this issue and rightly consider a national tax to be a temporary solution. Cf. B. Bonini, G. Galli, *La web tax italiana: prospettive e problemi*, in "Osservatorio CPI", 25 January 2020, https://osservatoriocpi.unicatt.it/cpi-archivio-studi-e-analisi-la-web-tax-italiana-prospettive-e-problemi [05/05/2020].

surplus value would have a much stronger chance of success than a tax on capital. For while the potential taxation of capital would make it fly to tax havens, the taxation of documedia surplus value would leverage a non-transferable capital, the collection of documents produced daily by mobilised Europeans.

Let's conceptualise the present using examples from the past, and think of Europe as a large factory made up of platforms, and Europeans as workers who produce value, i.e. documents, without being paid. Platforms would not be able to relocate their assets unless they gave up their users, and thus the possibility of further capital gains, because without the knowledge derived from users, the entire production system would be paralysed, just as would have happened in the case of a general strike in industrial times. Should the platforms then decide to resist to the bitter end in the face of the mere possibility of a general strike, it would be sufficient, though far from desirable, to nationalise them.

In this framework, the economic problem would be solved by understanding that what is perceived as taxation is in fact the remuneration of a workforce that is indispensable for the survival of the platforms themselves. Precisely because the users of free services are not the products, but the producers, these services are not really free, since they generate surplus value, so the platform must pay for them: not with a salary assigned to individual users (which would be practically unfeasible, theoretically problematic, and economically insignificant due to the fragmentation of the remuneration) but with a transfer of resources to the European Union for webfare purposes. On the one hand, this would solve a *de facto* problem, the circumstance that documents in themselves are worth little – no more than a passport or a banknote would be for Robinson Crusoe – and become significant and profitable only if they are aggregated and capitalised.[56] On the other hand, we would be able to address a *de iure* problem. Webfare would offer a double advantage compared to citizenship income: first, it would not be a subsidy, but a set of services paying for an activity that is actually carried out, namely the great mobilisation that produces documents, and therefore value, on the web. Second, the money would not come from the coffers of a heavily indebted

---

[56] It is true that 20 dollars a year, the profit that each user generates annually for Facebook, is not much: a low added value that generates capital only because of its large volume. But we must not forget an essential point, namely that capital, as a total volume, is generated by the network of meanings that is produced by all users. The real and fundamental added value therefore lies in the lives of all those who solve their practical problems by mobilising themselves on the web. This is priceless, and precisely for this reason it must be paid for. Reducing the annual work of those mobilised on Facebook to the sum of 20 dollars, therefore, not only fails to take into account that their mobilisation is not limited to Facebook, since there are many other documents that are created on other archives, but above all fails to consider that what is added is not just a part, but the whole, the overall meaning of the mobilisation, what sets the machine in motion and fuels it.

state, affecting future generations, but would be genuine new wealth that would circulate mainly in terms of services, thereby relaunching consumption. Obviously, the well-off would have nothing to claim, while those in need would be provided with housing, services and education, leaving room for individual initiative to achieve non-essential goals that are, nevertheless, important to human beings.

This leaves a moral problem. Am I proposing a not-so-updated version of the ethical state, which decides what is right for its citizens? No, because a webfare system would not prevent anyone from legally enriching themselves as much as they wished, or from breaking the law if they wanted to (they would have to deal with the police). Rather, it would simply offer opportunities to those who had fewer options than others. Also, the ethical state is also a closed commercial state, whereas webfare would be part of a supranational organisation that in turn would be part of a cosmopolitan perspective. Besides, any reform must be based on an idea of humanity, and in this case we would be dealing with its broadest and most cosmopolitan notion, that which conceives of humankind not as immutable, but as a becoming that leads from nature to culture. It follows that the good reasons for remunerating human mobilisation on platforms are also the guidelines that should inspire webfare. The surplus value generated by the lower costs resulting from automation and the possibility of targeted production and distribution enabled by mobilisation is such that, if redistributed, it can create consumption capacity that can in turn translate into new surplus value. Hence the Consumption-Capitalisation-Production-Consumption cycle. First of all we have the organic element that determines consumption; however, since consumption leaves traces and produces social objects, we see – and this is the novelty introduced by the documedia age – the production of social objects through consumption itself, which in turn will determine further consumption, and so on *ad infinitum*. Consumption thus generates new commodities, and the documedia revolution keeps pushing the threshold forward.

Having answered "Who's to pay?" and "How to pay?", where obviously politics and economics have the last word, we can shift to "What is one paying for?", where I think philosophy can offer some advice. One pays for reasons *de facto*, of course: otherwise one has to nationalise. And for reasons *de jure*: because the mobilised are human beings. So let's turn the perspective upside down in a revolutionary way, and consider all the elements which, according to political victimhood, make us slaves of technology: consumption, education and invention. I will show that they can – and should – rather be seen as reasons for the valorization and mastery of humankind over machines.

4.4.2. Consumption

Suppose someone said they invented a sushi-eating machine. No doubt, this initiative would appear implausible, not to mention worthless, as it would be the best approximation to the concept of a useless machine. If a machine for making sushi and a machine for distributing it make sense, it is because there are organisms that, for physiological reasons, need to be fed. If those organisms were to be replaced by automatons, the whole process would lose all meaning.[57] Therefore, to argue that fully-automated companies could do without humans even as consumers is nonsense.[58] A perfect machine, capable not only of producing but also of consuming, is theoretically conceivable but practically nonsensical: machines have an end in themselves, which is precisely to meet the needs of humans, whether good or bad, and the consumer machine would prove to be the most imperfect of machines. In reality, the Web is a complex set of automata in search of souls, a whole that needs to be nourished and mobilised by humans. Without embodiment, and without the human capital described in 2.3.2, neither value nor value production are conceivable. Conversely, the embodiment mentioned in 3.2.4 is why the possibility of increasing value, of maintaining value and of finalising value opens up. While highly complicated operations are within the reach of machines, that typical manifestation of embodiment, need, is within the reach of any newborn child.

Now, since antiquity philosophers have insisted on the values of consciousness, intelligence and desire. With respect to these, consumption has been seen as a subordinate phenomenon to be largely proscribed when it goes beyond the stage of satisfying basic needs. But this has failed to take into account that the humans who populate a shopping mall are fulfilling a spiritual function no different from that which takes place in religious rites, intellectual understanding, aesthetic enjoyment, and of course desire, of which consumption is a sublimated

---

[57] In *Grammatica della moltitudine*, DeriveApprodi, Rome 2002, P. Virno rightly writes: "The crucial point is, however, that while the material production of objects is delegated to the automated machine system, the performance of living labour increasingly resembles a linguistic-virtuosistic performance" (pp. 47–48). Yet this is only true if we consider labour as invention, which is a rare commodity. If, on the other hand, we consider consumption as labour, the matter changes completely.

[58] Hence the puzzling character of the following sentence: "However, it is far from certain that the future economy will need us even as consumers. Machines and computers could do that too. Theoretically, you can have an economy in which a mining corporation produces and sells iron to a robotics corporation, the robotics corporation produces and sells robots to the mining corporation, which mines more iron, which is used to produce more robots, and so on. These corporations can grow and expand to the far reaches of the galaxy,and all they need are robots and computers – they don't need humans even to buy their products." (Y.N. Harari, *21 Lessons for the 21st Century*, Random House, New York 2018, p. 37).

version, not unlike courtly love and mysticism. Consumption accompanies us until our last day, and even afterwards, if we have made onerous requests for our burial. In this framework, consumption is not the modern surrogate of the capital vices, but rather our organic need inserted in a socio-technical context. Hence all kinds of reinforcement, supplementation and metamorphosis of need, which is transformed into desire, ambition, luxury, waste, and culture. Above all, it should not be forgotten that consumption is a much broader phenomenon than the one captured by the criticism of consumerism: was Napoleon a consumerist because, when he came back from Russia, he threw away his books right after reading them? Was Leopardi a consumerist, having ruined his health with mad and desperate studies? Believe it if you like.

It is said that death makes us all equal. This is also true of a close relative of life, and therefore also of death, namely consumption, and shopping centres are the new cathedrals, soon to be replaced by platforms, where all walks of life meet. It seems difficult to draw a clear dividing line between consumption, religion and aesthetics, and one suspects that if one did, it would be to the advantage of consumption. The ideological differences between humans may be enormous, but consumption has a levelling function, overcoming ideologies and cultural barriers. Claiming that the French aristocracy had blue blood, the blood of the fleurs-de-lis, was the paradoxical answer to the fact that all humans have red blood, and a metabolism that binds them together, with or without titles of nobility. The *Grundphänomene* of humanity therefore include not only work, normativity, love and play, but also, and decisively, consumption. It is not true that only humans have a world or that only humans are aware of dying: some humans are poor in world and some are poor in death. But surely only humans have consumption and produce rubbish, which is the combination of the biological element of need and the technological element of production. Without mechanisms we would be simple organisms, but without us as organisms the whole process would be meaningless.

Of course, one cannot speak of "consumption" in the context of a subsistence economy, but already in the case of our hunter-gatherer ancestors their activity was mediated by supplements. Furthermore, strictly speaking, one of the rare cases of satisfaction of physiological needs in which the human animal is not conditioned by the supplement would be cannibalism out of necessity, like Ugolino della Gherardesca or shipwrecked people adrift in the ocean, not ritual cannibalism: in the latter case, culture has already entered human practice. I can imagine a possible objection: is there not something tautological in defining consumption as a strictly human fact? Isn't it like the tautology by which Heidegger defined death as a phenomenon that belongs only to humans? In both cases, in fact, "consumption" would be identified with "human consumption" and "death" with "human death". thereby developing an argument that is nothing more than a *petitio principii*. Now, as far as death is concerned, I believe I introduced a principle that may be debatable, but not tautological: true death, that which befalls all organisms, is an irreversible process, whereas automata are generally capable of reversible processes, unless they are programmed to be used only once. As regards consumption, I would like to resort to another argument, more akin to tautology and a *petitio principii*, which I nevertheless find convincing: consumption is peculiar to the human form of life, which may have similarities with other non-human practices, but only in appearance – rather like smiling, which for dogs and wolves is a show of aggression and not of cheerfulness or politeness. At this point there are two options.

The first one, which is the one chosen decades ago, consists in cursing the devil and his stand-ins who make us work even at home.[59] It seems to me that this is not very useful today, because we have to consume anyway, and consuming while complaining is neither nicer nor smarter than consuming blissfully; nor can I find any nobility in the Pascalian reed which, bent by the wind, knows it is bent: this seems like useless suffering to me, which moreover does not bring any change. Not to mention that often, by condemning consumption and proposing new (or, more often, old) alternatives, people end up idealising some supposedly "normal" era, which usually coincides with the complainer's childhood and youth. Now, it is unclear why the advancement of consumption should be seen as negative: does it not rather reveal a character that is proper to the human being, preferable to anthropophagy, religious wars, and above all to destitution and lack of goods? Moreover, even the most unbridled consumerist would never submit to their purchases in the same way as a religious fanatic does to their idols. The more people develop, the more they consume, the more they evolve spiritually. The more dead labour, the more living labour.

The second approach is the one I propose, which consists in conceptualising consumption as labour, or rather as mobilisation, with the justifiable objection that consumption involves neither fatigue nor alienation, and is therefore not work in the classical sense. If there are machines, which structurally have their purpose outside themselves, it is because there are organisms that satisfy their vital needs with the help of machines. Nihilism laments the lack of purpose, but, again, it presupposes the narrative that humans are meant to be angels, full of merit and goals, but they fall into the world and become depressed. Also, there is no shortage of good examples of virtuous consumption: "Vas positum erat aceto plenum; spongiam ergo plenam aceto hyssopo circumponentes, obtulerunt ori

---

[59] Think of G. Anders, *Die Antiquiertheit des Menschen*, C.H. Beck Verlag, Munich, 1956; M.-A. Dujarier, *Le travail du consommateur*, La Découverte, Paris 2008.

eius. Cum ergo accepisset acetum, Iesus dixit: 'Consummatum est'. Et inclinato capite tradidit spiritum."[60] Christ died by consuming a glass of vinegar,[61] and the double meaning of "it is finished" (the vinegar, as well as Christ's life) is not surprising. The end (goal) and the end (death) are one and the same thing, and it seems natural that the conclusion of Christ's incarnation should begin with an act of consumption. Having a nightcap and kicking the bucket are the same thing, because consumption, the obvious symbol of our embodiment, is also the most patently obvious emblem of our humanity. As someone who drank everything but vinegar, Francis Scott Fitzgerald, put it: "Of course all life is a process of breaking down".[62]

4.4.3. Education

There is a second good reason for webfare. It is all too evident that after the demise of *homo faber* another guiding principle will be needed for humanity. It is not simply a question of revaluing the professional skills derived from humanistic education, nor of the necessity of humanism in the modern world. The point is not about decision-makers, who have always managed on their own, but about humankind as a whole, which needs an alternative form of growth and development to the often tiring, thankless and alienating one that was traditionally guaranteed by labour. In other words, we need to map out the path and the prerequisites for a work of the spirit in which education and culture can drive the difficult but promising and, above all, indispensable transition from *prod-humanity,* in which we were surrogates for machines, to *doc-humanity*, in which machines will entirely surrogate us, leaving us to the purely human activity of living well, or at least of trying to. A documedia capital, whose greatest value lies in accumulated and invested documents, is the ideal pole of a labour that does not consist of toil and alienation, but of document production through mobilisation. At this point, the central question obviously becomes: What is humankind when it is no longer identified by labour? The biggest mistake, from this point of view, would be to consider humans as fallen angels, perfect individuals who are now able to express themselves in their full glory. That, of course, is not the case. Once ex-

---

[60] Jn 19:29–30: Now there was set a vessel full of vinegar: and they filled a sponge with vinegar, and put it upon hyssop, and put it to his mouth. 30 When Jesus therefore had received the vinegar, he said, It is finished: and he bowed his head, and gave up the ghost (KJV).

[61] The legionnaires were not as bad as they are made out to be: they offered Christ what they drank themselves. It was most likely *pulca*, a mixture of two parts water and one part vinegar, which was the Coca-Cola of Roman soldiers.

[62] F.S. Fitzgerald, *The Crack-Up*, New Directions, New York 1945. The first edition was published in February, March and April 1936 in "Esquire".

ploitation is over, education begins, and that is the great promise offered to us by automation and capital.

At a time when humanity is no longer formed from the outside, by toil and alienation (which are often only a deformation), it must be shaped from within, in the knowledge that the forms of its consumption will decide its own future and the happiness of future generations. One can legitimately wonder whether a passive mass of consumers does not represent a degraded humanity. Indeed, it does, but it is certainly less degraded than the (truly) alienated mass under inhumane working conditions that Marx knew in his time. In order to get out of this degradation, or simply to improve, it does not seem like a great idea to revive production and pride in a job well done as in Levi's *The Wrench*, because we do not know what to do with production. Rather, human beings must be educated, especially when they have the technical means to express their opinions. For this very reason, we must be aware that every advance solves old problems and creates new ones. For example, today, for the first time in the history of the world, it is virtually possible for the whole of humanity to express their ideas. An unimaginable number, not only in the time of Aristotle, but also when, in 1948, with the Universal Declaration of Human Rights, it was said that everyone has the right "to seek, receive and impart information and ideas through any media and regardless of frontiers". The Universal Declaration was made lightly, since it sanctioned a purely virtual right that no one thought would become real in seventy years: which explains why it was not supplemented with the clause "provided that the opinion expressed is reasonable and does not offend the dignity of others". Hence the online cacophony of billions of people shouting their views, each convinced of being right.

We are surprised that with the meteoric rise in the ability to express one's ideas, it is the worst ideas that take over, but in fact we should be surprised if the opposite were true. There is no reason to think that, once humans achieve the ability to freely express their thoughts, these will necessarily be good or true; ideas are often racist, sexist, and violent. Indeed, to appreciate this, there is no need for every human being to speak out: it is enough to read the declarations of so many of its rulers and spiritual leaders. And there is no reason for things to be otherwise, since the human being is neither angel nor beast, but something in between, engaged in a process of infinite growth and refinement of which education is the fundamental instrument. This is where we find the affirmation of the first principle of the Enlightenment, "dare to think for yourself". It didn't use to be the case. However, through education, the human animal can also learn the other two principles of the Enlightenment (unfortunately, we do not know how soon): learn to think by putting yourself in the shoes of others, and learn to think in accordance with yourself, i.e. consistently.[63]

So far, these are common-sense considerations, but I would not want them to be mistaken for a happy ending, simply because there is no ending that is happy in every way. The most important task of any webfare, as well as of any good life, is to learn to live, postponing the end, like Sherazade, with inventiveness and culture, snatching meaningful days from the meaninglessness *par excellence*, just as, in Faust's hope, the cobots digging his grave were actually snatching new land from the sea. For one *can* learn to live, and the proof is that we are here and we are living, me writing these lines and you reading them. It is a process of habituation: the purpose of society is to breed an animal that is allowed to make promises. We need technology and training to enable us to make promises, and to give back and reciprocate. This is why education is so important: it enables the development of human capacities through culture. This notion was already at the centre of important philosophical and economic reflections in the 20th century but, in the absence of a theoretical framework motivating investment in education, such ponderings were destined to remain wishful thinking.[64] What webfare offers, instead, is a philosophical and economic system that would motivate not only the practical availability of resources, but also their political and economic necessity, since these resources derive from the investment of human mobilisation on the web.

Now, I do not deny it: learning to live means playing a game that can never be won. For a mortal, wanting to live at all costs is the most absurd claim that can be made. Frederick the Great was well reminded of this at the battle of Kolin, on 18 June 1757, when he exhorted his soldiers who were already retreating: "Rascals, would you live forever?" This is why "learning to live" is an expression that can be deployed in many ways, whether melancholic ("will I ever learn to live?") or threatening ("I'm going to teach you how to live"), and generally manifests an illusion or a presumption. This illusion and this presumption are the hyperbole hidden in philosophy, and they generate the paradox whereby the philosopher, unwisely conceived as a model of wisdom, actually cultivates the least wise purpose one can set oneself: learning to live, the necessary but impossible task, the goal that fortunately remains often hidden under minor, secondary, diversionary purposes. This task often only appears at the end, when it's over, when one has the somewhat liberating impression that not much living is left, and that the urgency is, if anything, to learn to die. This latter task is even more absurd and

---

[63] Kant, *Answering the Question: What is Enlightenment?* cit.

[64] M.C. Nussbaum, *Cultivating Humanity. A Classical Defense of Reform in Liberal Education*, Harvard University Press, Cambridge (MA) 1997.

unattainable, given that it is a single action, which does not permit repetition, and which, as far as we can see, often takes place in an accidental way, which leaves everyone – starting with the person concerned – breathless.

But what if that's what learning to live was about? "Tout casse, tout passe, tout lasse, il n'est rien, et tout se remplace". *Telos*, which in Greek means "completion, goal, purpose" has the same value as the Latin *finis*,[65] and is linked to the mortuary veil, as in the *Iliad*: "Even as he thus spake the end [*telos*] of death enfolded him, his eyes alike and his nostrils".[66] But, at the same time, and in the same contexts, we also have the meaning of purpose: *telos*, in the *Odyssey*, is marriage, and more generally it means fulfilment, the goal of life or of a single action. In all these cases, just as in the veil of death, the *telos* is represented by a band. Breaking the finish line ribbon, signifying that one has fulfilled one's *telos*; cutting the ribbon at an inauguration; wrapping the winner with a sash. Why do we do this? We now know, and we know that it has to do with death, even at the very peak of life. How did Anaximander put it? "Whence things have their origin, Thence also their destruction happens, According to necessity".

4.4.4. Invention

A final point, which represents, at least I hope, the greatest possible openness to the future: humans are capable of invention. By "invention" I mean four distinct events: the first is the *creation* of a new device, for example the mobile phone; the second is the *re-signification*, by users, of what has been invented; the third is *recreation*, the fact that the human is the ultimate goal of invention; the fourth is *autopoiesis* as a characteristic element of the human being.

Let us start with *creation*. Invention is tautologically a human production, being the response to needs that are rooted in a body structured in a certain way and in a society organised in a certain way. By definition, a machine cannot invent new purposes or new rules, it can only apply the purposes and rules assigned to it by humans. We can therefore easily imagine a machine that can win a game of chess, but not a machine that changes the rules of chess – let alone a machine that

---

[65] Cf. R.B. Onians, *The Origins of European Thought*, Cambridge University Press, Cambridge 1951, pp. 510–511.

[66] *Iliad*, XVI, vv. 502 ff. The veil is linked to the Greek custom of veiling the head of the deceased (is this not the first thing the police do at a crime scene? And isn't it the last thing Caesar does, with a piece of his toga, when he recognises Brutus among the conspirators?) In the *Kalevala*, the deceased enters the realm of death with their faces covered; in the Minoan civilisation, the eyes of the deceased were covered with gold bands depicting eyes; not to mention that, for reasons that are not entirely obvious (what difference does it make?), people were often hooded before being hanged, and blindfolded before being shot.

stops playing chess because it is bored. Invention and intelligence are a form of intermittence with respect to the regular functioning of the automaton. It is the introduction of a discontinuity that produces a purpose.

Think of painters, who for millennia have had a secured livelihood, guaranteed by manual skill in producing representations.[67] At a certain point, machines were introduced in an activity traditionally connected with the craftsman's living labour. Or, to put it in a better way, this craft left the mill, the loom, and the workshop and entered into everyone's life. In particular, photography was invented in the mid-19th century, and ever since that moment painters have been asking what they are there for. They all came up with their own strategy. Some say they do not paint what is there, but their impressions. Others paint things that are not seen in the world, and therefore cannot be photographed, because they are abstractions. Others yet exhibit urinals in art galleries. This may look like the end of art but, instead, it is the dawn of a new day, not only for painting but – this is the novelty – for every form of art, as well as for the social world in general and for industry, where repetitive tasks are left to machines and humans are required to behave like artists.[68]

---

[67] I have developed this point in Ferraris, *From Fountain to Moleskine* cit.

[68] Here, too, documents play a crucial role. When the production of objects is automated, the constitutive aspect of art emerges: it does not lie so much in the production of an object, as the latter does not necessarily have to be handmade by the artist, nor expressly produced. Rather, the constitutive nature of art lies in the explication of the contract of definition and enjoyment of the work underlying the relationship between producer and user. But documents can also be used in a stronger form, to literally *produce* acts: thanks to strong documents, Theodore Fu Wan contractually changed his name to Saskatche Wan, Alix Lambert married five different wives in six months, and Maria Eichhorn conceived of her artistic activity as drafting contracts to protect urban areas threatened by speculation. And the attributive power of documents is at the core of practices such as those of Stefan Brüggemann and Robert Barry, who have agreed that authorship of two of their artworks will be transferred back and forth between them at the end of every five-year period. Likewise, by exploiting copyright laws, Philippe Parreno and Peter Huyge acquired the rights of a Manga figure. The contract can go as far as the staging of a subversion of rules that are no longer those of art, but of the penal code, as for example when the artist gives the order to rob a supermarket or as in "Corruption Contract" by the Superflex group, whose buyers – in clear exception to the theory of beauty as a symbol of moral good – agree to engage in activities of extortion or corruption for the contract period. One can go even further and create artworks by mere contractual agreement. Already in 1958 Yves Klein had created "Le vide", an exhibition without works, in which a contract for the sale of an "immaterial pictorial sensitivity zone" was issued, and much later, in 2010, Étienne Chambaud realised an artwork consisting only of contracts, certificates and declarations of authenticity. Likewise, a contract can transform even the author into an artwork, as in the provision by which Jill Magid gave a specialised company the mandate to transform her charred remains into a diamond. But the peak was reached perhaps by Robert Morris in 1963, whose contract consists of two parts: on the left, a lead plate with a few lines engraved on it, on the right a declaration in

So, painters have come up with things to do, and the art market has never been as flourishing as it has been since the camera came into existence. Indeed, paradoxically (but not too much), it is photographers that have disappeared, not painters. The former, in fact, exercise a skill that is becoming increasingly easy, widespread and accessible, because it has become automated, and is ultimately available to everyone thanks to the universal camera we all carry in our pockets. Painters, on the other hand, have to think up new rules, or simply ways to break the rules, and this does not necessarily have a single inventor. The same happened with the mobile phone: to its producers' surprise, the use of the keyboard has prevailed over verbal communication. What is the human input here? Production? Of course not: text messages have a human origin, but they were intended for more modest and functional purposes, just like Duchamp's urinal. The human contribution consists in consumption, which is the manifestation of needs and desires, and therefore also in giving new purposes to objects. In other words, it is the creation of a teleology through technology, since the increase in means leads to a multiplication of ends, as well as a human flourishing infinitely greater than that dictated by scarcity.

Let us come to *re-signification*. As we know, the mobile phone was conceived as a talking machine, which was bound by servitude to a wire that kept it attached to a specific place. However, responding much more to the needs of users than to the intentions of programmers, it turned out to be a machine for writing and recording, and eventually became a universal machine. In this case, as in so many others (writing, gunpowder ...), invention arose from the responsive relationship between the needs of an organism and the resources of a mechanism. Since mechanisms do not possess responsiveness, no invention is to be expected from

---

which the artist withdraws the character of artwork from the artwork, transferring the artistic aura onto the withdrawal document. Kant said that the main character of art is to make people think. But what thoughts do these works arouse? Essentially juridical questions. For example: is issuing instructions enough to be called an author? It is certainly enough to be called despotic: Seth Siegelaub, for example, prescribes in the execution contract of his works that even the slightest change involves an irreversible alteration of the works themselves. One can be even more despotic, and almost perverse: this is the case with Daniel Buren, who strictly abstains from signing or authenticating his works. Or else: is the curator of an exhibition or a museum an author, given that their responsibility goes far beyond the management of an exhibition space? (For his part, an artist, Maurizio Cattelan, co-curated the Berlin Biennial in 2006 with Massimiliano Gioni). Is a performance really an immaterial artwork that escapes the market? So it was in its original ideology, but now the world is full of performance recordings. Indeed, the world is full of documents, as happens in the philosophical conversations with Ian Wilson, of which what's left is only a sheet of paper with a signature. And it is also full of "scripta", works that are assembled and dismantled and come with instructions for use. Or artworks that consist only of documents, such as the complaint presented by Cattelan to the police headquarters of Forlì, reporting the theft from his car of an invisible work of art.

them: and we can therefore be certain that invention is destined to remain a human characteristic that can never be automated. Automaton, then, cannot do without flesh and soul, it needs them more than anything else. Just as it needs the obvious consequence of flesh and soul that is invention.

That this is possible is shown by the fact that, contrary to what the prophets of homologation claimed, we have never been as individual and idiomatic as we are today, so much so that in order to grasp correlations it is of no use to refer to perishable concepts such as "class". Instead, one must refer to profiling based on the individual and their consumption and behaviour, which have no predictable rationale. Unlike when it was thought that the technical society coincided with dehumanisation (an interpretation that, moreover, stemmed from a misunderstanding of the notion of "technique", which is co-extensive with the human being), never has there been so much humanity, and never has so much attention been paid to humanity. And what happens on the Web is the most obvious manifestation of this: fashion, travel, lifestyle, food, answers to the most diverse and bizarre questions. Nostalgics of the closed commercial state would be the last to give up this wealth and find themselves driving a Trabant. And above all, nothing seems more inappropriate than the characterisations of humankind and technology given in the 20th century – notions like, the one-dimensional man, the lonely man in the age of technology. As if! Think of how lonely a person can be in a village; this loneliness shows us how much human noise surrounds us and how difficult it is to deal with, because it is the noise of very different individuals.

Let us now turn to *recreation*. Today, several theories defend the argument that a computer can be creative,[69] and I have no problem with this, if only because the notion of "creativity" conceals many misunderstandings. In fact, I have no objection at all to recognising that creativity has something automatic about it, because it demands the perfect mastery of a series of rules and is often manipulated into a seriality that is very much machine-like, as we see in giants like Simenon, Mozart, or Picasso. Conversely, some works, such as *Finnegans Wake*, seem to have been composed by a badly programmed automaton. Given these premises, I don't have any difficulty in imagining a Simenon, a Mozart, a Picasso, or a Rembrandt produced by an automaton which, starting from its knowledge of the author's previous works, would generate a new, completely original and aesthetically satisfying one.[70] After all, this is what forgers have been doing in painting for a long time and with excellent results, so that the problem will be, if anything,

---

[69] See, in particular, the excellent study by G. Ajani, *Contemporary Artificial Art and the Law*, Brill, Leiden 2020.

[70] In the case of the Rembrandt, you don't even need to use your imagination: it has already been done. I am referring to the Next Rembrandt Project, in which artificial intelligence creates new art that even critics cannot distinguish from the real thing. The painting is made by com-

to establish whether entrusting the production of a work to an automaton is not a rather clumsy attempt to spare oneself the accusation of forgery.

One can also imagine a machine-made Rauschenberg + Degas + Piero della Francesca, or a Proust + Bernhard + Collodi. I suspect, however, that the result would be the extension to every art form of the disappointment usually generated by the eclecticism of the villas described by Gadda in *Acquainted with Grief*. What is inconceivable, rather, is an artificial user: an automaton that needs to be distracted, to rest, and thus to enjoy a work of art, i.e. not in a creative attitude, but in a recreational one. Indeed, what machines cannot do is enjoy the fruits of invention, which is the ultimate impulse of the creative act, the response to a need that we feel in ourselves and project onto others, and not the simple execution of a rule, however complex.[71] If, therefore, on the surface the human being can be considered a creative animal, on a deeper level we realise that its essential characteristic consists in being a recreational animal, capable of deriving pleasure, distraction and stimulation from its own or other people's inventions, even if they are mechanical, just as it can derive enjoyment from a sunset. This is true in many areas. Artificial intelligence, as we know, can write articles,[72] but is not in the least interested in reading them. So too, I can build a machine that performs mathematical calculations without knowing what mathematics is, or I can make a lie detector that doesn't know what truth is. And of course I can devise a machine to mimic T.S. Eliot or Rembrandt: but without a human user, it's hard to see what the point would be, as art is a form of human life. Indeed, it doesn't apply to animals: a peacock doesn't consider its train as a sophisticated outfit, and I doubt that a machine that produces a fake Rembrandt will ever ask to be paid – we can be sure that the one who collects the cash, or ends up in jail, is the programmer.

Finally, let us come to *autopoiesis*, the fourth, crucial sense in which I would like to articulate the idea of "invention". By "*autopoiesis*" I mean that, in accordance with the analyses carried out in 2.2, humanity has proved to be inventive with respect to itself. Let us say a few words about this invention of humanity that constitutes one of the strongest arguments for webfare. We come from a history that is billions of years long, and at the same time we exist in this precise moment. This very moment sums up and incorporates the entire past of the entire universe, without responding to any necessity whatsoever, but simply because the present sits on the shoulders of the past. Who we are, as human beings, was not already

---

puters using algorithms that draw on Rembrandt's pictorial characteristics (use of light, patterns in rendering faces, composition of painting materials, etc.).

[71] A. Donise, *Critica della ragione empatica*, il Mulino, Bologna 2020.

[72] Cf. https://www.theguardian.com/commentisfree/2020/sep/08/robot-wrote-this-article-gpt-3 [08/09/2020].

written from the beginning. Rather, the opposite is true. What we are appears the moment we glimpse a direction, the place towards which we are going.

There was very little in Lucy that led to us, but we can only understand that by starting from where we are today: Lucy did not know that, and neither did her descendants, near or far, or perhaps very far away, for example the humans who, a few generations ago, believed that our origin depended on an act of divine creation that shaped us exactly as we are (or, to be precise, as they were). We are what we become, and we will not understand who our oldest ancestor is by looking back, but by looking forward. That is why teleology is the only resource that can shed light on archaeology, and outline the direction the future will take. Looking at a contemporary politician, an ornamental brooch, or a vase, we are generally well aware of their ends (teleology); and it is on this basis that we interpret the conduct of an Athenian politician, the possible use of an Etruscan brooch, as well as that of an Egyptian vase (archaeology). The same, in a much more obvious form, applies to the future. That is why, having looked into the transformations of the documedia revolution, we are, I believe, in the best position to grasp the revelations about who we are – our deep archaeology.

Bare life had no meaning outside of itself, while "clothed" life has far too much meaning. And this overabundance of teleology gives rise to the retrospective view of archaeology, of the foundation, of the principle of reason. If we look at the origin of the individual or the species, it is only because these beginnings are called upon to shed light on fulfilment. Why tell a psychoanalyst about our childhood, study the bones of our most remote ancestors, display vases and buckles belonging to humans who lived long ago and whom we have never met, if not as an indication of a possible destiny? Darwin's project itself would be inconceivable if not in the light of this interplay between archaeology and teleology. The Galapagos tortoises are interesting because they say something about us. And if those who imputed a secret providentialist intent to Darwin are completely wrong (what we are in no way corresponds to a hypothetical design of "what we were meant to be". because this design simply does not exist),[73] this does not change the fact that there is no contradiction between emergentism and the philosophy of history. Rather, these two disciplines work hand in hand.

There is no such thing as human nature in itself: there is only a historical revelation of its possibilities and characteristics, in which technology, as a constitutive element of history, plays a decisive role. One could therefore legitimately ask whether this perspective does not suffer from panglossianism and, what is worse, fundamental quietism. Ultimately, we could not but become what we are,

---

[73] F. Nietzsche, *"On Teleology" or "Teleology since Kant"* (1868), trans. Paul Smith, in *Nietzscheana*, 8, 2000, pp. 1–20.

whatever we have become, so there is no alternative for humankind: we are what we were meant to be, and any attempt to steer history in a different direction, be it the vague dream of an originally perfect human or the revolutionary promotion of a new man, would be nipped in the bud by an identification between the rational and the real; we are what we are and could not be otherwise. I would suggest looking at it from another angle. Evil is not the absence of good. It is a resistance without which good could never take place. As Kant taught us, one might say that without air the dove, the symbol of moral action, would fly better, but in fact this is not so: it would not fly at all because it is precisely the friction of reality that allows the dove to fly and the utopian to hope, bringing out things that we people wouldn't believe. Because if vagueness makes us dream, determinism also has its advantages: thought requires resistance, otherwise it goes round in circles. Resistance, friction, is a resource. Whichever way you look at it, it is wrong to simply blame Satan, washing away all our sins, first and foremost those of gluttony, lust and avarice. After all, Satan is the lord of this world and it is to him – much more than to the good Lord, who merely expelled us, and without tools, from paradise – that we owe the best things we have: technology, society, and of course knowledge, be it Adam's apple or Newton's.[74]

Rather than an identification between the rational and the real, we should speak of a temporal action of a providence that does not govern the future, but manifests itself retrospectively, in hindsight.[75] Is there anything more providential than the fact that humans do not live forever? That is, that our ultimate end is our end, which allows us to devise all the intermediate ends we call "civilisation"? These contrivances do not arise from a specific project: they emerge precisely to respond, as we have seen, to our inadequacies. If God had created us as we are, he would have pulled a prank in very bad taste. Fortunately for him and for us, he did nothing of the sort, and we created our own providence, not out of a surplus of ingenuity, but just the opposite, out of the urgency of our deficits – something a cat will never experience. In this context, humanity finds in technology the vehicle of an ontological freedom that resists the biological determinism we share with all other living beings.

---

[74] "The demands made today by employers such as Amazon in Italy or Pegatron in Taipei and Foxconn in Shenzhen would have been unthinkable not only in the late 1960s but also in companies run by Taylor or Ford in the early 1900s." (D. De Masi, *Il lavoro nel XXI secolo*, Einaudi, Turin 2018, p. 584). Are we sure? Of course, they will do everything they can to automate their processes, and then they will have the problem of sustaining purchasing power, unless (as seems to be the case) they confuse capitalism with the Romanovs' stupid autocracy.

[75] To quote the subtitle of J. De Maistre, *Les Soirées de Saint-Pétersbourg ou Entretiens sur le gouvernement temporel de la Providence, suivis d'un Traité sur les sacrifices*, ed. by R. de Maistre, J.B. Pélagaud et Cie, Lyon-Paris 1821, 2 voll.

The current technological configuration, with the teleology it suggests, is therefore the privileged access route to our archaeology, as it allows us to retrospectively understand our becoming, the most conspicuous evidence of which is, at the same time, constituted and handed down by technology. What history gives us – quite contingently, because things could always have gone differently or not happened at all – is in fact an essence, which appears at the end rather than at the beginning, but which allows us to retrospectively reconstruct the beginning. The necessity of certain physical endowments in order to develop certain practical performances, such as the opposable thumb or standing upright, only becomes apparent in retrospect, in the light of current endowments and their foreseeable developments; but the fact that the revelation is contingent does not change the fact that what is revealed is an absolute, a defining essence, in some cases grasped long ago, in others hidden for millennia. It is enough to consider how, until very recently, the notion of the human being as a mobilised animal was concealed by an easier interpretation, which traced human mobilisation back to reasons of subsistence through physical effort, which today have often disappeared without the mobilisation ceasing. Is there a moral to this story? Perhaps there is. It is difficult, after all, to claim that existence has meaning in itself; but the multitude of intermediate meanings – from the starry sky above us to the Michelin star restaurant in front of us – is more than enough to make life worth living.

Note to the Text

What you have in your hands is the fruit of several years of research, writing, rewriting and (appearances can be deceiving) cutting. However, the final draft of this book coincided with the pandemic. This not only provided me with an unexpected amount of time (to reverse-quote Neil Armstrong, "That's one small step for mankind, one giant leap for a man"), but also meant that unspoken or merely implicit aspects came to light, or took on a different hue. In short, the virus acted as a reagent, making many facets of the system more evident to me and, I hope, to my readers. This system had always been there, right in front of us, but without the crisis and the following shock it would have seemed blurred, confused with thousands of different symptoms and moods.

Doc-humanity completes the quadrilogy in which I would like to sum up my career: *History of Hermeneutics* (1988), *Estetica Razionale* (Rational Aesthetics, 1997) and *Documentality* (2009). The first volume, with a historical approach, answered the question: What does it mean to think? The second answered the question: What is there in the natural world? The third investigated the nature of the social world. Bringing to a close my reflections of the last ten years, the present book seeks to answer the question: What is to be done?

I wish to thank Stefania Achella, Gianmaria Ajani, Tiziana Andina, Alessandro Arbo, Alessandro Armando, Carola Barbero, Fiorella Battaglia, Jocelyn Benoist, Mauricio Beuchot, Petar Bojanić, Stefano Bonaga, Francesca Borrelli, Olivier Bouin, Elisabetta Brizio, Michael Buckland, Clementina Cantillo, Stefano Caputo, Elena Casetta, Wang Chengbing, Michele Cometa, Angela Condello, Emilio Corriero, Fausto Corvino, Gianluca Cuozzo, Davide Dal Sasso, Ronald Day, Luca De Biase, Juan Carlos De Martin, Sarah De Sanctis, Jacopo Domenicucci, Anna Donise, Giovanni Durbiano, Antonia Egel, Maria Eichhorn, Melanie Erspamer, Donato Ferri, Günter Figal, Paolo Flores d'Arcais, Markus Gabriel, Sergio Genovesi, Werner Gephart, Sara Guindani, Jimmy Hernandez-Marcelo, José Luis Jerez, Hyun Kang Kim, Andreas Kemmerling, Stefano Lanterna, Anna Marmodoro, Valeria Martino, Sanja Milutinović, Kevin Mulligan, Julian Nida-Rümelin, Valentina Pisanty, Alberto Romele, Jens Rometsch, Lamberto Rondoni, Vincenzo Santarcangelo, René Scheu, Claire Scopsi, Ernesto Sferrazza Papa, Barry Smith, John Searle, Enrico Terrone, Tiziano Toracca, Giuliano Torrengo, Vera Tripodi, Jan Voosholzt, who have had the patience to read, correct, discuss and enrich parts of this book at different times during its gestation.

As Leonard Cohen put it, "And everybody knows that you live forever / Ah, when you've done a line or two". He was being ironic, as he often was. It is therefore with the awareness of making an obscure and fragile gift that I dedicate this book to Gianni Vattimo, who taught me how to think, and to the memory of my mother, who taught me how to live.

Afterword

Webfare: Who, What, Where, When, Why

Thanks to Markus Gabriel, series editor, Sarah De Sanctis, translator, and publisher Mohr Siebeck, this book is appearing in English just over a year after the Italian edition. The most concrete way to express my gratitude is to add an afterword on the developments in my work since the original publication, which focus on the notion of "Webfare" – the capitalisation and redistribution of the data produced by humanity to humanity itself, as I'll try to show in following the Five Ws rule – that constituted the end and the aim of Doc-humanity.

Who

The subject of Webfare is humanity understood as the bearer of needs, desires and curiosity. No longer interesting as *homo faber*, and always ready to turn into *homo necans* in the wars that are once more breaking out in Europe, man appears outwardly as *homo otiosum*, who – willingly, by choice, or unwillingly, due to the decrease in available jobs – spends his time connected to the Web. In reality, man is *homo consumens*: through his own consumption (which cannot be automated), he manifests the human life form – this is decisive for machines, as they do not know what it means to be human and have to learn it every day from us. Now, if we manage to interpret correctly the notion of "consumption" – which does not consist simply in eating junk food, but is the manifestation of the whole range of human needs, both material and spiritual – *homo consumens* must be recognised as *homo valens*, because he is the one who establishes the scale of values regulating the whole Web apparatus.

What

What is produced by *homo consumens qua homo valens* is an entirely new kind of capital, as until recently the vast majority of human acts did not translate into capitalisable data. We are dealing with world heritage in a radical sense here. Ontologically, it has nothing to do with common goods, which are part of nature, whereas this is about culture and society. Epistemologically, it is a treasure trove of finally reliable knowledge about humankind. Economically, it is infinitely renewable capital, since one can use the data as often as one likes. Finally, from a political point of view, it is radically democratic, since it is produced in equal measure by the richest, most educated, poorest, and most uneducated human beings, provided they are connected to the Web. Indeed, this capital does not differentiate between rich and poor, because even those who have no money at all (provided they are connected) generate a wealth of data just as the richest man on earth. Instead of being a sign of the divine election of the individual, as in the Calvinist origin of bourgeois capitalism, this new humanistic capital is only valuable insofar as it is shared among all humans, regardless of wealth, intelligence, race or faith.

Where

The place where this world heritage is produced is the Web as the repository of human life forms. It was once thought that the real capital of the Web was the (supposed) collective intelligence produced in the infosphere, namely semantic capital. This was not the case, both because the infosphere is also the realm of mass imbecility and fake news, and, above all, because the infosphere is simply a more accessible library than traditional ones. It is a library of Babel, if you will, but one that must abide by the inescapable rule that language was given to us not only to reveal our thoughts (which, incidentally, can be wrong), but also to hide them. The real capital of the Web is syntactic capital, produced by the acts of humans: what we do, where we are, what we search for, what we like. These acts are always true, even if most of the time they only mean something when put in relation with millions of other bits of data. These acts constitute the docusphere, the authentic seat of humanity's heritage, which is nourished by the biosphere (the totality of human life forms) and embedded in the ecosphere, i.e. the environment favourable to human life. The latter, indeed, must be preserved in the knowledge that protecting the environment is a moral obligation towards humanity – the only animal species for which it makes sense to speak of "morality".

When

When should Webfare be implemented? Right now, to meet the needs of post-co-ronial humanity, especially as its implementation does not call for political deliberations but for innovative actions. It is necessary to develop a form of proactive thinking, based on the fact that Webfare is a third, humanistic way compared to the two kinds of data capitalisation that have been successfully established for a couple of decades now. The first is the liberal way of American commercial platforms, which privatise capital without returning it (except for the least valuable part of it, information in the public domain) to the humanity that produced it. The second is the communist one of the Chinese nationalised platforms, which collectivise capital but lead to the state's total control over its citizens. Hence the third, humanist way: thanks to the possibility, guaranteed by European laws, of obtaining their own data, users can entrust the use of such data (which requires large numbers) to philanthropic platforms. The latter will thus create a data market, which does not exist today, returning the cognitive profits to the whole of humanity and the financial ones to the part of humanity which, today, has no money, but still produces data. This will initiate a path of citizenship which is not only formal, but concrete.

Why

Why? In order to realise the (evangelical and Marxist) principle of "From each according to his ability, to each according to his needs". This principle has not been implemented so far because in the world of *homo faber*, ability was much more rewarding than need – a manifestation of indigence that could at best be reserved for humiliating charity, and was not seen as value production. Instead, the peculiar nature of this world heritage could trigger a process that, starting from the recognition of the value that humans produce on the Web, would enable an alternative capitalisation of this value through philanthropic platforms, as well as triggering capacitation procedures that would allow *homo consumens*, recognised as *homo valens*, to take a few steps forward on the infinite path that leads to *homo sapiens*.

Bibliography

Aaberge, R., Brandolini, A., *Multidimensional Poverty and Inequality*, in "Temi di discussione", Banca d'Italia Working Papers, 976, September 2014

Accornero, A., *Era il secolo del lavoro*, il Mulino, Bologna 1997

Acerbi, A., *Le trasmissioni della cultura*, in *Modelli della mente e processi di pensiero. Il dibattito antropologico contemporaneo*, ed. by A. Lutri, Ed.It, Catania 2008

Acerbi, A., Parisi, D., *Cultural Transmission Between and Within Generations*, in "Journal of Artificial Societies and Social Simulation", 9, 1, 2006

Adorno, Th. W., *Introduction to the Sociology of Music*, Continuum, London 1989

Aiyagari, S.R., Wallace, N., *Existence of Steady States with Positive Consumption in the Kiyotaki-Wright Model*, in "Review of Economic Studies", 58, 1991, pp. 901–916

Ajani, G., *Contemporary Artificial Art and the Law*, Brill, Leiden 2020

Ajani, G., *I diritti dell'arte contemporanea*, Allemandi, Turin 2011

Albert, R., Jeong, H., Barabási, A.L., *Diameter of the World-Wide Web*, in "Nature", 401, 6749, 1999

Alexander, S., *Space, Time and Deity*, Macmillan, London 1920

Allais, L., *Manifest Reality: Kant's Idealism and His Realism*, Oxford University Press, Oxford/New York 2015

Allegri, G., Ciccarelli, R., *Il Quinto Stato*, Ponte alle Grazie, Milan 2013

Althusser, L., *On Ideology* (1970), Verso, London 2007

Anders, G., *Die Antiquiertheit des Menschen*, C.H. Beck, Munich, 1956

Andina, T., *Ontologia sociale. Transgenerazionalità, potere, giustizia*, Carocci, Rome 2016

Andina, T., *Transgenerazionalità: una filosofia per le generazioni future*, Carocci, Rome 2020

Andrejevic, M., Gates, K., *Big Data Surveillance. Introduction*, in "Surveillance & Society", 12, 2, 2014, pp. 185–196

Anon. The Oldest Systematic Programme of German Idealism in F. Beiser (ed. by), *The Early Political Writings of the German Romantics*, Cambridge Texts in the History of Political Thought, Cambridge University Press, Cambridge 1996, pp. 1–6

Appadurai, *Globalization*, A., Duke University Press, Durham/London 2001

Arendt, H., *The Human Condition*, University of Chicago Press, Chicago 1958

Aristotle, *Aristotle in 23 Volumes*, Harvard University Press, Cambridge, (MA); William Heinemann, London 1944

Armando, A., Durbiano, G., *Teoria del progetto architettonico. Dai disegni agli effetti*, Carocci, Rome 2017

Armstrong, D.M., *A World of States of Affairs*, Cambridge University Press, Cambridge 1997

Asadi, E., Gameiro da Silva, M., Henggeler Antunes, C., Dias, L., Glicksman, L., *Multi-objective Optimization for Building Retrofit: A Model Using Genetic Algorithm and Artificial Neural Network and an Application*, in "Energy and Buildings", 81, 2014, pp. 444–456

Asimov, I., *I, Robot*, Doubleday, Garden City/New York 1950

Askonas, P., Stewart, A., *Social Inclusion: Possibilities and Tensions*, St. Martin's, New York 2000

Atkinson, A.B., *Public Economics in Action: The Basic Income/Flat Tax Proposal*, Clarendon, Oxford 1995

Aunger, R.A., *Darwinizing Culture*, Oxford University Press, Oxford/New York 2000

Austin, J.L., *How to Do Things with Words*, Oxford University Press, Oxford/New York 1962

Avital, E., Jablonka, E., *Animal Traditions: Behavioural Inheritance in Evolution*, Cambridge University Press, Cambridge 2000

Axelrod, R., *Evolution of Cooperation*, Basic Books, New York 1984

Bachimont, B., *Patrimoine et numérique. Technique et politique de la mémoire*, Institut National de l'Audiovisuel, Bry-sur-Marne 2017

Badia, A., *The Information Manifold. Why Computers Can't Solve Algorithmic Bias and Fake News*, MIT Press, Cambridge (MA) 2019

Baldassarre, G., *Cultural Evolution of "Guiding Criteria" and Behaviour in a Population of Neural-Network Agents*, in "Journal of Memetics", 4, 2001

Barkow, J.H., Cosmides, L., Tooby, J., (ed. by), *The Adapted Mind: Evolutionary Psychology and the Generation of Culture*, Oxford University Press, Oxford/New York 2002

Barone, G., Mocetti, S., *Intergenerational Mobility in the Very Long Run: Florence 1427–2011*, in "Temi di discussione", Banca d'Italia Working Papers, 1060, April 2016

Barthes, R., *L'ancienne rhétorique [Aide-mémoire]*. in *Communications, 16, 1970. Recherches rhétoriques*, pp. 172–223

Bastani, A., *Fully Automated Luxury Communism: A Manifesto*, Verso, London 2019

Bateson, G., *Mind and Nature – a Necessary Unity*, E. P. Dutton, New York 1979

Bateson, G., *Steps to an Ecology of Mind* (1972), University of Chicago Press, Chicago 2000

Baucon, A., *Leonardo da Vinci, the Founding Father of Ichnology*, in "Palaios", 25, 2010, pp. 361–367

Baudrillard, J., *Symbolic Exchange and Death* (1976), Sage, London 2017

Baudrillard, J., *The Consumer Society: Myths and Structures* (1970), Sage, London 1998

Baudrillard, J., *The Perfect Crime*, Verso, London 1996

Bauman, Z., *Liquid Modernity*, Polity, Cambridge 2000

Beck, U., *Risk Society: Towards a New Modernity*, Sage, London 1992

Beck, U., *Schöne neue Arbeitswelt*, Campus, Frankfurt a. M. 1999

Beck, U., *The Brave New World of Work*, Polity Press, Cambridge 2000

Bell, C., *The Hand: Its Mechanism and Vital Endowments as Evincing Design*, William Pickering, London 1833

Bentivogli, M., *Contrordine compagni. Manuale di resistenza alla tecnofobia per la riscossa del lavoro e dell'Italia*, Rizzoli, Milan 2019

Berger, P.L., Luckmann, T., *The Social Construction of Reality: A Treatise in the Sociology of Knowledge*, Anchor Books, Garden City (NY) 1966

Bergson, H., *Matière et mémoire. Essai sur la relation du corps à l'esprit* (1896), PUF, Paris 1965

Bergson, H., *The Two Sources of Morality and Religion* (1932), MacMillan, London 1935

Bernhard, T., *Alte Meister* (1985), Suhrkamp, Frankfurt a. M. 1993

Berns, T., Rouvroy, A., *Gouvernementalité algorithmique et perspective d'émancipation. Le disparate comme condition d'individuation par la relation?*, in "Réseaux", 177, 1, 2013, pp. 163–196

Beverley, R.M., *The Darwinian Theory of the Transmutation of Species Examined*, Nisbet, London 1867

Bickerton, D., *More than Nature Needs: Language, Mind, and Evolution*, Harvard University Press, Cambridge (MA) 2014

Bigi, M., Cousin, O., Méda, D., Sibaud, L., Wieviorka, M., *Travailler au XXIe siècle. Des salariés en quête de reconnaissance*, Laffont, Paris 2015

Bigoni, M., Camera, G., Casari, M., *Money Is More than Memory*, in "Journal of Monetary Economics", 110, 2020, pp. 99–115

Blackmore, S., *The Meme Machine*, Oxford University Press, Oxford/New York 1999

Bloch, M., "Reflections of a Historian on the False News of the War" (1921), translated in *Michigan War Studies Review*, Volume 2013, *20th Century, World War I*, 2013

Borges, J. L., "The Library of Babel" in Id., *Collected Fictions*, Penguin, New York 1998

Bostrom, N., Yudkowsky, E., *The Ethics of Artificial Intelligence*, in *The Cambridge Handbook of Artificial Intelligence*, Cambridge University Press, Cambridge 2014, pp. 316–334

Bourdieu, P., *La précarité est aujourd'hui partout*, in Id., *Contre-feux: propos pour servir à la résistance contre l'invasion néo-liberale*, Liber-Raisons d'Agir, Paris 1998, pp. 95–101

Boyd, R., Richerson, P.J., *Culture and the Evolutionary Process*, University of Chicago Press, Chicago 1985

Boyd, R., Richerson, P.J., *Not By Genes Alone: How Culture Transformed Human Evolution*, University of Chicago Press, Chicago 2006

Boyd, R., Richerson, P.J., *The Origin and Evolution of Cultures*, Oxford University Press, Oxford/New York 2005

Brandt, R., *D'Artagnan und die Urteilstafel. Über ein Ordnungsprinzip der europäischen Kulturgeschichte (1, 2, 3/4) (1991)*, Deutscher Taschenbuch Verlag, München 1998

Bratman, M., *Faces of Intention*, Cambridge University Press, Cambridge 1999

Bratman, M.E., *Shared Cooperative Activity*, in "The Philosophical Review", 101, 1992, pp. 327–341

Braver, L., *A Thing of This World: A History of Continental Anti-Realism*, Northwestern University Press, Evanston 2007

Bray, D., *Wetware: A Computer in Every Living Cell*, Yale University Press, New Haven 2009

Bregman, R., *Utopia for Realists. The Case for Universal Basic Income, Open Borders, and a 15-Hours Workweek*, The Correspondent, Amsterdam 2016

Bricmont, J., A. Sokal, *Fashionable Nonsense*, Picador, New York 1998

Briet, S., *Qu'est-ce que la documentation?*, Édit, Paris 1951

Broad, C.D., *The Mind and Its Place in Nature*, Routledge-Kegan Paul, London 1923

Bronzini, G., *Il diritto a un reddito di base. Il welfare nell'era dell'innovazione*, Abele Edizioni, Turin 2017

Brynjolfsson, E., McAfee, A., *The Second Machine Age: Work, Progress, and Prosperity in a Time of Brilliant Technologies*, Norton, New York 2014

Buckland, M.K., *What is a "document"?*, in "Journal of the American Society for Information Science" (1986–1998), 48, 9, September 1998

Bunz M., *The Silent Revolution. How Digitalization Transforms Knowledge, Work, Journalism and Politics without Making Too Much Noise*, Palgrave Macmillan, London/New York 2013

Burkhardt, A., (ed. by), *Speech Acts, Meaning and Intentions. Critical Approaches to the Philosophy of J.R. Searle*, De Gruyter, Berlin/New York 1990

Butterworth, G., *Neonatal Imitation: Existence, Mechanisms and Motives*, in *Imitation in Infancy*, ed. by J. Nadel and G. Butterworth, Cambridge University Press, Cambridge 1999, pp. 63–88

Bybee, J., *From Usage to Grammar: the Mind's Response to Repetition*, in "Language", 82, 4, 2006, pp. 711–733

Byler D., Sanchez Boe, C., *Tech-Enabled 'Terror Capitalism' Is Spreading Worldwide. The Surveillance Regimes Must Be Stopped*, in "The Guardian", 24 July 2020

Byrne, R.W., Whiten, A., *Machiavellian Intelligence*, Clarendon Press, Oxford 1988

Bytwerk, R.L., *Landmark Speeches of National Socialism*, Texas A&M University Press, College Station (TX) 2008

Cacciari, M., *Il lavoro dello spirito*, Adelphi, Milan 2020

Cappiello, B., *Cepet leges in legibus. Cryptoasset and Cryptocurrencies Private International Law and Regulatory Issues from the Perspective of EU and Its Member States*, in "Diritto del commercio internazionale", 3, 2019

Carchia, G., Ferraris, M., (ed. by) *Interpretazione ed emancipazione. Studi in onore di Gianni Vattimo*, edited by, Raffaello Cortina, Milan 1995

Cardona, G.R., *Storia universale della scrittura*, Mondadori, Milan 1986

Carpo, M., *The Second Digital Turn. Design beyond Intelligence*, MIT Press, Cambridge (MA) 2017

Carr, N., *The Shallows: What the Internet Is Doing to Our Brains*, Norton, New York 2010

Casilli, A., *En attendant les robots – Enquête sur le travail du clic*, SEUIL, Paris 2019

Castel, R., *From Manual Workers to Wage Laborers: Transformation of the Social Question* (1995), Transaction Publishers, New Brunswick (NJ) 2003

Castelvetri, L., *Il diritto del lavoro delle origini*, Giuffrè, Milan 1994

Chakravarty, S.R., *Inequality, Polarization and Poverty: Advances in Distributional Analysis*, Springer, New York 2009

Chamayou, G., *Petits conseils aux enseignants-chercheurs qui voudront réussir leur évaluation*, in "Revue du Mouvement Anti-utilitariste dans les Sciences Sociales", 33 (*L'Université en crise. Mort ou résurrection?*), 2009, pp. 146–165

Châtelet, G., *L'enchantement du virtuel*, Éditions Rue d'Ulm, Paris 2010

Chirumbolo, P., *Letteratura e lavoro*, Rubbettino, Soveria Mannelli 2013

Ciccarelli, R., *Forza lavoro. Il lato oscuro della rivoluzione digitale*, DeriveApprodi, Rome 2018

Ciccotti, G., Cini, M., de Maria, M., Jona-Lasinio, G., *L'ape e l'architetto. Paradigmi scientifici e materialismo storico*, Feltrinelli, Milan 1976

Cingolani, P., *La précarité*, PUF, Paris 2017

Cingolani, P., *Révolutions précaires. Essai sur l'avenir de l'émancipation*, La Découverte, Paris 2014

Cingolani, P., *Un travail sans limites? Subordination, tensions, résistances*, ERES, Toulouse 2012

Clanchy, M.T., *From Memory to Written Record: England 1066–1307*, Blackwell, Oxford 1993

Clark, A., *Being There: Putting Brain, Body, and World Together Again*, MIT Press, Cambridge (MA) 1997

Clark, A., Chalmers, D.J., *The Extended Mind*, in "Analysis", 58, 1998, pp. 7–19

Clark, A., *Natural-Born Cyborgs*, Oxford University Press, Oxford/New York 2003

Clark, A., *Supersizing the Mind: Embodiment, Action, and Cognitive Extension*, Oxford University Press, Oxford/New York 2008

Clark, A., *Surfing Uncertainty: Prediction, Action, and the Embodied Mind*, Oxford University Press, Oxford/New York 2015

Clark, A., *Whatever Next? Predictive Brains, Situated Agents, and the Future of Cognitive Science*, in "Behavioral and Brain Sciences", 36, 3, 2013, pp. 181–204

Colonna, M., Pugliese, E., (ed. by), *Il futuro del lavoro in Europa. Occupazione, diritti civili, diritti sociali*, Edizioni Scientifiche Italiane, Naples 2007

Cometa, M., *Perché le storie ci aiutano a vivere: la letteratura necessaria*, Raffaello Cortina, Milan 2017

Condello, A., Toracca, T., *Lavoro, identità: riflessioni tra letteratura e diritto*, in "Il Ponte", 2, March 2016, pp. 120–126

Contarini, S., Jansen, M., Ricciardi, S., (ed. by), *Le culture del precariato*, Ombre corte, Verona 2015

Contarini, S., Marsi, L., (ed. by), *Precarietà. Per una critica della società della precarietà*, Ombre corte, Verona 2015

Corballis, M.C., *From Hand to Mouth: The Origins of Language*, Princeton University Press, Princeton 2002

Couprie, D., L., *Heaven and Earth in Ancient Greek Cosmology*, Springer, New York 2011

Crane, T., *The Significance of Emergence*, in *Physicalism and Its Discontents*, ed. by B. Loewer and G. Gillett, Cambridge University Press, Cambridge 2001

Crary, J.M, *24/7: Late Capitalism and the Ends of Sleep*, Verso, London 2013

Crescenzi, V., *La rappresentazione dell'evento giuridico. Origini e struttura della funzione documentaria*, Carocci, Rome 2005

Cukier, A., *Qu'est-ce que le travail*, Vrin, Paris 2018

d'Errico, F., Backwell, L., *From Tools to Symbols: From Early Hominids to Modern Humans*, Wits University Press, Johannesburg 2005

Darwin, Ch., *The Expression of the Emotions in Man and Animals*, John Murray, London 1872

Darwin, F., (ed. by), *More Letters of Charles Darwin*, 2 vols., D. Appleton, New York 1908

Day, R., *The Modern Invention of Information: Discourse, History, and Power*, Southern Illinois University Press, Edwardsville 2001

De Maistre, J., *Les Soirées de Saint-Pétersbourg ou Entretiens sur le gouvernement temporel de la Providence, suivis d'un Traité sur les sacrifices*, ed. by R. de Maistre, J.B. Pélagaud et Cie, Lyon-Paris 1821, 2 voll.

De Masi, D., *Il lavoro nel XXI secolo*, Einaudi, Turin 2018

De Soto, H., *The Mystery of Capital: Why Capitalism Triumphs in the West and Fails Everywhere Else*, Basic Books, New York 2000

De Tocqueville, A., *Democracy in America* [1830], Saunders & Otley, London 1840

De Vos, M., *From Labour Market Crisis to Labour Market Reform: Lessons from and for a Divided European Union*, in *Labour Regulation in the 21st Century: In Search of Flexibility and Security*, Cambridge Scholars Publishing, Newcastle 2012, pp. 117–131

Debord, G., *The Society of the Spectacle* [1967], Black & Red, Detroit 1970

Dehaene, S., *Reading in the Brain* (2007), Penguin, New York 2009

Deleuze, G., Guattari, F., *Anti-Oedipus* (1972), Continuum, London 2004

Dennett, D.C., *Can Machines Think?*, in *How We Know: Nobel Conference XX*, ed by M.G. Shafto, Harper & Row, San Francisco 1985

Dennett, D.C., *Competition in the Brain*, in *What Have You Changed Your Mind About?*, ed. by J. Brockman, HarperCollins, New York 2008

Dennett, D.C., *Darwin's Dangerous Idea*, Penguin, New York 1995

Dennett, D.C., *Freedom Evolves*, Penguin, New York 2003

Dennett, D.C., *From Bacteria to Bach and Back*, Penguin, New York 2017

Dennett, D.C., *From Bacteria to Bach and Back. The Evolution of Minds*, Bradford Books – MIT Press, Cambridge (MA) 2017

Dennett, D.C., *From Typo to Thinko: When Evolution Graduated to Semantic Norms*, in *Evolution and Culture*, ed. by S. Levinson and P. Jaisson, MIT Press, Cambridge (MA) 2006

Dennett, D.C., *Kinds of Minds. Towards an Understanding of Consciousness*, Basic Books, New York 1997

Dennett, D.C., *My Body Has a Mind of Its Own*, in *Distributed Cognition and the Will: Individual Volition and Social Context*, ed. by D. Ross, D. Spurrett, H. Kincaid and G.L. Stephens, MIT Press, Cambridge (MA) 2007, pp. 93–100

Dennett, D.C., *The Evolution of Culture*, in "The Monist", 84, 3, 2001, pp. 305–324

Dennett, D.C., *The Intentional Stance*, MIT Press, Cambridge (MA) 1987

Dennett, D.C., *Who's on First? Heterophenomenology Explained*, in "Journal of Consciousness Studies", 10, 9–10, 2003, pp. 19–30

Derrida, J., "Ousia and Gramme: Note on a Note from Being and Time," in *Margins of Philosophy*, trans. by A. Bass, University of Chicago Press, Chicago 1982

Derrida, J., *De la grammatologie*, Les Éditions de minuit, Paris 1967

Derrida, J., "Différance" [1968] in *Margins of Philosophy*, Chicago University Press, Chicago 1982

Derrida, J., *Edmund Husserl's Origin of Geometry: An Introduction* (1962), N. Hays, New York 1978

Derrida, J., *Heidegger's Hand* (1985) in *Deconstruction and Philosophy: The Texts of Jacques Derrida*, University of Chicago Press, Chicago, 1987

Derrida, J., *History of the Lie: Prolegomena*, in Id., *Without Alibi*, Stanford University Press, Stanford 2002

Derrida, J., *La pharmacie de Platon* (1968), in *La dissémination*, Seuil, Paris 1972

Derrida, J., *La Voix et le Phénomène*, PUF, Paris 1967

Derrida, J., *Le Facteur de la Vérité*, Poetique, 21, 1975, pp. 96–147

Derrida, J., *Limited Inc*, Galilée, Paris 1990; Northwestern University Press, Evanston (IL) 1988

Derrida, J., *The Animal That Therefore I Am*, Fordham University Press, New York 2008.

Derrida, J., *The Pit and the Pyramid: Introduction to Hegel's Semiology* (1970), in Id., *Margins of Philosophy*, University of Chicago Press, Chicago 1982

Derrida, J., *Writing and Difference*, [1967] University of Chicago Press, Chicago 1978

Descartes, R., *Meditations and Other Metaphysical Writings*, trans. by D.M. Clarke, Penguin, London 1998.

Descartes, R., *Passions of the Soul*, in *The Philosophical Writings of Descartes*, Cambridge University Press, Cambridge 1985, pp. 325–404

Detrain, C., Deneubourg, J., *Self-organized Structures in a Superorganism: Do Ants "Behave" like Molecules?*, in "Physics of Life Reviews", 3, 3, 2006, pp. 162–187

Dewey, J., *The Middle Works, 1899–1924*, vol. 7, 1912–1914, Southern Illinois University, Carbondale 1979

Di Lampedusa, G., *The Leopard* (1911), Collins and Harvill Press, London 1960

Diamond, J., *The Tasmanians: The Longest Isolation, the Simplest Technology*, in "Nature", 273, 1978, pp. 185–186

Dodd, N., T*he Social Life of Money*, Princeton University Press, Princeton 2014

Domingos, P., *The Master Algorithm: How the Quest for the Ultimate Learning Machine Will Remake Our World*, Penguin, New York 2015

Donati, A., *Law and Art. Diritto e arte contemporanea*, Giuffrè, Milan 2012

Donise, A., *Critica della ragione empatica*, il Mulino, Bologna 2020

Donise, A., *Valore*, Guida, Naples 2008

Doueihi, M., Domenicucci, J., (ed. by), *La confiance à l'ère numérique*, Éditions Rue d'Ulm-Berger Levrault, Paris 2018

Dyson, F.J., *Infinite in All Directions*, Harper & Row, New York 1988

Eco, U., *Fake and Forgeries* (1987), in "Versus", 46, thematic issue *Fake, Identity and the Real Thing* (now in Id., *The Limits of Interpretation*, Indiana University Press, Bloomington 1994

Eco, U., *Five Moral Pieces*, Penguin Random House, New York 2003

Eddington, A.S., *The Nature of the Physical World*, Macmillan, New York 1928

Engel, P., Rorty, R., *What's the Use of Truth?*, Columbia University Press, New York 2007

Engels, F., Marx, K., *German Ideology* (1845), International Publishers, New York 2004

Epstein, B., *The Ant Trap: Rebuilding the Foundations of the Social Sciences*, Oxford University Press, Oxford/New York 2015

Epstein, J.M., Axtell, R., *Growing Artificial Societies: Social Science from the Bottom Up*, MIT Press, Cambridge (MA) 1996

Ertzscheid, O., *L'homme, un document comme les autres*, in "Hermès", 53, 1, 2009, pp. 33–40

Farrell, J.P., *Babylon's Banksters: The Alchemy of Deep Physics, High Finance and Ancient Religion*, Feral House, Port Townsend (WA) 2010

Ferraris, M., *Anima e iPad*, Guanda, Parma 2011

Ferraris, M., *Ding an sich*, in *Das neue Bedürfnis nach Metaphysik / The New Desire for Metaphysics*, ed. by M. Gabriel, W. Hogrebe and A. Speer, De Gruyter, Berlin/Boston 2015, pp. 119–132

Ferraris, M., *Documentality. Why it is Necessary to Leave Traces*, Fordham University Press, New York 2012

Ferraris, M., *Emergenza*, Einaudi, Turin 2016

Ferraris, M., *Estetica razionale*, Raffaello Cortina, Milan 1997

Ferraris, M., *From Fountain to Moleskine*, Brill, Leiden 2019

Ferraris, M., *Goodbye, Kant!: What Still Stands of the Critique of Pure Reason*, SUNY Press, Albany 2014

Ferraris, M., *Hysteresis*, University of Edinburgh Press, Edinburgh (forthcoming)

Ferraris, M., *If It Is Free, You Are the Producer*, 24° World Congress of Philosophy, Beijing, 14 August 2018

Ferraris, M., *Il mondo esterno*, Bompiani, Milan 2001

Ferraris, M., *J.R. Searle, Il denaro e i suoi inganni*, Einaudi, Turin 2018

Ferraris, M., *L'esplosione della registrazione*, in *Filosofia del digitale*, ed. by L. Taddio and G. Giacomini, Mimesis, Milan-Udine 2020

Ferraris, M., *La filosofia e lo spirito vivente*, Laterza, Rome-Bari 1991

Ferraris, M., *Metafisica de la Web*, Dykinson, Madrid 2020

Ferraris, M., *Mobilitazione totale*, Laterza, Rome-Bari 2015

Ferraris, M., *Positive Realism*, John Hunt Publishing, London 2015

Ferraris, M., *Postverità e altri enigmi*, il Mulino, Bologna 2017

Ferraris, M., *Where Are You? An Ontology of the Cell Phone*, Fordham University Press, New York 2014

Feyerabend, P.K., *Against Method* (1975), Verso, London 2002

Fitzgerald, F.S., *The Crack-Up*, New Directions, New York 1945

Flake, G., Lawrence, S., Giles, C., Coetzee, F., *Self-organization and Identification of Web Communities*, in "Computer", 35, 3, 2002, pp. 66–70

Florida, R., *The Rise of the Creative Class*, Basic Books, London 2002

Floridi, L., (ed. by), *The Onlife Manifesto. Being Human in a Hyperconnected Era*, Springer International, London 2015

Floridi, L., *Information: A Very Short Introduction*, Oxford University Press, Oxford/New York 2010

Floridi, L., *Infosfera. Etica e filosofia nell'età dell'informazione*, Giappichelli, Turin 2009

Floridi, L., *The Fourth Revolution. How the Infosphere is Reshaping Human Reality*, Oxford University Press, Oxford/New York 2014

Floridi, L., *The Logic of Information: A Theory of Philosophy as Conceptual Design*, Oxford University Press, Oxford/New York 2019

Floridi, *Semantic Conceptions of Information*, in *The Stanford Encyclopedia of Philosophy* (Spring 2017 Edition), ed. by E.N. Zalta, https://plato.stanford.edu/archives/spr2017/en tries/information-semantic/

Formenti, C., *Felici e sfruttati. Capitalismo digitale ed eclissi del lavoro*, Egea, Milan 2011

Foster, J., *A World for Us. The Case for Phenomenalistic Idealism*, Oxford University Press, Oxford/New York 2008

Foucault, M., *Abnormal* (1975), Verso, London 2003

Foucault, M., *Discipline and Punish*, Allen Lane, London 1977

Foucault, M., *History of Madness*, Routledge, London 2006

Frankfurt, H., *On Bullshit*, Princeton University Press, Princeton 2005

Franklin, F., *Stove's Discovery of the Worst Argument in the World*, in *Philosophy*, 77, 2002, pp. 615–662

Frege, G., *Der Gedanke. Eine Logische Untersuchung* (1918–1919), in "Beiträge zur Philosophie des deutschen Idealismus", I, pp. 58–77

Fregni, M.C., *Mercato unico digitale e tassazione: misure attuali e progetti di riforma*, in "Rivista di diritto finanziario e scienza delle finanze", LXXVI, 1, 2017, pp. 51–81

Freud, S., "A Note upon the Mystic Writing-Pad" (1924) in *The Standard Edition of the Complete Psychological Works of Sigmund Freud*, Vol XIX, The Ego and the Id, The Hogarth Press, London 1961, pp. 227–232

Freud, S., "Psychoanalytic notes on an autobiographical account of a case of paranoia (Dementia Paranoides)", in the *Penguin Freud Library*, Volume 9, Case Histories II, Penguin, New York 1979, pp. 131–226

Freud, S., *Project for a Scientific Psychology (1895)* in *The Standard Edition of the Complete Psychological Works of Sigmund Freud, Volume I (1886–1899): Pre-Psycho-Analytic Publications and Unpublished Drafts*, Stanford University Press, Stanford 1950, pp. 281–391

Frey, C.B., Osborne, M.A., *The Future of Employment: How Susceptible Are Jobs to Computerisation?*, 17 September 2013, http://www.oxfordmartin.ox.ac.uk/downloads/academic/The_Future_of_Employment.pdf.

Fuchs, C., *A Method for Taxing Online Advertising and Digital Value*, in Id., *The Online Advertising Tax as the Foundation of a Public Service Internet: A CAMRI Extended Policy Report*, University of Westminster Press, London 2018, pp. 57–68

Fukuyama, F., *The End of History and the Last Man*, Free Press, New York 1992

Gallino, L., *Il lavoro non è una merce*, Laterza, Rome-Bari 2007

Gangemi, A., *Norms and Plans as Unification Criteria for Social Collectives*, in "Journal of Autonomous Agents and Multi-Agent Systems", 16, 3, 2008, pp. 70–112

García Martínez, A., *No, Data Is Not the New Oil*, in "Wired", 26 February 2019, https://www.wired.com/story/no-data-is-not-the-new-oil/.

Gazzaniga, M.S., *Human. Quel che ci rende unici*, Raffaello Cortina, Milan 2009

Gehlen, A., *Man in the Age of Technology* (1957), Columbia University Press, New York 1980

Gehlen, A., *Urmensch und Spätkultur. Philosophische Ergebnisse und Aussagen* (1956), Klostermann GmbH, Frankfurt a. M. 2004

Gelb, I.A., *A Study of Writing: The Foundations of Grammatology* (1952), 2a ed., The University of Chicago Press, Chicago 1963

Gibson, J.J., in *The Ecological Approach to Visual Perception*, Houghton Mifflin, Boston 1979

Gibson, K.R., Ingold, T., *Tools, Language, and Cognition in Human Evolution*, Cambridge University Press, Cambridge-New York 1993

Gilbert, M., *Group Membership and Political Obligation*, in "The Monist", 76, 1993, pp. 119–131

Gilbert, M., *Living Together. Rationality, Sociality, and Obligation*, Rowman & Littlefield, Lanham 1996

Gilbert, M., *On Social Facts*, Routledge, New York 1989

Gilbert, M., *Walking Together: A Paradigmatic Social Phenomenon*, in "Midwest Studies in Philosophy", 15, 1990, pp. 1–14

Gilbert, S., *Sur la technique* (1953–1983), Presses Universitaires de France, Paris, 2014

Gitelman, L., *Paper Knowledge: Toward a Media History of Documents*, Duke University Press, Durham 2014

Godfrey-Smith, P., *Conditions for Evolution by Natural Selection*, in "Journal of Philosophy", 104, 2007, pp. 489–516

Godfrey-Smith, P., *Darwinian Populations and Natural Selection*, Oxford University Press, Oxford/New York 2009

Godfrey-Smith, P., *Postscript on the Baldwin Effect and Niche Construction*, in *Evolution and Learning: The Baldwin Effect Reconsidered*, ed. by B.H. Weber and D.J. Depew, MIT Press, Cambridge (MA) 2003, pp. 210–223

Goethe, J.W., *Faust*, Princeton University Press, Princeton 1994

Goldman, A., *A Theory of Human Action*, Princeton University Press, Princeton 1970

Goldsmith, K., *Wasting Time on the Internet*, HarperCollins, New York 2016

Goody, J., *The Logic of Writing and the Organization of Society. Studies in Literacy, the Family, Culture and the State*, Cambridge University Press, Cambridge 1986

Gordon, D.M., *Ant Encounters: Interaction Networks and Colony Behavior*, Princeton University Press, Princeton 2010

Gori, R., Del Volgo, M.J., *L'idéologie de l'évaluation: un nouveau dispositif de servitude volontaire*, in "Nouvelle revue de psychosociologie", 2, 8, 2009, pp. 11–26

Gorz, A., *Critique of Economic Reason* (1988), Verso, London 2010

Gorz, A., *The Immaterial: Knowledge, Value and Capital*, Seagull, London 2010

Graeber, D., *Debt: The First 5000 Years*, Penguin, London 2011

Graham, M., *The Rise of the Planetary Labour Market – and What It Means for the Future of Work*, in "The New Statesman", 9 January 2018

Granaglia, E., Bolzoni, M., *Il reddito di base*, Ediesse, Rome 2016

Green, R.F., *A Crisis of Truth*, Pennsylvania University Press, Philadelphia 1999

Gregory of Nyssa, *On the Making of Man*, Christian Literature Publishing, New York 1892

Grice, H.P., *Logic and Conversation*, in: *Syntax and Semantics*, Vol. 3, Speech Acts, ed. by P. Cole and J.L. Morgan, Academic Press, New York 1975

Grimal, N., *A History of Ancient Egypt*, Wiley, Hoboken (NJ) 1988

Habermas, J., *The Structural Transformation of the Public Sphere: An Inquiry into a Category of Bourgeois Society*, MIT Press, Cambridge (MA) 1989

Han, B.C., *Noi, schiavi felici della pandemia digitale*, in "la Repubblica", 31 October 2020

Hanna, R., *et al.*, *In Defense of Intuitions: A New Rationalist Manifesto*, Palgrave Macmillan, London 2013

Harari, Y.N., *21 Lessons for the 21st Century*, Random House, New York 2018

Harari, Y.N., *Sapiens: A Brief History of Humankind*, Random House, New York 2014

Hardt, M., Negri, A., *Assembly*, Oxford University Press, Oxford/New York 2017

Hardt, M., Negri, A., *Empire*, Harvard University Press, Cambridge (MA) 2000

Harman, G., *Fear of Reality: On Realism and Infra-Realism*, in "The Monist", 98, 2, 2015, pp. 126–144

Harman, G., *Guerrilla Metaphysics. Phenomenology and the Carpentry of Things*, Open Court, Chicago 2005

Harman, G., *Object-Oriented Ontology: A New Theory of Everything*, Pelican, London 2018

Harman, G., *The Quadruple Object*, Zero Books, Alresford 2010

Harman, G., *Tool-Being: Heidegger and the Metaphysics of Objects*, Open Court, Chicago 2002

Hartmann, N., *Grundzüge einer Metaphysik der Erkenntnis*, Vereinigung wissenschaftlicher Verleger, Berlin 1921

Havelock, E.A., *Preface to Plato*, Harvard University Press, Cambridge (MA) 1963

Havelock, E.A., *The Muse Learns to Write. Reflections on Orality and Literacy from Antiquity to the Present*, Yale University Press, New Haven/London 1986

Hegel, G.W.F., *Aesthetics* (1836–1838), vol. 1, Clarendon Press, Oxford 1988

Hegel, G.W.F., *Encyclopedia of the Philosophical Sciences* (1817–1830), Clarendon Press, Oxford 1971

Heidegger, M., *Being and Time*, Wiley-Blackwell, Hoboken (NJ) 1978

Heidegger, M., *Grundbegriffe der Metaphysik*, in Id., *Gesamtausgabe*, Klostermann, Frankfurt a. M. 1929–1930

Heidegger, M., *Off the Beaten Track*, Cambridge University Press, Cambridge 2002

Heidegger, M., *The Question Concerning Technology and Other Essays*, Garland Publishing, New York/London 1977

Heidegger, M., *What is Called Thinking?* (1954), Harper & Row, London 1968

Herodotus, *Histories*, with an English translation by A.D. Godley, Harvard University Press, Cambridge (MA) 1920.

Herrenschmidt, C., *Les trois écritures. Langue, nombre, code*, Gallimard, Paris 2007

Hölldobler, B., Wilson, E.O., *The Superorganism: The Beauty, Elegance, and Strangeness of Insect Societies*, Norton, New York 2010

Honneth, A., *Axel Honneth* (2011), Wiley, Hoboken (NJ) 2014

Honneth, A., *The Struggle for Recognition*, Wiley, Hoboken (NJ) 1995

Horgan, T., *From Supervenience to Superdupervenience: Meeting the Demands of a Material World*, in "Mind", 102, 408, 1993, pp. 555–586

Hurley, S., *Consciousness in Action*, Harvard University Press, Cambridge (MA) 1998

Ifrah, G., *From One to Zero. A Universal History of Numbers*, Penguin, New York 1985

Jacomuzzi, A., *et al.*, *Molyneux's Question Redux*, in "Phenomenology and the Cognitive Sciences", 2, 2003, pp. 255–280

Johanson, D., Edgar, B., *From Lucy to Language*, Simon & Schuster, New York 1996

Johnson, S., *Emergence: The Connected Lives of Ants, Brains, Cities, and Software*, Scribner, New York 2001

Joullié, J-E, Spillane, R., *Philosophy of Leadership: The Power of Authority*, Palgrave MacMillan, London 2015

Jünger, E. (1930) 'Total Mobilization', in *The Heidegger Controversy*, ed. by R. Wolin, MIT Press, Cambridge (MA) 1993

Jünger, E., *The Forest Passage* (1951), Telos Press Publishing, Candor (NY) 2013.

Kafka, F., *The Penal Colony* (1919), Transaction Publishers, Livingston (NJ) 2000

Kafka, B., *The Demon of Writing: Powers and Failures of Paperwork*, Zone Books, New York 2012

Kafka, F., "The Cares of a Family Man" (1917) in *Franz Kafka: The Complete Stories*, Schocken, New York, 1971

Kallir, A., *Sign and Design: The Psychogenetic Source of the Alphabet*, Clarke, Cambridge 1961

Kant, I., *Answering the Question: What Is Enlightenment?* (1784), Penguin, New York 2013

Kant, I., *Critique of Judgment* (1790), Macmillan, London 1892

Kant, I., *Critique of Pure Reason*, Cambridge University Press, Cambridge 1998

Kant, I., *Idea for a Universal History with a Cosmopolitan Aim* (1784), Cambridge University Press, Cambridge 2009

Kelly, K., *What Technology Wants*, Penguin, New York 2010

Keynes, J. M., "Economic Possibilities for our Grandchildren (1930)," in *Essays in Persuasion*, Harcourt Brace, New York 1932, pp. 358–373

Keynes, J.M., *The General Theory of Employment, Interest and Money*, Macmillan Cambridge University Press, for Royal Economic Society, Cambridge 1936

Kittler, F., *Gramophone Film Typewriter* (1986), Stanford University Press, Stanford 1999

Klages, L., George, S., *L'anima e la forma* (1902), Fazi, Rome 1995

Klein, N., *How Big Tech Plans To Profit from the Pandemic*, in "The Guardian", 13 May 2020

Klein, N., *No Logo: Taking Aim at the Brand Bullies*, Picador, London 1999

Kocherlakota, N.R., *Money is Memory*, Federal Reserve Bank of Minneapolis, Research Department Staff Report 218, October 1996

Lafargue, P., *Le Droit à la paresse: réfutation du "Droit au travail" de 1848* (1883), Bureau d'éditions, Paris 1929

Lanier, J., *Tanti auguri, internet. Ma ora paga per i nostri dati*, in "La Repubblica", 29 October 2019.

Lanier, J., *Who Owns the Future?*, Simon & Schuster, New York 2013

Lanier, J., *You Are Not a Gadget*, Allen Lane, New York 2010

Laplace, J., *Introduction*, in Grégoire de Nysse, *La création de l'homme*, Éditions du Cerf, Paris 1944

Latour, B., *et al.*, *Enquêtes sur les modes d'existence. Une anthropologie des modernes*, La Découverte, Paris 2012

Laudani, R., *Disobbedienza*, il Mulino, Bologna 2010

Lavelli, M., *Intersoggettività*, Raffaello Cortina, Milan 2007

Lawson, T., *Debt as Money*, in "Cambridge Journal of Economics", 42, 4, 2018, pp. 1165–1181

Le Goff, J., *Documento/Monumento*, in *Enciclopedia Einaudi*, vol. V, Einaudi, Turin 1978

Leader, D., *Hands: What We Do with Them – and Why*, Penguin, New York 2016

Lehrer, S., *Hitler Sites: A City-by-City Guidebook (Austria, Germany, France, United States)*, McFarland, Jefferson 2002

Leibniz, G.W., *A New System of the Nature and the Communication of Substances, as well as the Union Between the Soul and the Body* in Loemker L.E. (ed. by) *Philosophical Papers and Letters. The New Synthese Historical Library (Texts and Studies in the History of Philosophy)*, vol 2. Springer, Dordrecht 1989

Leibniz, G.W., *Discourse on Metaphysics* (1686), Oxford University Press, Oxford/New York 2020

Leibniz, G.W., *New Essays on Human Understanding* (1705), II, Cambridge University Press, Cambridge 1996

Leistert, O., *Resistance against Cyber-Surveillance within Social Movements and how Surveillance Adapts*, in "Surveillance & Society", 9, 4, 2012, pp. 441–456

Leopardi, G., *Dialogo di Federico Ruysch e delle sue mummie*, in *Operette morali* (1827), Rizzoli, Milan 2008

Leroi-Gourhan, A., *Gesture and Program* (1964–65), MIT Press, Cambridge (MA) 1993

Lévi-Strauss, C., *A World on the Wane* (1955), Criterion Books, New York 1961

Lévinas, E., *Totality and Infinity: An Essay on Exteriority* (1961), Duquesne University Press, Pittsburgh 1969

Levy, P., *Collective Intelligence*, Basic Books, New York 1997

Libet, B., *Mind Time. The Temporal Factor in Consciousness*, Harvard University Press, Cambridge (MA) 2004

Lloyd Morgan, C., *Emergent Evolution*, Williams & Norgate, London 1923

Locke, J., *An Essay Concerning Human Understanding* (1689), Hackett Publishing Company, Indianapolis 1996

London, J., *John Barleycorn*, The Century Company, New York 1913

Lorusso, A.M., *Postverità*, Laterza, Bari-Rome 2018

Lovink, G., *Social Media Abyss: Critical Internet Cultures and the Force of Negation*, Polity, Cambridge 2016

Lucas, R. *Equilibrium in a Pure Currency Economy*, in *Models of Monetary Economies*, ed. by J.H. Kareken and N. Wallace, Federal Reserve Bank of Minneapolis, Minneapolis 1980

Ludlow, P., Stoljar, D., Nagasawa, Y., (eds.), *There's Something About Mary*, MIT Press, Cambridge (MA) 2004

Lyotard, J.F., *The Postmodern Condition* (1979), Manchester University Press, Manchester 1984

Mach, E., *The Analysis of Sensations and the Relation of the Physical to the Mental* (1903), Open Court, Chicago/London 1914

Maddalena, G., Gili, G., *Chi ha paura della post-verità? Effetti collaterali di una parabola*, Marietti, Bologna 2017

Maddalena, G., *The Philosophy of Gesture. Completing Pragmatists' Incomplete Revolution*, McGill-Queen's University Press, Montréal/New York 2015

Malthus, T.R., *Principles of Political Economy Considered with a View to their Applications* (1820), Pickering, London 1836

Mann, Th., *Buddenbrooks* (1902), Martin Secker, London 1922

Marazzi, C., *Capitale e linguaggio. Dalla New Economy all'economia di guerra*, DeriveApprodi, Rome 2002

Marconi, D., *Contro la mente estesa*, in "Sistemi intelligenti", 17, 3, 2005, pp. 389–398

Marconi, D., *Per la verità*, Einaudi, Turin 2007

Mari, G., "Il lavoro 4.0 come atto linguistico performativo. Per una svolta linguistica nell'analisi delle trasformazioni del lavoro", in *Il lavoro 4.0. La quarta rivoluzione industriale e le trasformazioni delle attività lavorative*, ed. by A. Cipriani, A. Gramolati and G. Mari, Firenze University Press, Firenze 2018, pp. 321–335

Marres, N., *Digital Sociology*, Polity Press, Cambridge 2017

Marsh, E., *A Number of People: A Book of Reminiscences*, Harper & Brothers, New York/London 1939

Marsh, L., Onof, C., *Stigmergic Epistemology, Stigmergic Cognition*, in "Cognitive Systems Research", 9, 1–2, 2008, pp. 136–149

Marx, K., *Capital*, Volume One, Penguin Classics edition, New York 1976

Marx, K., *Grundrisse, Foundations of the Critique of Political Economy*, Penguin, New York, 1973

Mason, P., *PostCapitalism: A Guide to Our Future*, Allen Lane, London 2015

Mayer-Schönberger, V., Cukier, K., *Big Data: A Revolution That Will Transform How We Live, Work, and Think*, Eamon Dolan, Boston/New York 2013

Mayer-Schönberger, V., Ramge, T., *Reinventing Capitalism in the Age of Big Data*, John Murray, London 2018

Mayergoyz, I.D., Bertotti, G., (ed. by), *The Science of Hysteresis*, 3 voll., Academic Press, Cambridge (MA) 2005

McDowell, J., *Mind and World*, Harvard University Press, Cambridge (MA) 1994

McGinn, C., *Prehension: The Hand and the Emergence of Humanity*, MIT Press, Cambridge (MA) 2015

McLaughlin, B., *The Rise and Fall of British Emergentism*, in *Emergence or Reduction?*, ed. by A. Beckerman, H. Flohr and J. Kim, De Gruyter, Berlin 1992

McLuhan, M., *The Gutenberg Galaxy: The Making of Typographic Man*, University of Toronto Press Toronto 1962

Meillassoux, Q., *After Finitude: An Essay on the Necessity of Contingency*, Continuum, London 2008

Menger, P. M., *Portrait de l'artiste en travailleur. Métamorphoses du capitalisme*, Seuil, Paris 2003

Merlini, F., *Ubicumque*, Quodlibet, Macerata 2015

Meyer, R., *The Cataclysmic Break That (Maybe) Occurred in 1950*, in "The Atlantic", 16 April 2019

Mill, J.S., *A System of Logic, Ratiocinative and Inductive*, Harper, New York 1875

MIT Work of the Future Task Force, *The Work of the Future: Building Better Jobs in an Age of Intelligent Machines*, MIT Press, Cambridge (MA) 2020

Montague, B., *A Virus is Haunting Europe – The Vector Is Capitalism*, in "The Ecologist", 18 March 2020

Mulligan, K., Simons, P., Smith, B., *Truth-Makers*, in "Philosophy and Phenomenological Research", 44, 1984, pp. 287–321

Munck, R., *Globalization and Labour. The New 'Great Transformation'*, Zed Books, London-New York 2002

Nagel, T., "Panpsychism", in *Nagel's Mortal Questions,* Cambridge University Press, Cambridge 1979, pp. 181–195

Nagel, T., *Consciousness and Content*, in "Proceedings of the British Academy", 74, 1988, pp. 225–245

Napier, J., *Hands*, Princeton University Press, Princeton 1980

Napier, J., *The Roots of Mankind*, Smithsonian Institution Press, New York 1970

Nietzsche, F., *"On Teleology" or "Teleology since Kant"* (1868), trans. Paul Smith, in *Nietzscheana*, 8, 2000, pp. 1–20

Nietzsche, F., *On the Genealogy of Morality* (1887), Cambridge University Press, Cambridge 2007

Nietzsche, F., *Philosophy in the Tragic Age of the Greeks* [1873], Regnery Gateway, Chicago 1962

Noë, A., *Action in Perception*, MIT Press, Cambridge (MA) 2004

Noë, A., *Out of Our Heads: Why You Are Not Your Brain, and Other Lessons from the Biology of Consciousness*, Farrar, Straus and Giroux, New York 2009

Nozick, R., *Invariances: The Structure of the Objective World*, Harvard University Press, Cambridge (MA) 2001

Nussbaum, M.C., *Cultivating Humanity. A Classical Defense of Reform in Liberal Education*, Harvard University Press, Cambridge (MA) 1997

OECD, *OECD Employment Outlook 2020: Worker Security and the COVID-19 Crisis*, OECD Publishing, Paris 2020, https://doi.org/10.1787/1686c758-en

Onians, R.B., *The Origins of European Thought*, Cambridge University Press, Cambridge, 1951

Oppenheimer, S., *Out of Eden: The Peopling of the World*, Constable & Robinson, London 2004

Ostroy, J.M., *The Informational Efficiency of Monetary Exchange*, in "American Economical Review", 63, 4, 1973, pp. 597–610

Otlet, P., *Traité de documentation. Le livre sur le livre. Théorie et pratique*, Mundaneum, Bruxelles 1934

Pagès, R., *Transformations documentaires et milieu culturel*, in "Review of Documentation", XV, 3, 1948

Pareyson, L., *Estetica. Teoria della formatività* [1954], Bompiani, Milan 1988

Pasquinelli, E., *Irresistibili schermi*, Mondadori, Milan 2012

Pettit, P., *The Common Mind: An Essay on Psychology, Society, and Politics*, Oxford University Press, Oxford/New York 1996

Phillips, L., Rozworski, M., *The People's Republic of Walmart: How the World's Biggest Corporations Are Laying the Foundation for Socialism*, Verso, London 2019

Piaget, J., *La formation du symbole chez l'enfant. Imitation, jeu et rêve, image et représentation*, Delachaux et Niestlé, Neuchâtel 1945

Pievani, T., *Finitudine*, Raffaello Cortina, Milan 2020

Pievani, T., *Imperfezione. Una storia naturale*, Raffaello Cortina, Milan 2019

Piketty, T., *Capital in the Twenty-First Century*, Harvard University Press, Cambridge (MA) 2014

Pinker, S., *Commentary on Daniel Dennett*, in "Mind, Brain, and Behavior", Lecture at Harvard University, 23 April 2009

Pinker, S., *How the Mind Works*, Norton, New York 1997

Pinker, S., Jackendoff, R., *The Faculty of Language: What's Special About It?*, in "Cognition", 95, 2, 2005, pp. 201–236

Pinker, S., *Language as an Adaptation to the Cognitive Niche*, in *Language Evolution*, ed. by M.H. Christiansen, S. Kirby, Oxford University Press, Oxford/New York 2003

Plato, *Platonis Opera*, ed. by J. Burnet, Oxford University Press, Oxford/New York 1903

Pouivet, R., *De van Inwagen à saint Athanase, une ontologie personnelle de la résurrection des corps*, in "Klesis, Revue philosophique", 17, 2010, pp. 98–123

Preston, J., *Is Your Mobile Part of Your Mind?*, in *Seeing, Understanding, Learning in the Mobile Age*, ed. by K. Nyíri, Passagen Verlag, Wien 2003, pp. 313–324

Proust, M., *In Search of Lost Time*, vol. IV, Time Regained, New York, Modern Library 1993

Putnam, H., *The Meaning of "Meaning"*, in Id., *Philosophical Papers*, vol. II, Mind, Language and Reality, Cambridge University Press, Cambridge 1975, pp. 215–271

Putnam, H., *The Threefold Cord: Mind, Body and World*, Columbia University Press, New York 1999

R. Boyd, P.J. Richerson, J. Henrich, *The Cultural Niche: Why Social Learning Is Essential for Human Adaptation*, in "Proceedings of the National Academy of Sciences", 108 (suppl. 2), 2011, pp. 10918–10925

Radman, Z., (ed. by), *The Hand, an Organ of the Mind. What the Manual Tells the Mental*, MIT Press, Cambridge (MA) 2013

Raventós, D., *Basic Income: The Material Conditions of Freedom*, Pluto Press, London 2007

Recanati, F., *Mental Files*, Oxford University Press, Oxford/New York 2012

Rendell, L., Boyd, R., Cownden, D., Enquist, M., Eriksson, K., Feldman, M.W., Fogarty, L., Ghirlanda, S., Lillicrap, T., Laland, K.N., *Why Copy Others? Insights from the Social Learning Strategies Tournament*, in "Science", 328, 5975, 2010, pp. 208–213

Resnick, M., *Turtles, Termites, and Traffic Jams: Explorations in Massively Parallel Microworlds*, MIT Press, Cambridge (MA) 1997

Ressi, A., *Dell'economia nella specie umana*, vol. III, in Stamperia e Libreria di Pietro Bizzoni, Pavia 1819

Rey, P.J., *Alienation, Exploitation, and Social Media*, in "American Behavioral Scientist", 56, 4, 2012, pp. 399–420

Rifkin, J., *The End of Work: The Decline of the Global Labor Force and the Dawn of the Post-Market Era*, Putnam, New York 1995

Roberts, A., *Churchill: Walking with Destiny*, Penguin, New York 2018, p. 397

Robinson, J., *Economic Philosophy*, Watts, London 1962

Romele, A, Terrone, E, (eds.), *Towards a Philosophy of Digital Media*, Palgrave MacMillan, London 2018

Romele, A., Gallino, F., Emmenegger, C., Gorgone, D., *Panopticism Is not Enough: Social Media as Technologies of Voluntary Servitude*, in "Surveillance & Society", 15, 2, 2017

Romele, A., Severo, M., *The Economy of the Digital Gift: From Socialism to Sociality Online*, in "Theory, Culture & Society", 33, 5, 2016, pp. 43–63

Roncaglia, G., *La quarta rivoluzione. Sei lezioni sul futuro del libro*, Laterza, Rome-Bari 2010

Rorty, R., *Achieving Our Country. Leftist Thought in Twentieth-Century America*, Harvard University Press, Cambridge (MA) 1998

Rorty, R., *Charles Taylor on Truth*, in Id., *Philosophical Papers*, vol. III, *Truth and Progress*, Cambridge University Press, Cambridge 1998, pp. 84–97

Rorty, R., *Nineteenth-Century Idealism and Twentieth-Century Textualism*, in "The Monist", 64, 2, 1981, pp. 155–174

Rorty, R., *Solidarity or Objectivity?* (1990), in Id., *Philosophical Papers*, vol. I, *Objectivity, Relativism, and Truth*, Cambridge University Press, Cambridge 1991, pp. 21–34

Rossi, S., *Oro*, il Mulino, Bologna 2017

Rousseau, J.-J., *The Social Contract* (1762), J.M. Dent & Sons, Ltd, London; E.P. Dutton & Co., New York 1913

Rouvroy, A., *Face à la gouvernementalité algorithmique, repenser le sujet de droit comme puissance*, unpublished paper

Rovelli, C., *Helgoland*, Adelphi, Milan 2020

Rowthorn, R., Ramaswamy, R., *Deindustrialization: Causes and Implications*, in "IMF Working Papers", 42, 1997

Roy, D., *The Birth of a Word*, TED talk 2011, available at: http://www.ted.com/talks/deb_roy_the_birth_of_a_word

Ruben, D.H., *The Metaphysics of Social Reality*, The Free Press, New York 1985

Ryle, G., *The Concept of Mind*, Hutchinson & Co, London 1949

Sacco, R., *Diritto muto*, in "Rivista di diritto civile", I, 1993

Sadin, E., *La silicolonisation du monde. L'irrésistible expansion du libéralisme numérique*, L'échappée, Paris 2016

Sagan, D., Skoyles, J., *Up from Dragons: The Evolution of Human Intelligence*, McGraw-Hill, New York 2002

Salmon, C., *Storytelling: Bewitching the Modern Mind*, Verso, London 2008

Scheler, M., *Formalism in Ethics and Non-formal Ethics of Values* (1913–1916), Northwestern University Press, Evanston 1973

Schelling, F.W.J., *Philosophy of Revelation*, 95, SW II/3, in *The Grounding of Positive Philosophy* ed. and trans. B. Matthews, SUNY Press, Albany 2007

Schelling, F.W.J., *System der gesammten Philosophie und der Naturphilosophie insbesondere [1804])*

Schmitt, C., *Das Problem der Legalität* (1950), in *Verfassungsrechtliche Aufsätze*, Duncker & Humblot, Berlin 1958, pp. 440–451

Scholz, T., Schneider, N., *Ours to Hack and to Own. The Rise of Platform Cooperativism, a New Vision for the Future of Work and a Fairer Internet*, OR Books, New York/London 2016

Seabright, P., in *The Company of Strangers: A Natural History of Economic Life*, Princeton University Press, Princeton 2004

Searle, J.R., "Minds, Brains, and Programs" (1981), *Behavioral and Brain Sciences 3*, no 3 (1980), pp. 417–24

Searle, J.R., *Making the Social World: The Structure of Human Civilization*, Oxford University Press, Oxford/New York 2010

Searle, J.R., *The Construction of Social Reality*, Free Press, New York 1995

Searle, J.R., *The Rediscovery of the Mind*, MIT Press, Cambridge (MA) 1992

Sennet, R., *The Corrosion of Character: The Personal Consequences of Work in the New Capitalism*, Norton, New York 1998

Sennet, R., *The Fall of Public Man*, Penguin, New York 1976

Seung, H.S., *Learning in Spiking Neural Networks by Reinforcement of Stochastic Synaptic Transmission*, in "Neuron", 40, 6, 2003, pp. 1063–1073

Shannon, C.E., Weaver, W., *The Mathematical Theory of Communication*, University of Illinois Press, Urbana/Chicago 1949

Shapiro, S.A., *Massively Shared Agency*, in *Rational and Social Agency: Essays on the Philosophy of Michael Bratman*, ed. by M. Vargas and G. Yaffe, Oxford University Press, Oxford/New York 2014

Shepard, R.N., Metzler, J., *Mental Rotation of Three-Dimensional Objects*, in "Science", 171, 3972, 1971, pp. 701–703

Simmel, G., *The Philosophy of Money* (1900), Routledge, London 2004

Simondon, G., *Imagination et invention (1965–1966)*, Éditions de la Transparence, Chatou 2008

Sini, C., *Etica della scrittura* (1992), new ed., Mimesis, Milan 2009

Sini, C., *Filosofia e scrittura*, Laterza, Rome-Bari 1994

Smit, J.P., Buekens, F., Du Plessis, S., *Cigarettes, Dollars and Bitcoins. An Essay on the Ontology of Money*, in "Journal of Institutional Economics", 12, 2016, pp. 327–347

Smith, B., *Diagrams, Documents, and the Meshing of Plans*, in *How To Do Things With Pictures: Skill, Practice, Performance. Visual Learning*, ed. by A. Benedek and K. Nyíri, Lang, Frankfurt a. M. 2013, pp. 165–179

Smith, B., *Objects and Their Environments: From Aristotle to Ecological Ontology*, in *The Life and Motion of Socio-Economic Units*, ed. by A. Frank, J. Raper and J.P. Cheylan, Taylor and Francis, London 2001, pp. 79–97

Smith, B.,*Toward a Realistic Science of Environments*, in "Ecological Psychology", 21, 2, 2009, pp. 121–130

Sohn-Rethel, A., *Das Geld, die bare Münze des Apriori*, Wagenbach, Berlin 1976

Sohn-Rethel, A., *Intellectual and Manual Labour* (1970), Brill, Boston 1978

Sohn-Rethel, A., *Warenform und Denkform*, Suhrkamp, Frankfurt a. M. 1971

Spengler, O., *The Decline of the West* (1918), Oxford University Press, Oxford/New York 1932

Sperber, D., *Explaining Culture: A Naturalistic Approach*, Blackwell, Oxford 1996

Srnicek, N., Williams, A., *ACCELERATE MANIFESTO. For an Accelerationist Politics* (2013), https://criticallegalthinking.com/2013/05/14/accelerate-manifesto-for-an-accelerationist-politics

Srnicek, N., Williams, A., *Inventing the Future: Postcapitalism and a World Without Work*, Verso, London 2015

Stiegler, B., *L'emploi est mort, vive le travail! Entretiens avec Ariel Kyrou*, Fayard, Paris 2015

Stiegler, B., *La société automatique*, Fayard, Paris 2015

Stove, D., *Idealism: a Victorian Horror Story (Part Two)*, in Id., *The Plato Cult and Other Philosophical Follies*, Blackwell, Oxford 1991, pp. 135–178

Supiot, A., (dir.), *Au-delà de l'emploi*, Flammarion, Paris 2016

Tallis, R., *The Hand: A Philosophical Inquiry into Human Being*, Edinburgh University Press, Edinburgh 2003

Tarde, G., *Les lois de l'imitation* (1895), new ed. Les Empêcheurs de penser en rond, Paris 2001

Teilhard de Chardin, P., *Hominization*, in Id., *The Vision of the Past*, Harper & Row, New York 1923

The Future of Work, in "Nature", vol. 550, 19 October 2017 (https://doi.org/10.1038/550315a)

Theraulaz, G., Bonabeau, E., *A Brief History of Stigmergy*, in "Artificial Life", 5, 1999, pp. 97–116

Thomasson, A., *Fiction and Metaphysics*, Cambridge University Press, Cambridge 1998

Thompson, E., *Mind in Life. Biology, Phenomenology and the Sciences of Mind*, Harvard University Press, Cambridge (MA) 2008

Thomson, J.J., *The Right to Privacy*, in "Philosophy and Public Affairs", 4, 1975, pp. 295–314

Toffler, A., *The Third Wave*, William Morrow, New York 1980

Tomasello, M., *A Natural History of Human Thinking*, Harvard University Press, Cambridge (MA) 2014

Tomasello, M., *The Cultural Origins of Human Cognition*, Harvard University Press, Cambridge (MA) 1999

Toracca, T., *Labour Between Law and Literature: Historical Similarities and Critical Propositions on the Present*, in "Pólemos", 11, 2, 2017, pp. 361–377

Toscano, M.A., (ed. by), *Homo instabilis. Sociologia della precarietà*, Jaca Book, Milan 2007

Tranquilli, V., *Il concetto di lavoro da Aristotele a Calvino*, Ricciardi, Milan-Naples 1979

Trevisan, V., *Works*, Einaudi, Turin 2016

Trible, W., Kronauer, D.J.C., *Caste Development and Evolution in Ants: It's All About Size*, in "Journal of Experimental Biology", 220, 2007, pp. 53–62

Tripet, F., Nonacs, P., *Foraging for Work and Age-Based Polyethism: The Roles of Age and Previous Experience on Task Choice in Ants*, in "Ethology", 110, 11, 2004, pp. 863–877

Tullio, P., *La forma naturale delle lettere dell'alfabeto*, in "Bollettino della Accademia Italiana di Stenografi 1931

Tuomela, R., *The Importance of Us*, Stanford University Press, Stanford 1995

Tuomela, R., *The Philosophy of Social Practices*, Cambridge University Press, Cambridge 2002

Türcke, C., *Erregte Gesellschaft. Philosophie der Sensation*, C.H. Beck, Munich 2002

Turkle, S., *Alone Together: Why We Expect More From Technology And Less From Each Other*, Basic Books, New York 2011.

Turner, S.P., in *Explaining the Normative*, Polity Press, Cambridge 2010

Van Cleve, J., *Emergence or Panpsychism: Magic or Mind-dust?*, in "Philosophical Perspectives", 4, 1990, pp. 215–226

Van Den Eede, Y., *In Between Us: On the Transparency and Opacity of Technological Mediation*, in "Foundations of Science", 16, 2, 2011, pp. 139–159

Van Inwagen, P., *The Possibility of Resurrection and Other Essays in Christian Apologetics*, Westview Press, Boulder 1998

Van Parijs, P., Vanderborght, Y., *Basic Income: A Radical Proposal for a Free Society and a Sane Economy*, Harvard University Press, Cambridge (MA) 2017

van Romburgh, M., *Facebook Makes More Profit per Employee than Other Fortune 500 Tech Companies*, in "Bizwomen", 8 July 2019, https://www.bizjournals.com/bizwomen/news/latest-news/2019/07/facebook-makes-more-profit-per- employee-than-other.html?page=all

Vattimo, G., *A Farewell to Truth*, Columbia University Press, New York 2011

Vattimo, G., *The Transparent Society*, Johns Hopkins University Press, Baltimore 1992

Vecchi, B., *Il capitalismo delle piattaforme*, Manifestolibri, Rome 2017

Vercellone, C., (ed. by), *Capitalismo cognitivo*, Manifestolibri, Rome 2006

Verdenius, W.J., *Mimesis. Plato's Doctrine of Artistic Imitation and Its Meaning to Us*, Brill, Leiden 1949

Vico, G.B., *La scienza nuova* (1744), in *The New Science of Giambattista Vico*, Cornell University Press, Ithaca 1948

Virilio, P., *Vitesse et politique*, Galilée, Paris 1977

Virno, P., *General Intellect, Exodus, Multitude*, in "Fine Art Critical Studies", 1, June 2002

Virno, P., *Grammatica della moltitudine*, DeriveApprodi, Rome 2002

Volponi, P., *Le mosche del capitale*, Einaudi, Turin 1989

von Mises, L., *The Theory of Money and Credit* (1912), Jonathan Cape Ltd., London 1934

von Moltke, H., *Über Strategie*, in C. von Clausewitz, H. von Moltke, *Kriegstheorie und Kriegsgeschichte*, Deutscher Klassiker Verlag, Frankfurt a.M. 1993

Vonnegut, K., *Player Piano*, Schribner & Sons, New York 1952

W. James, "Two English Critics", in *The Meaning of Truth*, Longman Green and Co, New York 1911, pp. 272–286

W. James, *Pragmatism: A New Name for Some Old Ways of Thinking*, Longmans, Green & Co, New York-London 1907

Walton, K., *Mimesis as Make-Believe: On the Foundations of the Representational Arts*, Harvard University Press, Cambridge (MA) 1990

Warburton, W., *The Divine Legation of Moses Demonstrated* (1741), Garland, New York/London 1978

Westbury, C., Dennett, D.C., *Mining the Past to Construct the Future: Memory and Belief as Forms of Knowledge*, in *Memory, Brain, and Belief*, ed. by D.L. Schacter and E. Scarry, Harvard University Press, Cambridge (MA) 2000, pp. 11–32

Whitehead, A.N., *Process and Reality. An Essay in Cosmology. Gifford Lectures Delivered in the University of Edinburgh During the Session 1927–1928*, Macmillan, New York, Cambridge University Press, Cambridge 1929

Whiten, A., Ham, R., *On the Nature and Evolution of Imitation in the Animal Kingdom. Reappraisal of a Century of Research*, in "Advances in the Study of Behavior", 21, 1992, pp. 239–283

Widerquist, K., Lewis, M.A., Pressman, S., (ed. by), *The Ethics and Economics of the Basic Income Guarantee*, Routledge, New York 2005

Wieviorka, M., *L'Impératif numérique*, CNRS Éditions, Paris 2013

Winner, L., *The Whale and the Reactor: A Search for Limits in an Age of High Technology*, Chicago University Press, Chicago 1986

Wittgenstein, L., *Philosophical Investigations* (1953), Cambridge University Press, Cambridge 2010

Žižek, S., *The Sublime Object of Ideology*, Verso, London 1989

Zuazua, M., *The Fourth Industrial Revolution Will Change Production Forever. Here's How*, in "World Economic Forum", 18 January 2019

Zuboff, S., *The Age of Surveillance Capitalism: The Fight for a Human Future at the New Frontier of Power*, Profile Books, London 2019

Index